T0329735

**Nonparametric Statistics with Applications
to Science and Engineering with R**

WILEY SERIES IN PROBABILITY AND STATISTICS

Established by *Walter A. Shewhart and Samuel S. Wilks*

Editors: *David J. Balding, Noel A. C. Cressie, Garrett M. Fitzmaurice, Geof H. Givens, Geert Molenberghs, David W. Scott, Adrian F. M. Smith, Ruey S. Tsay*

Editors Emeriti: *Harvey Goldstein, J. Stuart Hunter, Iain M. Johnstone, Joseph B. Kadane, Jozef L. Teugels*

The **Wiley Series in Probability and Statistics** is well established and authoritative. It covers many topics of current research interest in both pure and applied statistics and probability theory. Written by leading statisticians and institutions, the titles span both state-of-the-art developments in the field and classical methods.

Reflecting the wide range of current research in statistics, the series encompasses applied, methodological and theoretical statistics, ranging from applications and new techniques made possible by advances in computerized practice to rigorous treatment of theoretical approaches.

This series provides essential and invaluable reading for all statisticians, whether in academia, industry, government, or research.

A complete list of titles in this series can be found at http://www.wiley.com/go/wsps

Nonparametric Statistics with Applications to Science and Engineering with R

Second Edition

Paul Kvam
University of Richmond
Richmond, Virginia, USA

Brani Vidakovic
Texas A&M University
College Station, Texas, USA

Seong-joon Kim
Chosun University
Gwangju, South Korea

Registered Office
John Wiley & Sons, Inc., 111 River Street, Hoboken, NJ 07030, USA

Editorial Office
John Wiley & Sons, Inc., 111 River Street, Hoboken, NJ 07030, USA

For details of our global editorial offices, customer services, and more information about Wiley products visit us at www.wiley.com.

Wiley also publishes its books in a variety of electronic formats and by print-on-demand. Some content that appears in standard print versions of this book may not be available in other formats.

Library of Congress Cataloging-in-Publication Data applied for

Hardback ISBN: 9781119268130

Cover image: © Aleksandr Semenov/Shutterstock
Cover design by Wiley

Set in 9.5/12.5pt STIXTwoText by Straive, Chennai, India

Contents

Preface

Danger lies not in what we don't know, but in what we think we know that just ain't so.

Mark Twain (1835–1910)

This textbook is a substantial revision of a previous textbook written in 2007 by Kvam and Vidakovic. The biggest difference in this version is the adoption of the R programming language as a supplementary learning tool for the purpose of teaching concepts, illustrating examples, and completing computational homework assignments. In the original book, the authors relied on Matlab.

There has been plenty of change in the world of nonparametric statistics since we finished the first edition of this book. While the statistics community had already adapted to a modern framework for data analysis that relies increasingly on nonparametric procedures (not to mention Bayesian alternatives to traditional inference), we sense more adapters in engineering, medical research, chemistry, biology, and especially the behavioral sciences with each passing year. However, the field of nonparametric statistics has also receded toward the periphery of the statistics curriculum in the wake of data science, which continues to encroach on graduate curriculums associated with statistics, causing more programs to replace traditional statistics courses with the trendier versions involving data structures.

There are quality monographs/texts dealing with nonparametric statistics, such as the encyclopedic book by Hollander and Wolfe, *Nonparametric Statistical Methods*, or the excellent book by Conover, *Practical Nonparametric Statistics*, which has served as a staple for a generation of professors tasked to teach a course in this subject. Before engaging in writing the first version of this textbook, we taught several iterations of a graduate course on nonparametric statistics at Georgia Tech. The audience consisted of MS and PhD students in Engineering Statistics, Electrical Engineering, Bioengineering, Management, Logistics, Applied Mathematics, and Physics. While comprising a nonhomogeneous group, all of the students had solid mathematical, programming, and statistical training needed to benefit from the course.

In our course, we relied on the third edition of Conover's book, which is mainly concerned with what most of us think of as traditional nonparametric statistics: proportions, ranks, categorical data, goodness of fit, and so on, with the understanding that the text would be supplemented by the instructor's handouts. We ended up supplying an increasing number of handouts every year, for units such as density and function estimation, wavelets, Bayesian approaches to nonparametric problems, EM algorithm, splines, machine learning, and other arguably modern nonparametric topics. Later on, we decided to merge the handouts and fill the gaps.

With this new edition, we adhere to the traditional form one expects in an academic textbook, but we aim to provide more informal discussion and commentary to balance with the regimen of lessons that help the student progress through a statistics methods course. Unlike newer books that focus on data science, we want to help the student learn more than just how to implement a statistical procedure. We want them to understand, to a higher degree, what they are doing (or what R is doing for them).

We hope the book provides all of the tools and motivation for a student to study methods of nonparametric statistics, but we also aim to keep a conversational tone in our writing. Reading math-infused textbooks can be challenging, but it need not be a drudgery. For that reason, we remind the reader of the bigger picture, including the historical and cultural aspects linked to the development and application of nonparametric procedures. We think it is important to acknowledge the fundamental contributions to the field of nonparametric statistics by not only our field's pioneers, such as Karl Pearson, Nathan Mantel, or Brad Efron, but also others in our vanguard, including François-Marie Arouet (Voltaire), Karl Popper, and Baron Von Munchausen.

Computing. The book is integrated with R, and for many procedures covered in this book, we feature subroutines and packages (free libraries of code) of R code. The choice of software was natural: engineers, scientists, and increasingly statisticians are communicating in the "R language." R is an open-source language for statistical computing and quickly emerging environment as the standard for research and development. R provides a wide variety of packages that allow to perform various kinds of analyses and powerful graphic components. For Bayesian calculation we previously relied on WinBUGS, a free software from Cambridge's Biostatistics Research Unit. Both R and WinBUGS are briefly covered in two appendices for readers less familiar with them. For R-programmers who want to see a variety of programming modules for nonparametric inference in the

R language, we refer you to the R-series guide *Nonparametric Statistical Methods Using R* by Kloke and McKean.

Outline of Chapters. For a typical graduate student to cover the full breadth of this textbook, two semesters would be required. For a one-semester course, the instructor should necessarily cover Chapters 1–3 and 5–9 to start. Depending on the scope of the class, the last part of the course can include different chapter selections.

Chapters 2–4 contain important background material the student needs to understand to effectively learn and apply the methods taught in a nonparametric analysis course. Because the ranks of observations have special importance in a nonparametric analysis, Chapter 5 presents basic results for order statistics and includes statistical methods to create tolerance intervals.

Traditional topics in estimation and testing are presented in Chapters 7–10 and should receive emphasis even to students who are most curious about advanced topics such as density estimation (Chapter 11), curve fitting (Chapter 13), and wavelets (Chapter 14). These topics include a core of rank tests that are analogous to common parametric procedures (e.g. *t*-tests, analysis of variance).

Basic methods of categorical data analysis are contained in Chapter 9. Although most students in the biological sciences are exposed to a wide variety of statistical methods for categorical data, engineering students and other students in the physical sciences typically receive less schooling in this quintessential branch of statistics. Topics include methods based on tabled data, chi-square tests, and the introduction of general linear models. Also included in the first part of the book is the topic of "goodness of fit" (Chapter 6), which refers to testing data not in terms of some unknown parameters, but the unknown distribution that generated it. In a way, goodness of fit represents an interface between distribution-free methods and traditional parametric methods of inference, and both analytical and graphical procedures are presented. Chapter 10 presents the nonparametric alternative to maximum likelihood estimation and likelihood ratio-based confidence intervals.

The term "regression" is familiar from your previous course that introduced you to statistical methods. Nonparametric regression provides an alternative method of analysis that requires fewer assumptions of the response variable. In Chapter 12, we use the regression platform to introduce other important topics that build on linear regression, including isotonic (constrained) regression, robust regression, and generalized linear models. In Chapter 13, we introduce more general curve fitting methods. Regression models based on wavelets (Chapter 14) are presented in a separate chapter.

In the latter part of the book, emphasis is placed on nonparametric procedures that are becoming more relevant to engineering researchers and practitioners. Beyond the conspicuous rank tests, this text includes many of the newest nonparametric tools available to experimenters for data analysis. Chapter 17 introduces fundamental topics of statistical learning as a basis for data mining and pattern recognition and includes discriminant analysis, nearest-neighbor classifiers, neural networks, and binary classification trees. Computational tools needed for nonparametric analysis include bootstrap resampling (Chapter 15) and the EM algorithm (Chapter 16). Bootstrap methods, in particular, have become indispensable for uncertainty analysis with large data sets and elaborate stochastic models.

The textbook also unabashedly includes a review of Bayesian statistics and an overview of nonparametric Bayesian estimation. If you are familiar with Bayesian methods, you might wonder what role they play in nonparametric statistics. Admittedly, the connection is not obvious, but in fact nonparametric Bayesian methods (Chapter 18) represent an important set of tools for complicated problems in statistical modeling and learning, where many of the models are nonparametric in nature.

The book is intended both as a reference text and a text for a graduate course. We hope the reader will find this book useful. All comments, suggestions, updates, and critiques will be appreciated.

April 2022

Paul Kvam
Department of Mathematics
University of Richmond

Brani Vidakovic
Department of Statistics
Texas A & M University

Seong-joon Kim
Department of Industrial Engineering
Chosun University

Acknowledgments

We would like to thank Lori Kvam, Draga Vidakovic, and the rest of our families.

1

Introduction

For every complex question, there is a simple answer… and it is wrong.

H. L. Mencken

Jacob Wolfowitz first coined the term *nonparametric*, saying "We shall refer to this situation [*where a distribution is completely determined by the knowledge of its finite parameter set*] as the parametric case, and denote the opposite case, where the functional forms of the distributions are unknown, as the non-parametric case" (Wolfowitz, 1942). From that point on, nonparametric statistics was defined by what it is not: traditional statistics based on known distributions with unknown parameters. Randles, Hettmansperger, and Casella (2004) extended this notion by stating that "nonparametric statistics can and should be broadly defined to include all methodology that does not use a model based on a single parametric family."

Traditional statistical methods are based on parametric assumptions; that is, the data can be assumed to be generated by some well-known family of distributions, such as normal, exponential, Poisson, and so on. Each of these distributions has one or more parameters (e.g. the normal distribution has μ and σ^2), at least one of which is presumed unknown and must be inferred. The emphasis on the normal distribution in linear model theory is often justified by the central limit theorem, which guarantees *approximate normality* of sample means provided the sample sizes are large enough. Other distributions also play an important role in science and engineering. Physical failure mechanisms often characterize the lifetime distribution of industrial components (e.g. Weibull or lognormal), so parametric methods are important in reliability engineering.

However, with complex experiments and messy sampling plans, the generated data might not be attributed to any well-known distribution. Analysts limited to basic statistical methods can be trapped into making parametric assumptions about the data that are not apparent in the experiment or the data. In the case where the experimenter is not sure about the underlying distribution of the

Nonparametric Statistics with Applications to Science and Engineering with R, Second Edition. Paul Kvam, Brani Vidakovic, and Seong-joon Kim.
© 2023 John Wiley & Sons, Inc. Published 2023 by John Wiley & Sons, Inc.

data, statistical techniques are needed that can be applied regardless of the true distribution of the data. These techniques are called *nonparametric methods*, or *distribution-free methods*.

> The terms nonparametric and distribution-free are not synonymous... Popular usage, however, has equated the terms... Roughly speaking, a nonparametric test is one which makes no hypothesis about the value of a parameter in a statistical density function, whereas a distribution-free test is one which makes no assumptions about the precise form of the sampled population.
>
> J. V. Bradley (1968)

It can be confusing to understand what is implied by the word "nonparametric." What is termed *modern nonparametrics* includes statistical models that are quite refined, except the distribution for error is left unspecified. Wasserman's recent book *All Things Nonparametric* (Wasserman, 2005) emphasizes only modern topics in nonparametric statistics, such as curve fitting, density estimation, and wavelets. Conover's *Practical Nonparametric Statistics* (Conover, 1999), on the other hand, is a classic nonparametrics textbook but mostly limited to traditional binomial and rank tests, contingency tables, and tests for goodness of fit. Topics that are not really under the distribution-free umbrella, such as robust analysis, Bayesian analysis, and statistical learning also have important connections to nonparametric statistics and are all featured in this book. Perhaps this text could have been titled *A Bit Less of Parametric Statistics with Applications in Science and Engineering*, but it surely would have sold fewer copies. On the other hand, if sales were the primary objective, we would have titled this *Nonparametric Statistics for Data Science* or maybe *Nonparametric Statistics with Pictures of Naked People*.

1.1 Efficiency of Nonparametric Methods

> Doubt is not a pleasant condition, but certainty is absurd.
> Francois Marie Voltaire (1694–1778)

It would be a mistake to think that nonparametric procedures are simpler than their parametric counterparts. On the contrary, a primary criticism of using parametric methods in statistical analysis is that they oversimplify the population or process we are observing. Indeed, parametric families are not more useful because they are perfectly appropriate, rather because they are perfectly convenient.

Table 1.1 Asymptotic relative efficiency (ARE) of some basic nonparametric tests.

	Parametric test	Nonparametric test	ARE (normal)	ARE (double exponential)
Two-sample test	t-test	Mann–Whitney	0.955	1.50
Three-sample test	One-way layout	Kruskal–Wallis	0.864	1.50
Variances test	F-test	Conover	0.760	1.08

Nonparametric methods are inherently less powerful than parametric methods. This must be true because the parametric methods are assuming more information to construct inferences about the data. In these cases the estimators are inefficient, where the efficiencies of two estimators are assessed by comparing their variances for the same sample size. This inefficiency of one method relative to another is measured in power in hypothesis testing, for example.

However, even when the parametric assumptions hold perfectly true, we will see that nonparametric methods are only slightly less powerful than the more presumptuous statistical methods. Furthermore, if the parametric assumptions about the data fail to hold, only the nonparametric method is valid. A t-test between the means of two normal populations can be dangerously misleading if the underlying data are not actually normally distributed. Some examples of the relative efficiency of nonparametric tests are listed in Table 1.1, where asymptotic relative efficiency (ARE) is used to compare parametric procedures (second column) with their nonparametric counterparts (third column). ARE describes the relative efficiency of two estimators of a parameter as the sample size approaches infinity and is listed for the normal distribution, where parametric assumptions are justified, and the double-exponential distribution. For example, if the underlying data are normally distributed, the t-test requires 955 observations to have the same power of the Wilcoxon signed-rank test based on 1000 observations.

Parametric assumptions allow us to extrapolate away from the data. For example, it is hardly uncommon for an experimenter to make inferences about a population's extreme upper percentile (say, 99th percentile) with a sample so small that none of the observations would be expected to exceed that percentile. If the assumptions are not justified, this is grossly unscientific.

Nonparametric methods are seldom used to extrapolate outside the range of observed data. In a typical nonparametric analysis, little or nothing can be said about the probability of obtaining future data beyond the largest sampled observation or less than the smallest one. For this reason, the actual measurements of a sample item means less than its rank within the sample. In fact, nonparametric methods are typically based on *ranks* of the data, and properties of the population are deduced using *order statistics* (Chapter 5). The measurement scales for typical data are as follows:

Nominal scale: numbers used only to categorize outcomes (e.g. we might define a random variable to equal one in the event a coin flips heads and zero if it flips tails).

Ordinal scale: numbers can be used to order outcomes (e.g. the event X is greater than the event Y if X = *medium* and Y = *small*).

Interval scale: order between numbers and distances between numbers are used to compare outcomes.

Only interval scale measurements can be used by parametric methods. Nonparametric methods based on ranks can use ordinal scale measurements, and simpler nonparametric techniques can be used with nominal scale measurements.

The binomial distribution is characterized by counting the number of independent observations that are classified into a particular category. Binomial data can be formed from measurements based on a *nominal scale* of measurements; thus binomial models are most encountered models in nonparametric analysis. For this reason, Chapter 3 includes a special emphasis on statistical estimation and testing associated with binomial samples.

1.2 Overconfidence Bias

Be slow to believe what you worst want to be true

Samual Pepys (1633–1703)

Confirmation Bias or *Overconfidence Bias* describes our tendency to search for or interpret information in a way that confirms our preconceptions. Business and finance has shown interest in this psychological phenomenon (Tversky and Kahneman, 1974) because it has proven to have a significant effect on personal and corporate financial decisions where the decision maker will actively seek out and give extra weight to evidence that confirms a hypothesis they already favor. At the same time, the decision maker tends to ignore evidence that contradicts or disconfirms their hypothesis.

Overconfidence bias has a natural tendency to affect an experimenter's data analysis for the same reasons. While the dictates of the experiment and the data sampling should reduce the possibility of this problem, one of the clear pathways open to such bias is the infusion of parametric assumptions into the data analysis. After all, if the assumptions seem plausible, the researcher has much to gain from the extra certainty that comes from the assumptions in terms of narrower confidence intervals and more powerful statistical tests.

Nonparametric procedures serve as a buffer against this human tendency of looking for the evidence that best supports the researcher's underlying hypothesis. Given the subjective interests behind many corporate research findings, nonparametric methods can help alleviate doubt to their validity in cases when these procedures give statistical significance to the corporation's claims.

> If everything isn't black and white, I say...
> Why the hell not?
>
> John Wayne (1907–1979)

1.3 Computing with R

Because a typical nonparametric analysis can be computationally intensive, computer support is essential to understand both theory and applications. Numerous software products can be used to complete exercises and run nonparametric analysis in this textbook, including SAS, SPSS, MINITAB, MATLAB, StatXact, and JMP (to name a few). A student familiar with one of these platforms can incorporate it with the lessons provided here, and without too much extra work.

It must be stressed, however, that demonstrations in this book rely mainly on a single software called R (maintained by R Foundation). R is a "GNU"-(free) programming environment for statistical computing and graphics. Today, the R is one of the fastest growing software programs with over 5000 packages that enable us to perform various kinds of statistical analysis. Because of its open source and extensible nature, it has been widely used in research and engineering practice and is rapidly becoming the dominant software tool for data manipulation, modeling, analysis, and graphical display. R is available on Unix systems, Microsoft Windows, and Apple Macintosh. If you are unfamiliar with R, in the first appendix, we present a brief tutorial along with a short description of some R procedures that are used to solve analytical problems and demonstrate nonparametric methods in this book. For a more comprehensive guide, we recommend the book *An Introduction to R* (Venables, Smith, and the R Core Team, 2014). For more detail information, visit

http://www.r-project.org

A user-friendly computing platform for R is provided by R-Studio, which can be downloaded for free at

https://www.rstudio.com

RStudio Cloud allows students a convenient way of accessing the RStudio development environment without having to worry about installation problems associated with R and RStudio. Classroom instructors and students can easily share work spaces using R-Markdown files, for example,

http://rstudio.cloud

We hope that many students of statistics will find this book useful, but it was written primarily with the scientist and engineer in mind. With nothing against statisticians (some of our acquaintances know statisticians), our approach emphasizes the application of the method over its mathematical theory. We have intentionally made the text less heavy with theory and instead emphasized applications and examples. If you come into this course thinking the history of nonparametric statistics is dry and unexciting, you are probably right, at least compared with the history of ancient Rome, the British monarchy, or maybe even Wayne Newton.[1] Nonetheless, we made efforts to convince you otherwise by noting the interesting historical context of the research and the personalities behind its development. For example, we will learn more about Karl Pearson (1857–1936) and R. A. Fisher (1890–1962), legendary scientists and competitive archrivals, who both contributed greatly to the foundation of nonparametric statistics through their separate research directions.

In short, this book features techniques of data analysis that rely less on the assumptions of the data's good behavior – the very assumptions that can get researchers in trouble. Science's gravitation toward distribution-free techniques is due to both a deeper awareness of experimental uncertainty and the availability of ever-increasing computational abilities to deal with the implied ambiguities in the experimental outcome.

1.4 Exercises

1.1 Describe a potential data analysis in your field of study where parametric methods are appropriate. How would you defend this assumption?

1.2 Describe another potential data analysis in your field of study where parametric methods may not be appropriate. What might prevent you from using parametric assumptions in this case?

1.3 Describe three ways in which overconfidence bias can affect the statistical analysis of experimental data. How can this problem be overcome?

---

1 Strangely popular Las Vegas entertainer.

1.4 For an analysis of variance involving three treatment groups, the traditional one-way layout is more efficient than Kruskal and Wallis's nonparametric test. If the Kruskal and Wallis test requires 400 observations to achieve the required test power that is desired, how many samples would the parametric test need to achieve the same power?

1.5 Find an example of data from your field of study that is considered ordinal.

References

Bradley, J. V. (1968), *Distribution Free Statistical Tests*, Englewood Cliffs, NJ: Prentice Hall.

Conover, W. J. (1999), *Practical Nonparametric Statistics*, New York: Wiley.

Randles, R. H., Hettmansperger, T.P., and Casella, G. (2004), "Introduction to the Special Issue Nonparametric Statistics," *Statistical Science*, 19, 561–562.

Tversky, A., and Kahneman, D. (1974), "Judgment Under Uncertainty: Heuristics and Biases," *Science*, 185, 1124–1131.

Venables, W. N., Smith, D. M., and the R Core Team (2014), *An Introduction to R, version 3.1.0.*, Technical Report, The Comprehensive R Archive Network(CRAN).

Wasserman, L. (2005), *All of Nonparametric Statistics*, New York: Springer-Verlag.

Wolfowitz, J. (1942), "Additive Partition Functions and a Class of Statistical Hypotheses," *Annals of Statistics*, 13, 247–279.

2

Probability Basics

Probability theory is nothing but common sense reduced to calculation.
Pierre Simon Laplace (1749–1827)

In Chapters 2 and 3, we review some fundamental concepts of elementary probability and statistics. If you think you can use these chapters to catch up on all the statistics you forgot since you passed "Introductory Statistics" in your college sophomore year, you are acutely mistaken. What is offered here is an abbreviated reference list of definitions and formulas that have applications to nonparametric statistical theory. Some parametric distributions, useful for models in both parametric and nonparametric procedures, are listed, but the discussion is abridged.

2.1 Helpful Functions

- *Permutations*: The number of arrangements of n distinct objects is $n! = n(n-1)\cdots(2)(1)$. In R: `factorial(n)`.
- *Combinations*: The number of distinct ways of choosing k items from a set of n is

$$\binom{n}{k} = \frac{n!}{k!(n-k)!}.$$

  In R: `choose(n,k)`. Note that all possible ways of choosing k items from a set of n can be obtained by `combn(n,k)`.
- $\Gamma(t) = \int_0^\infty x^{t-1} e^{-x}\, dx, \ t > 0$ is called the gamma function. If t is a positive integer, $\Gamma(t) = (t-1)!$. In R: `gamma(t)`.

Nonparametric Statistics with Applications to Science and Engineering with R, Second Edition.
Paul Kvam, Brani Vidakovic, and Seong-joon Kim.

- Incomplete gamma is defined as $\gamma(t, z) = \int_0^z x^{t-1} e^{-x} dx$. In R: `pgamma(t,z,1)`. The upper tail incomplete gamma is defined as $\Gamma(t, z) = \int_z^\infty x^{t-1} e^{-x} dx$, in R: `1-pgamma(t,z,1)`. If t is an integer,

$$\Gamma(t, z) = (t - 1)! \ e^{-z} \sum_{i=0}^{t-1} z^i / i!.$$

Note that `pgamma` is a cumulative distribution function (CDF) of the gamma distribution. By letting the scale parameter λ set to 1, `pgamma` reduced to the incomplete gamma.

- *Beta function:* $B(a, b) = \int_0^1 t^{a-1} (1 - t)^{b-1} \ dt = \Gamma(a)\Gamma(b)/\Gamma(a + b)$. In R: `beta(a,b)`.

- *Incomplete beta:* $B(x, a, b) = \int_0^x t^{a-1} (1 - t)^{b-1} dt$, $0 \le x \le 1$. In R: `pbeta(x,a,b)` represents normalized incomplete beta defined as $I_x(a, b) = B(x, a, b)/B(a, b)$.

- *Summations of powers of integers:*

$$\sum_{i=1}^n i = \frac{n(n + 1)}{2}, \quad \sum_{i=1}^n i^2 = \frac{n(n + 1)(2n + 1)}{6}, \quad \sum_{i=1}^n i^3 = \left(\frac{n(n + 1)}{2} \right)^2.$$

- *Floor function:* $\lfloor a \rfloor$ denotes the greatest integer $\le a$. In R: `floor(a)`.

- *Geometric series:*

$$\sum_{j=0}^n p^j = \frac{1 - p^{n+1}}{1 - p}, \text{ so that for } |p| < 1, \sum_{j=0}^\infty p^j = \frac{1}{1 - p}.$$

- *Stirling's formula:* To approximate the value of a large factorial,

$$n! \approx \sqrt{2\pi} \ e^{-n} n^{n+1/2}.$$

- *Common limit for e:* For a constant α,

$$\lim_{x \to 0} (1 + \alpha x)^{1/x} = e^\alpha.$$

This can also be expressed as $(1 + \alpha/n)^n \longrightarrow e^\alpha$ as $n \longrightarrow \infty$.

- *Newton's formula:* For a positive integer n,

$$(a + b)^n = \sum_{j=0}^n \binom{n}{j} a^j b^{n-j}.$$

- *Taylor Series expansion:* For a function $f(x)$, its Taylor series expansion about $x = a$ is defined as

$$f(x) = f(a) + f'(a)(x - a) + f''(a) \frac{(x - a)^2}{2!} + \cdots + f^{(k)}(a) \frac{(x - a)^k}{k!} + R_k,$$

where $f^{(m)}(a)$ denotes mth derivative of f evaluated at a and, for some $\bar{a}$ between a and x,

$$R_k = f^{(k+1)}(\bar{a}) \frac{(x - a)^{k+1}}{(k + 1)!}.$$

- *Convex function*: A function h is *convex* if for any $0 \leq \alpha \leq 1$,

$$h(\alpha x + (1 - \alpha)y) \leq \alpha h(x) + (1 - \alpha)h(y),$$

for all values of x and y. If h is twice differentiable, then h is convex if $h''(x) \geq 0$. Also, if $-h$ is convex, then h is said to be *concave*.
- *Bessel function*: $J_n(x)$ is defined as the solution to the equation

$$x^2 \frac{\partial^2 y}{\partial x^2} + x \frac{\partial y}{\partial x} + (x^2 - n^2)y = 0.$$

In R: `besselJ(x,n)`.

2.2 Events, Probabilities, and Random Variables

- The *conditional probability* of event A occurring given that event B occurs is $P(A|B) = P(AB)/P(B)$, where AB represents the intersection of events A and B, and $P(B) > 0$.
- Events A and B are stochastically *independent* if and only if $P(A|B) = P(A)$ or equivalently, $P(AB) = P(A)P(B)$.
- *Law of total probability*: Let $A_1, \ldots, A_k$ be a partition of the sample space Ω, i.e. $A_1 \cup A_2 \cup \cdots \cup A_k = \Omega$ and $A_i A_j = \emptyset$ for $i \neq j$. For event B, $P(B) = \sum_i P(B|A_i)P(A_i)$.
- *Bayes formula*: For an event B where $P(B) \neq 0$ and partition $(A_1, \ldots, A_k)$ of Ω,

$$P(A_j|B) = \frac{P(B|A_j)P(A_j)}{\sum_i P(B|A_i)P(A_i)}.$$

- A function that assigns real numbers to points in the sample space of events is called a *random variable*.[1]
- For a random variable X, $F_X(x) = P(X \leq x)$ represents its (cumulative) *distribution function*, which is nondecreasing with $F(-\infty) = 0$ and $F(\infty) = 1$. In this book, it will often be denoted simply as CDF. The *survivor function* is defined as $S(x) = 1 - F(x)$.
- If the CDF's derivative exists, $f(x) = \partial F(x)/\partial x$ represents the *probability density function*, or PDF.
- A *discrete random variable* is one that can take on a countable set of values $X \in \{x_1, x_2, x_3, \ldots\}$ so that $F_X(x) = \sum_{t \leq x} P(X = t)$. Over the support X, the probability $P(X = x_i)$ is called the probability mass function, or PMF.

---

1 While writing their early textbooks in statistics, J. Doob and William Feller debated on whether to use this term. Doob said, "I had an argument with Feller. He asserted that everyone said *random variable* and I asserted that everyone said *chance variable*. We obviously had to use the same name in our books, so we decided the issue by a stochastic procedure. That is, we tossed for it and he won."

- A *continuous random variable* is one that takes on any real value in an interval, so $P(X \in A) = \int_A f(x)\, dx$, where $f(x)$ is the density function of X.
- For two random variables X and Y, their *joint distribution function* is $F_{X,Y}(x,y) = P(X \leq x, Y \leq y)$. If the variables are continuous, one can define joint density function $f_{X,Y}(x,y)$ as $\frac{\partial^2}{\partial x\, \partial y} F_{X,Y}(x,y)$. The conditional density of X, given $Y = y$ is $f(x|y) = f_{X,Y}(x,y)/f_Y(y)$, where $f_Y(y)$ is the density of Y.
- Two random variables X and Y, with distributions F_X and F_Y, are *independent* if the joint distribution $F_{X,Y}$ of (X, Y) is such that $F_{X,Y}(x,y) = F_X(x)F_Y(y)$. For any sequence of random variables $X_1, \ldots, X_n$ that are independent with the same (identical) marginal distribution, we will denote this using *i.i.d.*

2.3 Numerical Characteristics of Random Variables

- For a random variable X with distribution function F_X, the *expected value* of some function $\phi(X)$ is defined as $\mathbb{E}(\phi(X)) = \int \phi(x)\, dF_X(x)$. If F_X is continuous with density $f_X(x)$, then $\mathbb{E}(\phi(X)) = \int \phi(x)f_X(x)\, dx$. If X is discrete, then $\mathbb{E}(\phi(X)) = \sum_x \phi(x)P(X = x)$.
- The kth *moment* of X is denoted as $\mathbb{E}X^k$. The kth moment about the mean or kth central moment of X is defined as $\mathbb{E}(X - \mu)^k$, where $\mu = \mathbb{E}X$.
- The *variance* of a random variable X is the second central moment, $\mathbb{V}\mathrm{ar}X = \mathbb{E}(X - \mu)^2 = \mathbb{E}X^2 - (\mathbb{E}X)^2$. Often, the variance is denoted by σ_X^2, or simply by σ^2 when it is clear which random variable is involved. The square root of variance, $\sigma_X = \sqrt{\mathbb{V}\mathrm{ar}X}$, is called the standard deviation of X.
- With $0 \leq p \leq 1$, the pth *quantile* of F, denoted x_p is the value x such that $P(X \leq x) \geq p$ and $P(X \geq x) \geq 1 - p$. If the CDF F is invertible, then $x_p = F^{-1}(p)$. The 0.5th quantile is called the *median* of F.
- For two random variables X and Y, the *covariance* of X and Y is defined as $\mathbb{C}\mathrm{ov}(X, Y) = \mathbb{E}[(X - \mu_X)(Y - \mu_Y)]$, where μ_X and μ_Y are the respective expectations of X and Y.
- For two random variables X and Y with covariance $\mathbb{C}\mathrm{ov}(X, Y)$, the *correlation coefficient* is defined as

$$\mathrm{Corr}(X, Y) = \frac{\mathbb{C}\mathrm{ov}(X, Y)}{\sigma_X \sigma_Y},$$

where σ_X and σ_Y are the respective standard deviations of X and Y. Note that $-1 \leq \rho \leq 1$ is a consequence of the Cauchy–Schwartz inequality (Section 2.8).
- The *characteristic function* of a random variable X is defined as

$$\varphi_X(t) = \mathbb{E}e^{itX} = \int e^{itx}\, dF(x).$$

The *moment generating function* of a random variable X is defined as

$$m_X(t) = \mathbb{E}e^{tX} = \int e^{tx} \, dF(x),$$

whenever the integral exists. By differentiating r times and letting $t \to 0$, we have that

$$\frac{d^r}{dt^r} m_X(0) = \mathbb{E}X^r.$$

- The *conditional expectation* of a random variable X is given $Y = y$ is defined as

$$\mathbb{E}(X|Y = y) = \int xf(x|y) \, dx,$$

where $f(x|y)$ is a conditional density of X given Y.
- For random variables X and Y with finite means and variances, we can obtain moments of X through its conditional distribution:

$$\mathbb{E}X = \mathbb{E}(\mathbb{E}(X|Y)),$$

$$\mathbb{V}\text{ar}X = \mathbb{V}\text{ar}(\mathbb{E}(X|Y)) + \mathbb{E}(\mathbb{V}\text{ar}(X|Y)).$$

These two equations are commonly referred to as Adam and Eve's rules.

2.4 Discrete Distributions

Ironically, parametric distributions have an important role to play in the development of nonparametric methods. Even if we are analyzing data without making assumptions about the distributions that generate the data, these parametric families appear nonetheless. In counting trials, for example, we can generate well-known discrete distributions (e.g. binomial, geometric) assuming only that the counts are independent and probabilities remain the same from trial to trial.

2.4.1 Binomial Distribution

A simple Bernoulli random variable Y is dichotomous with $P(Y = 1) = p$ and $P(Y = 0) = 1 - p$ for some $0 \le p \le 1$. It is denoted as $Y \sim Ber(p)$. Suppose an experiment consists of n independent trials $(Y_1, \ldots, Y_n)$ in which two outcomes are possible (e.g. success or failure), with $P(\text{success}) = P(Y = 1) = p$ for each trial. If $X = x$ is defined as the number of successes (out of n), then $X = Y_1 + Y_2 + \cdots + Y_n$, and there are $\binom{n}{x}$ arrangements of x successes and $n - x$

failures, each having the same probability $p^x(1-p)^{n-x}$. X is a *binomial* random variable with PMF:

$$p_X(x) = \binom{n}{x} p^x(1-p)^{n-x}, \quad x = 0, 1, \ldots, n.$$

This is denoted by $X \sim Bin(n, p)$. From the moment generating function $m_X(t) = (pe^t + (1-p))^n$, we obtain $\mu = \mathbb{E}X = np$ and $\sigma^2 = \mathbb{V}arX = np(1-p)$.

The cumulative distribution for a binomial random variable is not simplified beyond the sum; i.e. $F(x) = \sum_{i \leq x} p_X(i)$. However, interval probabilities can be computed in R using pbinom(x,n,p), which computes the CDF at value x. The PMF is also computed in R using dbinom(x,n,p).

2.4.2 Poisson Distribution

A Poisson random variable may characterize the number of events occurring in some fixed interval of time, so that the events occur with constant rate and independently of the time since any previous event. The PMF for the Poisson distribution is

$$p_X(x) = \frac{\lambda^x}{x!} e^{-\lambda}, \quad x = 0, 1, 2, \ldots$$

This is denoted by $X \sim \mathcal{P}(\lambda)$. From $m_X(t) = \exp\{\lambda(e^t - 1)\}$, we have $\mathbb{E}X = \lambda$ and $\mathbb{V}arX = \lambda$; the mean and the variance coincide.

The sum of a finite independent set of Poisson variables also has a Poisson distribution. Specifically, if $X_i \sim \mathcal{P}(\lambda_i)$, then $Y = X_1 + \cdots + X_k$ is distributed as $\mathcal{P}(\lambda_1 + \cdots + \lambda_k)$. Furthermore, the Poisson distribution is a limiting form for a binomial model, i.e.

$$\lim_{n, np \to \infty, \lambda} \binom{n}{x} p^x(1-p)^{n-x} = \frac{1}{x!} \lambda^x e^{-\lambda}. \tag{2.1}$$

R commands for Poisson CDF, PDF, quantile, and a random number are ppois, dpois, qpois, and rpois.

2.4.3 Negative Binomial Distribution

Suppose we are dealing with i.i.d. trials again, this time counting the number of successes observed until a fixed number of failures (k) occur. If we observe k consecutive failures at the start of the experiment, for example, the count is $X = 0$ and $P_X(0) = p^k$, where p is the probability of failure. If $X = x$, we have observed x successes and k failures in $x + k$ trials. There are $\binom{x+k}{x}$ different ways of arranging those $x + k$ trials, but we can only be concerned with the arrangements in which the last trial ended in a failure. So there are really only $\binom{x+k-1}{x}$ arrangements, each equal in probability. With this in mind, the PMF is

$$p_X(x) = \binom{k+x-1}{x} p^k(1-p)^x, \quad x = 0, 1, 2, \ldots$$

This is denoted by $X \sim \mathcal{N}B(k, p)$. From its moment generating function

$$m(t) = \left(\frac{p}{1 - (1 - p)e^t} \right)^k,$$

the expectation of a negative binomial random variable is $\mathbb{E}X = k(1 - p)/p$ and variance $\mathbb{V}\mathrm{ar}X = k(1 - p)/p^2$. R commands for negative binomial CDF, PDF, quantile, and a random number are `pnbinom`, `dnbinom`, `qnbinom`, and `rnbinom`.

2.4.4 Geometric Distribution

Random events characterized by counting the number of independent trials until a specific event occurs will have a geometric distribution. As such, the geometric is a special case of negative binomial for $k = 1$ called the geometric distribution. Random variable X has geometric $\mathcal{G}(p)$ distribution if its PMF is

$$p_X(x) = p(1 - p)^x, \quad x = 0, 1, 2, \ldots$$

If X has geometric $\mathcal{G}(p)$ distribution, its expected value is $\mathbb{E}X = (1 - p)/p$ and variance $\mathbb{V}\mathrm{ar}X = (1 - p)/p^2$.

The geometric random variable can be considered as the discrete analog to the (continuous) exponential random variable because it possesses a "memoryless" property. That is, if we condition on $X \geq m$ for some nonnegative integer m, then for $n \geq m$, $P(X \geq n | X \geq m) = P(X \geq n - m)$.

R commands for geometric CDF, PDF, quantile, and a random number are `pgeom`, `dgeom`, `qgeom`, and `rgeom`.

2.4.5 Hypergeometric Distribution

The hypergeometric distribution is a natural byproduct of counting principles. Suppose a box contains m balls, k of which are white and $m - k$ of which are gold. Suppose we randomly select and remove n balls from the box *without replacement*, so that when we finish, there are only $m - n$ balls left. If X is the number of white balls chosen (without replacement) from n, then

$$p_X(x) = \frac{\binom{k}{x} \binom{m-k}{n-x}}{\binom{m}{n}}, \quad x \in \{0, 1, \ldots, \min\{n, k\}\}.$$

This PMF can be deduced with counting rules. There are $\binom{m}{n}$ different ways of selecting the n balls from a box of m. From these (each equally likely), there are $\binom{k}{x}$ ways of selecting x white balls from the k white balls in the box and similarly $\binom{m-k}{n-x}$ ways of choosing the gold balls.

It can be shown that the mean and variance for the hypergeometric distribution are, respectively,

$$\mathbb{E}(X) = \mu = \frac{nk}{m} \quad \text{and} \quad \mathbb{V}\text{ar}(X) = \sigma^2 = \left(\frac{nk}{m}\right)\left(\frac{m-k}{m}\right)\left(\frac{m-n}{m-1}\right).$$

R commands for hypergeometric CDF, PDF, quantile, and a random number are `phyper`, `dhyper`, `qhyper`, and `rhyper`.

Example 2.1 Games such as poker based on a deal of x cards from a deck of m cards resemble the hypergeometric distribution. With the hypergeometric PMF

$$p_X(x) = \frac{\binom{k}{x}\binom{m-k}{n-x}}{\binom{m}{n}}, \tag{2.2}$$

if $k \to \infty$ and $m \to \infty$ in such a way that $k/m \to p$ for some $0 < p < 1$, then (2.2) converges to the binomial probability

$$\binom{n}{x}p^x(1-p)^{n-x}.$$

To a gambler, this results makes card-counting a futile effort if the casino uses multiple card packs in a game of 21 or blackjack. The player's advantage in such games relies on the hypergeometric distribution's *without replacement* properties of the card deal, which slowly disappear with multiple decks. The game eventually resembles a *with replacement* counting problem, which is characterized by the binomial distribution.

2.4.6 Multinomial Distribution

The binomial distribution is based on dichotomizing event outcomes. If the outcomes can be classified into $k \geq 2$ categories, then out of n trials, we have X_i outcomes falling in the category i, $i = 1, \ldots, k$. The PMF for the vector $(X_1, \ldots, X_k)$ is

$$p_{X_1,\ldots,X_k}(x_1, \ldots, x_k) = \frac{n!}{x_1! \cdots x_k!}p_1^{x_1} \cdots p_k^{x_k},$$

where $p_1 + \cdots + p_k = 1$, so there are $k - 1$ free probability parameters to characterize the multivariate distribution. This is denoted by $\mathbf{X} = (X_1, \ldots, X_k) \sim \mathcal{M}n(n, p_1, \ldots, p_k)$. R commands for multinomial PDF and a random number are `dmultinom` and `rmultinom`.

The mean and variance of X_i is the same as a binomial because this is the marginal distribution of X_i, i.e. $\mathbb{E}(X_i) = np_i$, $\mathbb{V}\text{ar}(X_i) = np_i(1 - p_i)$. The covariance between X_i and X_j is $\mathbb{C}\text{ov}(X_i, X_j) = -np_ip_j$ because $\mathbb{E}(X_iX_j) = \mathbb{E}(\mathbb{E}(X_iX_j|X_j)) = \mathbb{E}(X_j\mathbb{E}(X_i|X_j))$ and conditional on $X_j = x_j$, X_i is binomial $Bin(n - x_j, p_i/(1 - p_j))$. Thus, $\mathbb{E}(X_iX_j) = \mathbb{E}(X_j(n - X_j))p_i/(1 - p_j)$, and the covariance follows from this.

2.5 Continuous Distributions

> I can only recognize the occurrence of the normal curve - the Laplacian curve of errors - as a very abnormal phenomenon. It is roughly approximated to in certain distributions; for this reason, and on account for its beautiful simplicity, we may, perhaps, use it as a first approximation, particularly in theoretical investigations.
>
> Karl Pearson (1857–1936)

Discrete distributions are often associated with nonparametric procedures, but continuous distributions will play a role in how we learn about nonparametric methods. The normal distribution, of course, can be produced in a sample mean when the sample size is large, as long as the underlying distribution of the data has finite mean and variance. Many other distributions will be referenced throughout the text book.

2.5.1 Exponential Distribution

The PDF for an exponential random variable is

$$f_X(x) = \lambda\, e^{-\lambda x}, \quad x > 0,\ \lambda > 0.$$

An exponentially distributed random variable X is denoted by $X \sim E(\lambda)$. Its moment generating function is $m(t) = \lambda/(\lambda - t)$ for $t < \lambda$, and the mean and variance are $1/\lambda$ and $1/\lambda^2$, respectively. This distribution has several interesting features; for example, its *failure rate*, defined as

$$r_X(x) = \frac{f_X(x)}{1 - F_X(x)},$$

is constant and equal to λ.

The exponential distribution has an important connection to the Poisson distribution. Suppose we measure i.i.d. exponential outcomes $(X_1, X_2, \ldots)$ and define $S_n = X_1 + \cdots + X_n$. For any positive value t, it can be shown that $P(S_n < t < S_{n+1}) = p_Y(n)$, where $p_Y(n)$ is the PMF for a Poisson random variable Y with parameter λt. Similar to a geometric random variable, an exponential random variable has the *memoryless property* because for $t > x$, $P(X \geq t | X \geq x) = P(X \geq t - x)$.

The median value, representing a typical observation, is roughly 70% of the mean, showing how extreme values can affect the population mean. This is easily shown because of the ease at which the inverse CDF is computed:

$$p \equiv F_X(x; \lambda) = 1 - e^{-\lambda x} \quad \Longleftrightarrow \quad F_X^{-1}(p) \equiv x_p = -\frac{1}{\lambda} \log(1 - p).$$

R commands for exponential CDF, PDF, quantile, and a random number are pexp, dexp, qexp, and rexp. For example, the CDF of random variable $X \sim E(3)$ distribution evaluated at $x = 2$ is calculated in R as pexp(2, 3).

2.5.2 Gamma Distribution

The gamma distribution is an extension of the exponential distribution. Random variable X has gamma $Gamma(r, \lambda)$ distribution if its PDF is given by

$$f_X(x) = \frac{\lambda^r}{\Gamma(r)} x^{r-1} e^{-\lambda x}, \quad x > 0, \ r > 0, \ \lambda > 0.$$

The moment generating function is $m(t) = (\lambda/(\lambda - t))^r$, so in the case $r = 1$, gamma is precisely the exponential distribution. From $m(t)$ we have $\mathbb{E}X = r/\lambda$ and $\mathbb{V}arX = r/\lambda^2$.

If $X_1, \ldots, X_n$ are generated from an exponential distribution with (rate) parameter λ, it follows from $m(t)$ that $Y = X_1 + \cdots + X_n$ is distributed gamma with parameters λ and n; that is, $Y \sim Gamma(n, \lambda)$. The CDF in R is pgamma(x, r, lambda), and the PDF is dgamma(x, r, lambda). The function qgamma(p, r, lambda) computes the pth quantile of the gamma.

2.5.3 Normal Distribution

The PDF for a normal random variable with mean $\mathbb{E}X = \mu$ and variance $\mathbb{V}arX = \sigma^2$ is

$$f_X(x) = \frac{1}{\sqrt{2\pi\sigma^2}} e^{-\frac{1}{2\sigma^2}(x-\mu)^2}, \quad \infty < x < \infty.$$

The distribution function is computed using integral approximation because no closed form exists for the anti-derivative; this is generally not a problem for practitioners because most software packages will compute interval probabilities numerically. For example, in R, pnorm(x, mu, sigma) and dnorm(x, mu, sigma) find the CDF and PDF at x, and qnorm(p, mu, sigma) computes the inverse CDF with quantile probability p. A random variable X with the normal distribution will be denoted $X \sim \mathcal{N}(\mu, \sigma^2)$.

The central limit theorem (CLT) (formulated in a later section of this chapter) elevates the status of the normal distribution above other distributions. Despite its difficult formulation, the normal is one of the most important distributions in all science, and it has a critical role to play in nonparametric statistics. Any linear combination of normal random variables (independent or with simple covariance structures) is also normally distributed. In such sums, then, we need only keep track of the mean and variance, because these two parameters completely characterize the distribution. For example, if $X_1, \ldots, X_n$ are i.i.d. $\mathcal{N}(\mu, \sigma^2)$, then the sample mean $\overline{X} = (X_1 + \cdots + X_n)/n \sim \mathcal{N}(\mu, \sigma^2/n)$ distribution.

2.5.4 Chi-square Distribution

The PDF for an chi-square random variable with the parameter k, called the *degrees of freedom*, is

$$f_X(x) = \frac{2^{-k/2}}{\Gamma(k/2)} x^{k/2-1} e^{-x/2}, -\infty < x < \infty.$$

The chi-square distribution (χ^2) is a special case of the gamma distribution with parameters $r = k/2$ and $\lambda = 1/2$. Its mean and variance are $\mathbb{E}X = \mu = k$ and $\mathbb{V}\text{ar}X = \sigma^2 = 2k$.

If $Z \sim \mathcal{N}(0,1)$, then $Z^2 \sim \chi_1^2$, that is, a chi-square random variable with one degree of freedom. Furthermore, if $U \sim \chi_m^2$ and $V \sim \chi_n^2$ are independent, then $U + V \sim \chi_{m+n}^2$.

From these results, it can be shown that if $X_1, \ldots, X_n \sim \mathcal{N}(\mu, \sigma^2)$ and $\overline{X}$ is the sample mean, then the *sample variance* $S^2 = \sum_i (X_i - \overline{X})^2 / (n - 1)$ is proportional to a chi-square random variable with $n - 1$ degrees of freedom:

$$\frac{(n-1)S^2}{\sigma^2} \sim \chi_{n-1}^2.$$

In R, the CDF and PDF for a χ_k^2 are `pchisq(x,k)` and `dchisq(x,k)`. The *p*th quantile of the χ_k^2 distribution is `qchisq(p,k)`.

2.5.5 (Student) *t*-Distribution

Random variable X has Student's *t*-distribution with k degrees of freedom, $X \sim t_k$, if its PDF is

$$f_X(x) = \frac{\Gamma\left(\frac{k+1}{2}\right)}{\sqrt{k\pi}\,\Gamma(k/2)} \left(1 + \frac{x^2}{k}\right)^{-\frac{k+1}{2}}, -\infty < x < \infty.$$

The *t*-distribution[2] is similar in shape to the standard normal distribution except for the fatter tails. If $X \sim t_k$, $\mathbb{E}X = 0$, $k > 1$; and $\mathbb{V}\text{ar}X = k/(k-2)$, $k > 2$. For $k = 1$, the *t*-distribution coincides with the Cauchy distribution.

The *t*-distribution has an important role to play in statistical inference. With a set of i.i.d. $X_1, \ldots, X_n \sim \mathcal{N}(\mu, \sigma^2)$, we can standardize the sample mean using the simple transformation of $Z = (\overline{X} - \mu)/\sigma_{\overline{X}} = \sqrt{n}(\overline{X} - \mu)/\sigma$. However, if the variance is unknown, by using the same transformation except substituting the sample standard deviation S for σ, we arrive at a *t*-distribution with $n - 1$ degrees of freedom:

$$T = \frac{(\overline{X} - \mu)}{S/\sqrt{n}} \sim t_{n-1}.$$

---

2 William Sealy Gosset derived the *t*-distribution in 1908 under the pen name "Student" (Gosset, 1908). He was a researcher for Guinness Brewery, which forbids any of their workers to publish "company secrets."

More technically, if $Z \sim \mathcal{N}(0,1)$ and $Y \sim \chi_k^2$ are independent, then $T = Z/\sqrt{Y/k} \sim t_k$. In R, the CDF at x for a t-distribution with k degrees of freedom is calculated as pt (x, k), and the PDF is computed as dt (x, k). The pth percentile is computed with qt (p, k).

2.5.6 Beta Distribution

The density function for a beta random variable is

$$f_X(x) = \frac{1}{B(a, b)} x^{a-1}(1-x)^{b-1}, \quad 0 < x < 1, \ a > 0, \ b > 0,$$

and B is the beta function. Because X is defined only in $(0,1)$, the beta distribution is useful in describing uncertainty or randomness in proportions or probabilities. A beta-distributed random variable is denoted by $X \sim Be(a, b)$. The *uniform distribution* on $(0,1)$, denoted as $\mathcal{U}(0,1)$, serves as a special case with $(a, b) = (1,1)$. The beta distribution has moments

$$\mathbb{E}X^k = \frac{\Gamma(a+k)\Gamma(a+b)}{\Gamma(a)\Gamma(a+b+k)} = \frac{a(a+1)\cdots(a+k-1)}{(a+b)(a+b+1)\cdots(a+b+k-1)}$$

so that $\mathbb{E}(X) = a/(a+b)$ and $\mathbb{V}\mathrm{ar}X = ab/[(a+b)^2(a+b+1)]$.

In R, the CDF for a beta random variable (at $x \in (0,1)$) is computed with pbeta (x, a, b), and the PDF is computed with dbeta (x, a, b). The pth percentile is computed qbeta (p, a, b). If the mean μ and variance σ^2 for a beta random variable are known, then the basic parameters (a, b) can be determined as

$$a = \mu \left(\frac{\mu(1-\mu)}{\sigma^2} - 1 \right), \quad \text{and} \quad b = (1-\mu) \left(\frac{\mu(1-\mu)}{\sigma^2} - 1 \right). \tag{2.3}$$

2.5.7 Double-Exponential Distribution

Random variable X has double-exponential $\mathcal{DE}(\mu, \lambda)$ distribution if its density is given by

$$f_X(x) = \frac{\lambda}{2} e^{-\lambda|x-\mu|}, \ -\infty < x < \infty, \ \lambda > 0.$$

The expectation of X is $\mathbb{E}X = \mu$, and the variance is $\mathbb{V}\mathrm{ar}X = 2/\lambda^2$. The moment-generating function for the double-exponential distribution is

$$m(t) = \frac{\lambda^2 e^{\mu t}}{\lambda^2 - t^2}, \quad |t| < \lambda.$$

Double exponential is also called *Laplace distribution*. If X_1 and X_2 are independent $E(\lambda)$, then $X_1 - X_2$ is distributed as $\mathcal{DE}(0, \lambda)$. Also, if $X \sim DE(0, \lambda)$, then $|X| \sim E(\lambda)$.

2.5.8 Cauchy Distribution

The Cauchy distribution is symmetric and bell-shaped like the normal distribution, but with much heavier tails. For this reason, it is a popular distribution to use in nonparametric procedures to represent non-normality. Because the distribution is so spread out, it has no mean and variance (none of the Cauchy moments exist). Physicists know this as the *Lorentz distribution*. If $X \sim Ca(a, b)$, then X has density

$$f_X(x) = \frac{1}{\pi} \frac{b}{b^2 + (x - a)^2}, \quad -\infty < x < \infty.$$

The moment-generating function for Cauchy distribution does not exist, but its characteristic function is $\mathbb{E}e^{iX} = \exp\{iat - b|t|\}$. The $Ca(0,1)$ coincides with t-distribution with one degree of freedom.

The Cauchy is also related to the normal distribution. If Z_1 and Z_2 are two independent $\mathcal{N}(0,1)$ random variables, then $C = Z_1/Z_2 \sim Ca(0,1)$. Finally, if $C_i \sim Ca(a_i, b_i)$ for $i = 1, \ldots, n$, then $S_n = C_1 + \cdots + C_n$ is distributed Cauchy with parameters $a_S = \sum_i a_i$ and $b_S = \sum_i b_i$.

2.5.9 Inverse Gamma Distribution

Random variable X is said to have an inverse gamma $\mathcal{IG}(r, \lambda)$ distribution with parameters $r > 0$ and $\lambda > 0$ if its density is given by

$$f_X(x) = \frac{\lambda^r}{\Gamma(r)x^{r+1}} e^{-\lambda/x}, \quad x \geq 0.$$

The mean and variance of X are $\mathbb{E}X = \lambda^k/(r - 1)$ and $\mathbb{V}\text{ar}X = \lambda^2/((r - 1)^2(r - 2))$, respectively. If $X \sim \mathcal{G}amma(r, \lambda)$, then its reciprocal X^{-1} is $\mathcal{IG}(r, \lambda)$ distributed.

2.5.10 Dirichlet Distribution

The Dirichlet distribution is a multivariate version of the beta distribution in the same way the multinomial distribution is a multivariate extension of the binomial. A random variable $X = (X_1, \ldots, X_k)$ with a Dirichlet distribution $(X \sim Dir(a_1, \ldots, a_k))$ has PDF

$$f(x_1, \ldots, x_k) = \frac{\Gamma(A)}{\prod_{i=1}^{k} \Gamma(a_i)} \prod_{i=1}^{k} x_i^{a_i - 1},$$

where $A = \sum a_i$ and $x = (x_1, \ldots, x_k) \geq 0$ is defined on the simplex $x_1 + \cdots + x_k = 1$. Then

$$\mathbb{E}(X_i) = \frac{a_i}{A}, \quad \mathbb{V}\text{ar}(X_i) = \frac{a_i(A - a_i)}{A^2(A + 1)}, \quad \text{and} \quad \mathbb{C}\text{ov}(X_i, X_j) = -\frac{a_i a_j}{A^2(A + 1)}.$$

The Dirichlet random variable can be generated from gamma random variables $Y_1, \ldots, Y_k \sim Gamma(a, b)$ as $X_i = Y_i/S_Y$, $i = 1, \ldots, k$ where $S_Y = \sum_i Y_i$. Obviously, the marginal distribution of a component X_i is $Be(a_i, A - a_i)$.

2.5.11 F Distribution

Random variable X has F distribution with m and n degrees of freedom, denoted as $F_{m,n}$, if its density is given by

$$f_X(x) = \frac{m^{m/2}n^{n/2}\, x^{m/2-1}}{B(m/2, n/2)\,(n + mx)^{-(m+n)/2}}, \quad x > 0.$$

The CDF of the F distribution has no closed form, but it can be expressed in terms of an incomplete beta function.

The mean is given by $\mathbb{E}X = n/(n - 2), n > 2$, and the variance by $\mathbb{V}\text{ar}X = [2n^2(m + n - 2)]/[m(n - 2)^2(n - 4)], n > 4$. If $X \sim \chi_m^2$ and $Y \sim \chi_n^2$ are independent, then $(X/m)/(Y/n) \sim F_{m,n}$. If $X \sim Be(a, b)$, then $bX/[a(1 - X)] \sim F_{2a,2b}$. Also, if $X \sim F_{m,n}$ then $mX/(n + mX) \sim Be(m/2, n/2)$.

The F-distribution is one of the most important distributions for statistical inference; in introductory statistical courses, test of equality of variances and ANOVA are based on the F-distribution. For example, if S_1^2 and S_2^2 are sample variances of two independent normal samples with variances σ_1^2 and σ_2^2 and sizes m and n, respectively, the ratio $(S_1^2/\sigma_1^2)/(S_2^2/\sigma_2^2)$ is distributed as $F_{m-1,n-1}$.

In R, the CDF at x for a F distribution with m, n degrees of freedom is calculated as pf(x,m,n), and the PDF is computed as df(x,m,n). The pth percentile is computed with qf(p,m,n).

2.5.12 Pareto Distribution

> The greater part of human actions have their origin not in logical reasoning but in sentiment.
>
> Vilfredo Pareto (1848–1923)

The Pareto distribution is named after the Italian economist/sociologist Vilfredo Pareto. Some examples in which the Pareto distribution provides a good-fitting model include wealth distribution, sizes of human settlements, visits to encyclopedia pages, and file size distribution of Internet traffic. Random variable X has a Pareto $Pa(x_0, \alpha)$ distribution with parameters $0 < x_0 < \infty$ and $\alpha > 0$ if its density is given by

$$f(x) = \frac{\alpha}{x_0} \left(\frac{x_0}{x}\right)^{\alpha+1}, \quad x \geq x_0, \ \alpha > 0.$$

The mean and variance of X are $\mathbb{E}X = \alpha x_0/(\alpha - 1)$ and $\mathbb{V}\text{ar}X = \alpha x_0^2/((\alpha - 1)^2(\alpha - 2))$. If $X_1, \ldots, X_n \sim Pa(x_0, \alpha)$, then $Y = 2x_0 \sum \ln(X_i) \sim \chi_{2n}^2$.

2.5.13 Weibull Distribution

Weibull (1939) first published what is now known as the Weibull distribution in his research characterizing material strength. Weibull models continue to serve as baselines distribution for a variety of engineering problems involving material strength, component lifetime, manufacturing rates, and even wind speeds. The random variable $X \sim \mathcal{W}(a, b)$ has a Weibull distribution with shape parameter $(a > 0)$, and scale parameter $(b > 0)$ has density

$$f(x) = \frac{ax^{a-1}}{b^a}\, e^{(x/b)^a}, \quad X > 0.$$

The mean and variance for $X \sim \mathcal{W}(a, b)$ are $\mathbb{E}X = b\Gamma((a+1)/a))$ and $\mathbb{V}\mathrm{ar}X = b^2\left(\Gamma(1 + \frac{2}{1}) - \Gamma(1 + \frac{1}{a})^2\right)$.

2.6 Mixture Distributions

Mixture distributions occur when the population consists of heterogeneous subgroups, each of which is represented by a different probability distribution. If the sub-distributions cannot be identified with the observation, the observer is left with an unsorted mixture. For example, a finite mixture of k distributions has PDF

$$f_X(x) = \sum_{i=1}^{k} p_i f_i(x),$$

where f_i is a density and the weights $(p_i \geq 0, \ i = 1, \ldots, k)$ are such that $\sum_i p_i = 1$. Here, p_i can be interpreted as the probability that an observation will be generated from the subpopulation with PDF f_i.

In addition to applications where different types of random variables are mixed together in the population, mixture distributions can also be used to characterize extra variability (dispersion) in a population. A more general continuous mixture is defined via a *mixing distribution* $g(\theta)$ and the corresponding mixture distribution

$$f_X(x) = \int_\theta f(t; \theta)g(\theta)d\theta.$$

Along with the mixing distribution, $f(t; \theta)$ is called the *kernel distribution*.

Example 2.2 Suppose an observed count is distributed $Bin(n, p)$, and overdispersion is modeled by treating p as a mixing parameter. In this case, the binomial distribution is the kernel of the mixture. If we allow $g_P(p)$ to follow a beta distribution with parameters (a, b), then the resulting mixture distribution

$$p_X(x) = \int_0^1 p_{X|p}(t; p)g_P(p; a, b)\, dp = \binom{n}{x}\frac{B(a + x, n + b - x)}{B(a, b)},$$

is the *beta-binomial* distribution with parameters (n, a, b) and B is the beta function.

Example 2.3   In 1 MB dynamic random access memory (DRAM) chips, the distribution of defect frequency is approximately exponential with $\mu = 0.5/\text{cm}^2$. The 16 MB chip defect frequency, on the other hand, is exponential with $\mu = 0.1/\text{cm}^2$. If a company produces 20 times as many 1 MB chips as they produce 16 MB chips, the overall defect frequency is a mixture of exponentials:

$$f_X(x) = \frac{1}{21}10e^{-10x} + \frac{20}{21}2e^{-2x}.$$

In R, we can produce a graph (see Figure 2.1) of this mixture using the following code:

```
> x <- seq(0,1,by=0.01)
> y <- (10/21)*exp(-x*10)+(40/21)*exp(-x*2)
> z <- 2*exp(-2*x)
> p <- ggplot()+geom_line(aes(x=x,y=y),lwd=0.7)
> p <- p + geom_line(aes(x=x,y=z),lty=2,col="red",lwd=0.7)
> p <- p + xlim(c(0,1))+ylab(c(0,2.5))+xlab("Chip area (cm^2)")
> p <- p + ylab("Probability density")
> print(p)
```

Estimation problems involving mixtures are notoriously difficult, especially if the mixing parameter is unknown. In Section 16.3, the Expectation–Maximization (EM) algorithm is used to aid in statistical estimation.

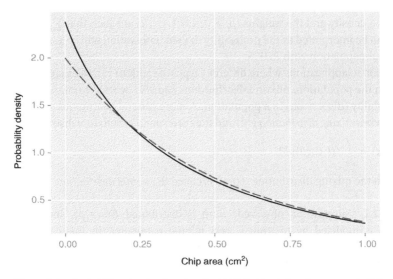

Figure 2.1   Probability density function for DRAM chip defect frequency (*solid*) against exponential PDF (*dashed*).

2.7 Exponential Family of Distributions

We say that y_i is from the exponential family if its distribution is of the form

$$f(y|\theta, \phi) = \exp\left\{ \frac{y\theta - b(\theta)}{\phi} + c(y, \phi) \right\}, \tag{2.4}$$

for some given functions b and c. Parameter θ is called *canonical parameter*, and ϕ dispersion parameter.

Example 2.4   We can write the normal density as

$$\frac{1}{\sqrt{2\pi}\sigma} \exp\left\{ -\frac{(y-\mu)^2}{2\sigma^2} \right\} = \exp\left\{ \frac{y\mu - \mu^2/2}{\sigma^2} - 1/2[y^2/\sigma^2 + \log(2\pi\sigma^2)] \right\}.$$

Thus, it belongs to the exponential family, with $\theta = \mu$, $\phi = \sigma^2$, $b(\theta) = \theta^2/2$, and $c(y, \phi) = -1/2[y^2/\phi + \log(2\pi\phi)]$.

2.8 Stochastic Inequalities

The following four simple inequalities are often used in probability proofs:

1. *Markov inequality*: If $X \geq 0$ and $\mu = \mathbb{E}(X)$ are finite, then

   $$P(X > t) \leq \mu/t.$$

2. *Chebyshev's inequality*: If $\mu = \mathbb{E}(X)$ and $\sigma^2 = \mathbb{V}\text{ar}(X)$, then

   $$P(|X - \mu| \geq t) \leq \frac{\sigma^2}{t^2}.$$

3. *Cauchy–Schwartz inequality*: For random variables X and Y with finite variances,

   $$\mathbb{E}|XY| \leq \sqrt{\mathbb{E}(X^2)\mathbb{E}(Y^2)}.$$

4. *Jensen's inequality*: Let $h(x)$ be a convex function. Then,

   $$h(\mathbb{E}(X)) \leq \mathbb{E}(h(X)).$$

   For example, $h(x) = x^2$ is a convex function, and Jensen's inequality implies $[\mathbb{E}(X)]^2 \leq \mathbb{E}(X^2)$.

Most comparisons between two populations rely on direct inequalities of specific parameters such as the mean or median. We are more limited if no parameters are specified. If $F_X(x)$ and $G_Y(y)$ represent two distributions (for random variables X and Y, respectively), there are several direct inequalities used to describe how one distribution is larger or smaller than another. They are stochastic ordering, failure rate ordering, uniform stochastic ordering, and likelihood ratio ordering.

Stochastic ordering: X is smaller than Y in stochastic order ($X \leq_{ST} Y$) iff $F_X(t) \geq G_Y(t)$ $\forall t$. Some texts use stochastic ordering to describe any general ordering of distributions, and this case is referred to as *ordinary stochastic ordering*.

Failure rate ordering: Suppose F_X and G_Y are differentiable and have PDFs f_X and g_Y, respectively. Let $r_X(t) = f_X(t)/(1 - F_X(t))$, which is called the *failure rate* or *hazard rate* of X. X is smaller than Y in failure rate order ($X \leq_{HR} Y$) iff $r_X(t) \geq r_Y(t)$ $\forall t$.

Uniform stochastic ordering: X is smaller than Y in uniform stochastic order ($X \leq_{US} Y$) if the ratio $(1 - F_X(t))/(1 - G_Y(t))$ is decreasing in t.

Likelihood ratio ordering. Suppose F_X and G_Y are differentiable and have PDFs f_X and g_Y, respectively. X is smaller than Y in likelihood ratio order ($X \leq_{LR} Y$) if the ratio $f_X(t)/g_Y(t)$ is decreasing in t.

It can be shown that uniform stochastic ordering is equivalent to failure rate ordering. Furthermore, there is a natural ordering to the three different inequalities:

$$X \leq_{LR} Y \Rightarrow X \leq_{HR} Y \Rightarrow X \leq_{ST} Y.$$

That is, stochastic ordering is the weakest of the three. Figure 2.2 shows how these orders relate two different beta distributions. The R code below plots the ratios $(1 - F(x))/(1 - G(x))$ and $f(x)/g(x)$ for two beta random variables that have the same mean but different variances. Figure 2.2a shows that they do not have uniform stochastic ordering because $(1 - F(x))/(1 - G(x))$ is not monotone. This also

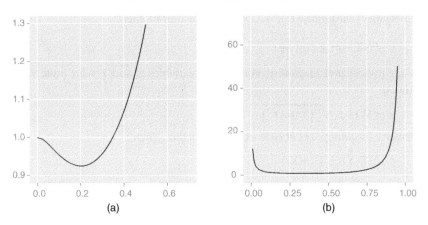

Figure 2.2 Distribution functions F ($Be(2,4)$) and G ($Be(3,6)$): (a) plot of $(1 - F(x))/(1 - G(x))$ and (b) plot of $f(x)/g(x)$.

assures us that the distributions do not have likelihood ratio ordering, which is illustrated in Figure 2.2b:

```
> x1 <- seq(0,0.7,by=0.01)
> r1 <- (1-pbeta(x1,2,4))/(1-pbeta(x1,3,6))
> ggplot()+geom_line(aes(x=x1,y=r1))+xlim(c(0,0.7))+ylim(c(0.9,1.3))
>
> x2 <- seq(0,0.99,by=0.01)
> r2 <- dbeta(x2,2,4)/dbeta(x2,3,6)
> ggplot()+geom_line(aes(x=x2,y=r2))+xlim(c(0,1))+ylim(c(0,70))
```

2.9 Convergence of Random Variables

> The most important questions of life are, for the most part, really only problems of probability.
>
> Pierre Simon Laplace (1749–1827)

Unlike number sequences for which the convergence has a unique definition, sequences of random variables can converge in many different ways. In statistics, convergence refers to an estimator's tendency to look like what it is estimating as the sample size increases.

For general limits, we will say that $g(n)$ is *small "o" of n* and write $g_n = o(n)$ if and only if $g_n/n \to 0$ when $n \to \infty$. Then if $g_n = o(1)$, $g_n \to 0$. The *"big O" notation* concerns equiconvergence. Define $g_n = O(n)$ if there exist constants $0 < C_1 < C_2$ and integer n_0 so that $C_1 < |g_n/n| < C_2 \quad \forall n > n_0$. By examining how an estimator behaves as the sample size grows to infinity (its *asymptotic limit*), we gain a valuable insight as to whether estimation for small- or medium-sized samples make sense. Four basic measure of convergence are as follows.

Convergence in distribution: A sequence of random variables $X_1, \ldots, X_n$ converges in distribution to a random variable X if $P(X_n \leq x) \to P(X \leq x)$. This is also called *weak convergence* and is written $X_n \Longrightarrow X$ or $X_n \to_d X$.

Convergence in probability: A sequence of random variables $X_1, \ldots, X_n$ converges in probability to a random variable X if, for every $\varepsilon > 0$, we have $P(|X_n - X| > \varepsilon) \to 0$ as $n \to \infty$. This is symbolized as $X_n \overset{P}{\to} X$.

Almost-sure convergence: A sequence of random variables $X_1, \ldots, X_n$ converges almost surely (a.s.) to a random variable X (symbolized $X_n \overset{a.s.}{\to} X$) if $P(\lim_{n \to \infty} |X_n - X| = 0) = 1$.

Convergence in mean square: A sequence of random variables $X_1, \ldots, X_n$ converges in mean square to a random variable X if $\mathbb{E}|X_n - X|^2 \to 0$. This is also called *convergence in* $\mathbb{L}_2$ and is written as $X_n \xrightarrow{\mathbb{L}_2} X$.

Convergence in distribution, probability, and almost sure can be ordered; i.e.

$$X_n \xrightarrow{a.s.} X \Rightarrow X_n \xrightarrow{P} X \Rightarrow X_n \Longrightarrow X.$$

The $\mathbb{L}_2$-convergence implies convergence in probability and in distribution, but it is not comparable with the almost-sure convergence.

If $h(x)$ is a continuous mapping, then the convergence of X_n to X guarantees the same kind of convergence of $h(X_n)$ to $h(X)$. For example, if $X_n \xrightarrow{a.s.} X$ and $h(x)$ is continuous, then $h(X_n) \xrightarrow{a.s.} h(X)$, which further implies that $h(X_n) \xrightarrow{P} h(X)$ and $h(X_n) \Longrightarrow h(X)$.

Laws of large numbers (LLN): For i.i.d. random variables $X_1, X_2, \ldots$ with finite expectation $\mathbb{E}X_1 = \mu$, the sample mean converges to μ in the almost-sure sense, that is, $S_n/n \xrightarrow{a.s.} \mu$, for $S_n = X_1 + \cdots + X_n$. This is termed the *strong law of large numbers* (SLLN). Finite variance makes the proof easier, but it is not a necessary condition for the SLLN to hold. If, under more general conditions, $S_n/n = \overline{X}$ converges to μ in probability, we say that the *weak law of large numbers* (WLLN) holds. LLN are important in statistics for investigating the consistency of estimators.

Slutsky's Theorem: Let $\{X_n\}$ and $\{Y_n\}$ be two sequences of random variables on some probability space. If $X_n - Y_n \xrightarrow{P} 0$, and $Y_n \Longrightarrow X$, then $X_n \Longrightarrow X$.

Corollary to Slutsky's Theorem: In some texts, this is sometimes called Slutsky's Theorem. If $X_n \Longrightarrow X$, $Y_n \xrightarrow{P} a$, and $Z_n \xrightarrow{P} b$, then $X_n Y_n + Z_n \Longrightarrow aX + b$.

Delta method: If $\mathbb{E}X_i = \mu$ and $\mathbb{V}\mathrm{ar}X_i = \sigma^2$, and if h is a differentiable function in the neighborhood of μ with $h'(\mu) \neq 0$, then $\sqrt{n}\left(h(X_n) - h(\mu)\right) \Longrightarrow W$, where $W \sim \mathcal{N}(0, [h'(\mu)]^2\sigma^2)$.

Central limit theorem (CLT): Let $X_1, X_2, \ldots$ be i.i.d. random variables with $\mathbb{E}X_1 = \mu$ and $\mathbb{V}\mathrm{ar}X_1 = \sigma^2 < \infty$. Let $S_n = X_1 + \cdots + X_n$. Then

$$\frac{S_n - n\mu}{\sqrt{n\sigma^2}} \Longrightarrow Z,$$

where $Z \sim \mathcal{N}(0,1)$. For example, if $X_1, \ldots, X_n$ is a sample from population with the mean μ and finite variance σ^2, by the CLT, the sample mean $\overline{X} = (X_1 + \cdots + X_n)/n$ is approximately normally distributed, $\overline{X} \overset{appr}{\sim} \mathcal{N}(\mu, \sigma^2/n)$, or equivalently, $(\sqrt{n}(\overline{X} - \mu))/\sigma \overset{appr}{\sim} \mathcal{N}(0,1)$. In many cases, usable approximations are achieved for n as low as 20 or 30.

Example 2.5 We illustrate the CLT with R simulations. A single sample of size $n = 300$ from Poisson $\mathcal{P}(1/2)$ distribution is generated with the first line of code: `sample <- rpois(300, 1/2)`. According to the CLT, the sum $S_{300} = X_1 + \cdots + X_{300}$ should be approximately normal $\mathcal{N}(300 \times 1/2, 300 \times 1/2)$. The histogram of the original sample is depicted in Figure 2.3a.

Next, we generated $N = 5000$ similar samples, each of size $n = 300$ from the same distribution, and for each we found the sum S_{300}:

```
> sample <- rpois(300,0.5)
> p <- ggplot()+geom_histogram(aes(x=sample),col="black",fill="gray",
+ binwidth=1)
> p <- p + xlim(c(0,6)) + xlab("") + ylab("")
> print(p)
>
> S300 <- vector(mode="integer",length=5000)
> for(i in 1:5000){
+     S300[i]<- sum(rpois(300,0.5))
+ }
> p <- ggplot()+geom_histogram(aes(x=S300),col="black",fill="gray",
+ binwidth=2)
> p <- p + xlim(c(100,200)) + ylim(c(0,400)) + xlab("") + ylab("")
> print(p)
```

The histograms of a single sample data generated from $\mathcal{P}(1/2)$ distribution and 5000 realizations of S_{300} are shown in Figure 2.3a,b, respectively. Notice that the histogram of sums is bell-shaped and normal-like, as predicted by the CLT. It is centered near $300 \times 1/2 = 150$.

A more general CLT can be obtained by relaxing the assumption that the random variables are identically distributed. Let $X_1, X_2, \ldots$ be independent random

(a) (b)

Figure 2.3 (a) Histogram of single sample generated from Poisson $\mathcal{P}(1/2)$ distribution. (b) Histogram of S_n calculated from 5000 independent samples of size $n = 300$ generated from Poisson $\mathcal{P}(1/2)$ distribution.

variables with $\mathbb{E}(X_i) = \mu_i$ and $\mathbb{V}ar(X_i) = \sigma_i^2 < \infty$. Assume that the following limit (called *Lindeberg's condition*) is satisfied.
For $\varepsilon > 0$,

$$(D_n^2)^{-1} \sum_{i=1}^{n} \mathbb{E}[(X_i - \mu_i)^2] \mathbf{1}_{\{|X_i - \mu_i| \geq \varepsilon D_n\}} \to 0, \quad \text{as } n \to \infty, \tag{2.5}$$

where

$$D_n^2 = \sum_{i=1}^{n} \sigma_i^2.$$

Extended CLT: Let $X_1, X_2, \ldots$ be independent (not necessarily identically distributed) random variables with $\mathbb{E}X_i = \mu_i$ and $\mathbb{V}arX_i = \sigma_i^2 < \infty$. If condition (2.5) holds, then

$$\frac{S_n - \mathbb{E}S_n}{D_n} \Longrightarrow Z,$$

where $Z \sim \mathcal{N}(0,1)$ and $S_n = X_1 + \cdots + X_n$.

Continuity theorem: Let $F_n(x)$ and $F(x)$ be distribution functions that have characteristic functions $\varphi_n(t)$ and $\varphi(t)$, respectively. If $F_n(x) \Longrightarrow F(x)$, then $\varphi_n(t) \longrightarrow \varphi(t)$. Furthermore, let $F_n(x)$ and $F(x)$ have characteristic functions $\varphi_n(t)$ and $\varphi(t)$, respectively. If $\varphi_n(t) \longrightarrow \varphi(t)$ and $\varphi(t)$ is continuous at 0, then $F_n(x) \Longrightarrow F(x)$.

Example 2.6   Consider the following array of independent random variables

$$
\begin{array}{llll}
X_{11} \\
X_{21} & X_{22} \\
X_{31} & X_{32} & X_{33} \\
\vdots & \vdots & \vdots & \ddots
\end{array}
$$

where $X_{nk} \sim Ber(p_n)$ for $k = 1, \ldots, n$. The X_{nk} have characteristic functions

$$\varphi_{X_{nk}}(t) = p_n e^{it} + q_n$$

where $q_n = 1 - p_n$. Suppose $p_n \to 0$ in such a way that $np_n \to \lambda$, and let $S_n = \sum_{k=1}^{n} X_{nk}$. Then

$$
\begin{aligned}
\varphi_{S_n}(t) &= \prod_{k=1}^{n} \varphi_{X_{nk}}(t) &&= (p_n e^{it} + q_n)^n \\
&= (1 + p_n e^{it} - p_n)^n &&= [1 + p_n(e^{it} - 1)]^n \\
&\approx [1 + \tfrac{\lambda}{n}(e^{it} - 1)]^n &&\to \exp[\lambda(e^{it} - 1)],
\end{aligned}
$$

which is the characteristic function of a Poisson random variable. So, by the continuity theorem, $S_n \Longrightarrow P(\lambda)$.

2.10 Exercises

2.1 For the characteristic function of a random variable X, prove the three following properties:

(i) $\varphi_{aX+b}(t) = e^{ib}\varphi_X(at)$.

(ii) If $X = c$, then $\varphi_X(t) = e^{ict}$.

(iii) If $X_1, X_2, \ldots X_n$ are independent, then $S_n = X_1 + X_2 + \cdots + X_n$ has characteristic function $\varphi_{S_m}(t) = \prod_{i=1}^{n} \varphi_{X_i}(t)$.

2.2 Let $U_1, U_2, \ldots$ be independent uniform $\mathcal{U}(0,1)$ random variables. Let $M_n = \min\{U_1, \ldots, U_n\}$. Prove $nM_n \Longrightarrow X \sim E(1)$, the exponential distribution with rate parameter $\lambda = 1$.

2.3 Let $X_1, X_2, \ldots$ be independent geometric random variables with parameters $p_1, p_2, \ldots$ Prove, if $p_n \to 0$, then $p_n X_n \Longrightarrow E(1)$.

2.4 Use Newton's formula from Section 2.1 to derive the moment-generating function for the binomial distribution:

$$m_X(t) = \mathbb{E}e^{tX} = (1 - p + p\,e^t)^n$$

For $-\infty < t < \infty$.

2.5 Show that for continuous distributions that have continuous density functions, failure rate ordering is equivalent to uniform stochastic ordering. Then show that it is weaker than likelihood ratio ordering and stronger than stochastic ordering.

2.6 Derive the mean and variance for a Poisson distribution using (a) just the PMF and (b) the moment-generating function.

2.7 Show that the Poisson distribution is a limiting form for a binomial model, as given in Eq. (2.1) on page 14.

2.8 Show that, for the exponential distribution, the median is less than 70% of the mean.

2.9 Prove the memoryless property of the exponential distribution: if $X \sim E(\lambda)$, then for $t > x$, $P(X \geq t | X \geq x) = P(X \geq t - x)$.

2.10 Use a Taylor series expansion to show the following:

(i) $e^{-ax} = 1 - ax + (ax)^2/2! - (ax)^3/3! + \cdots$

(ii) $\log(1 + x) = x - x^2/2 + x^3/3 - \cdots$

2.11 Show that if X is stochastically smaller than Y, it follows that $P(X \leq Y) \geq 1/2$.

2.12 Show that failure rate ordering implies uniform stochastic ordering.

2.13 Use R to plot a mixture density of two normal distributions with mean and variance parameters (3,6) and (10,5). Plot using weight function $(p_1, p_2) = (0.5, 0.5)$.

2.14 For the beta-binomial distribution in Example 2.2, derive the mean and variance using Adam's and Eve's rules.

2.15 Write a R function to compute, in table form, the following quantiles for a χ^2 distribution with v degrees of freedom, where v is a function (user) input:

$$\{0.005, 0.01, 0.025, 0.05, 0.10, 0.90, 0.95, 0.975, 0.99, 0.995\}.$$

2.16 Which is the more likely outcome, obtaining at least one six in four rolls of a single (balanced) six-sided die or obtaining at least one "double six" in 24 rolls of a pair of such dice?

2.17 Suppose an urn contains six balls: two red, two white, and two black. If we select three balls at random with replacement, what is the probability of getting one of each color? How does the probability change if we select the balls without replacement?

2.18 Suppose two six-sided dice are rolled, and let X be the (absolute) difference in the outcomes. Find the PMF for X and compute its mean and variance.

2.19 Suppose that X is distributed $Bin(n, p)$. Show that $\mathbb{E}(2^X) = (1 + p)^n$.

2.20 Suppose $\sqrt{n}(X_n - \lambda) \Longrightarrow Z$, where $Z \sim N(0,1)$. Use the delta method to find the asymptotic distribution of $Y_n = \sqrt{T}$.

2.21 **The coupon collector problem.** Suppose that there is an urn of n different coupons that are randomly drawn with replacement. What is the probability that more than m draws are needed to collect all n coupons?

2.22 **The Monty Hall problem.** A television game show called "Let's Make a Deal" aired in the 1960s and 1970s across the United States, hosted by actor, producer, and sportscaster Monte Halparin, better known as Monty Hall.

Steve Selvin (1975) first posed the Monty Hall problem. You are given the choice of three doors. Behind one door is a car; behind the others, goats. You pick a door, say, No. 1, and the host, who knows what is behind the doors, opens another door, say, No. 3, that has a goat. He then says to you, "Do you want to pick door No. 2?" What should you do?

References

Gosset, W. S. (1908), "The Probable Error of a Mean," *Biometrika*, 6, 1–25.

Selvin, S. (1975), "A problem in probability (letter to the editor)," *American Statistician*, 29 (1), 67.

Weibull, W. (1939), "A statistical theory of the strength of materials," *Proceedings of the Royal Swedish Institute for Engineering Research*, No. 149.

3

Statistics Basics

Daddy's rifle in my hand felt reassurin',
he told me "Red means run, son. Numbers add up to nothin'."
But when the first shot hit the dog, I saw it comin'...
Neil Young (from the song Powderfinger)

In this chapter, we review fundamental methods of statistics.

Most students experience basic statistical estimation by learning, assuming the random outcomes are generated by a familiar distribution (such as the normal). In nonparametric statistics, we may have familiar goals of estimating means and variances, but the applied methods rely less on those underlying assumptions.

We emphasize some statistical methods that are important for nonparametric inference. Specifically, tests and confidence intervals for the binomial parameter p are described in detail and serve as building blocks to many nonparametric procedures. The empirical distribution function, a nonparametric estimator for the underlying cumulative distribution, is introduced in the first part of the chapter. We also introduce the likelihood ratio, which is a fundamental statistic for nonparametric inference.

3.1 Estimation

For distributions with unknown parameters (say, θ), we form a point estimate $\hat{\theta}_n$ as a function of the sample $X_1, \ldots, X_n$. Because $\hat{\theta}_n$ is a function of random variables, it has a distribution itself, called the *sampling distribution*. If we sample randomly from the same population, then the sample is said to be independently and identically distributed, or i.i.d.

Nonparametric Statistics with Applications to Science and Engineering with R, Second Edition.
Paul Kvam, Brani Vidakovic, and Seong-joon Kim.
© 2023 John Wiley & Sons, Inc. Published 2023 by John Wiley & Sons, Inc.

An *unbiased estimator* is a statistic $\hat{\theta}_n = \hat{\theta}_n(X_1, \ldots, X_n)$ whose expected value is the parameter it is meant to estimate; i.e. $\mathbb{E}(\hat{\theta}_n) = \theta$. An estimator is weakly *consistent* if, for any $\varepsilon > 0$, $P(|\hat{\theta}_n - \theta| > \varepsilon) \to 0$ as $n \to \infty$ (i.e. $\hat{\theta}_n$ converges to θ in probability). In compact notation, $\hat{\theta}_n \overset{P}{\longrightarrow} \theta$. A parameter θ is said to be *estimable* if there exists an estimator $\hat{\theta}(X_1, \ldots, X_n)$ that is unbiased for θ.

Unbiasedness and consistency are desirable qualities in an estimator, but there are other ways to judge an estimate's efficacy. To compare estimators, one might seek the one with smaller mean squared error (MSE), defined as

$$\text{MSE}(\hat{\theta}_n) = \mathbb{E}(\hat{\theta}_n - \theta)^2 = \mathbb{V}\text{ar}(\hat{\theta}_n) + [\text{Bias}(\hat{\theta}_n)]^2,$$

where $\text{Bias}(\hat{\theta}_n) = \mathbb{E}(\hat{\theta}_n - \theta)$. If the bias and variance of the estimator have limit 0 as $n \to \infty$ (or, equivalently, $\text{MSE}(\hat{\theta}_n) \to 0$), the estimator is consistent. An estimator is defined as *strongly* consistent if, as $n \to \infty$, $\hat{\theta}_n \overset{\text{a.s.}}{\longrightarrow} \theta$.

Example 3.1   Suppose $X \sim Bin(n, p)$. If p is an unknown parameter, $\hat{p} = X/n$ is unbiased and strongly consistent for p. This is because the strong law of large numbers (SLLN) holds for i.i.d. $Ber(p)$ random variables and X coincides with S_n for the Bernoulli case; see laws of large numbers on p. 28.

3.2 Empirical Distribution Function

Let $X_1, X_2, \ldots, X_n$ be a sample from a population with continuous cumulative distribution function (CDF) F. An *empirical (cumulative) distribution function* (EDF) based on a random sample is defined as

$$F_n(x) = \frac{1}{n} \sum_{i=1}^{n} \mathbf{1}(X_i \leq x), \tag{3.1}$$

where $\mathbf{1}(\rho)$ is called the *indicator function* of ρ and is equal to 1 if the relation ρ is true and 0 if it is false. In terms of ordered observations $X_{1:n} \leq X_{2:n} \leq \cdots \leq X_{n:n}$, the empirical distribution function can be expressed as

$$F_n(x) = \begin{cases} 0, & \text{if } x < X_{1:n}, \\ k/n, & \text{if } X_{k:n} \leq x < X_{k+1:n}, \\ 1, & \text{if } x \geq X_{n:n}. \end{cases}$$

We can treat the empirical distribution function as a random variable with a sampling distribution, because it is a function of the sample. Depending on the argument x, it equals one of $n + 1$ discrete values, $\{0/n, 1/n, \ldots, (n-1)/n, 1\}$. It is easy to see that, for any fixed x, $nF_n(x) \sim Bin(n, F(x))$, where $F(x)$ is the true CDF of the sample items.

Indeed, for $F_n(x)$ to take value k/n, $k = 0,1,\ldots,n$, k observations from $X_1,\ldots,X_n$ should be less than or equal to x, and $n - k$ observations larger than x. The probability of an observation being less than or equal to x is $F(x)$. Also, the k observations less than or equal to x can be selected from the sample in $\binom{n}{k}$ different ways. Thus,

$$P\left(F_n(x) = \frac{k}{n}\right) = \binom{n}{k}(F(x))^k(1 - F(x))^{n-k}, \quad k = 0,1,\ldots,n.$$

From this it follows that $\mathbb{E}F_n(x) = F(x)$ and $\mathbb{V}\mathrm{ar}F_n(x) = F(x)(1 - F(x))/n$.

A simple graph of the EDF is available in R the following codes. For example, the code below creates Figure 3.1 that shows how the EDF becomes more refined as the sample size increases:

```
> y1 <- rnorm(20)
> y2 <- rnorm(200)
> x <- seq(-3,3,0.01)
> y <- pnorm(x,0,1)
> p <- ggplot()+geom_line(aes(x=x,y=y),lwd=0.8)
> p <- p + geom_step(aes(x=sort(y1),y=seq(0,1,length=20)))
> p <- p + geom_step(aes(x=sort(y2),y=seq(0,1,length=200)))
> print(p)
```

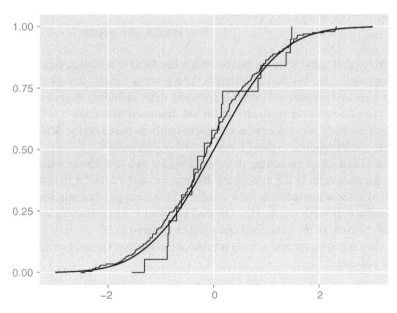

Figure 3.1 Empirical distribution function based on normal samples (sizes 20 and 200) plotted along with the true CDF.

3.2.1 Convergence for EDF

The MSE is defined for F_n as $\mathbb{E}(F_n(x) - F(x))^2$. Because $F_n(x)$ is unbiased for $F(x)$, the MSE reduces to $\mathbb{V}\mathrm{ar}F_n(x) = F(x)(1 - F(x))/n$, and as $n \to \infty$, $\mathrm{MSE}(F_n(x)) \to 0$, so that $F_n(x) \overset{P}{\longrightarrow} F(x)$.

There are a number of convergence properties for F_n that are of limited use in this book and will not be discussed. However, one fundamental limit theorem in probability theory that is the Glivenko–Cantelli theorem is worthy of mention.

Theorem 3.1 *(Glivenko–Cantelli)* *If $F_n(x)$ is the empirical distribution function based on an i.i.d. sample $X_1, \dots, X_n$ generated from $F(x)$,*

$$\sup_x |F_n(x) - F(x)| \overset{\text{a.s.}}{\longrightarrow} 0.$$

3.3 Statistical Tests

> I shall not require of a scientific system that it shall be capable of being singled out, once and for all, in a positive sense; but I shall require that its logical form shall be such that it can be singled out, by means of empirical tests, in a negative sense: it must be possible for an empirical scientific system to be refuted by experience.
>
> Karl Popper, Philosopher (1902–1994)

Uncertainty associated with the estimator is a key focus of statistics, especially *tests of hypothesis* and *confidence intervals*. There are a variety of methods to construct tests and confidence intervals from the data, including Bayesian (see Chapter 4) and frequentist methods, which are discussed in Section 3.5. Of the two general methods adopted in research today, methods based on the *likelihood ratio* are generally superior to those based on *Fisher information*.

In a traditional setup for testing data, we consider two hypotheses regarding an unknown parameter in the underlying distribution of the data. Experimenters usually plan to show new or alternative results, which are typically conjectured in the *alternative hypothesis* (H_1 or H_a). The *null hypothesis*, designated H_0, usually consists of the parts of the parameter space not considered in H_1.

When a test is conducted and a claim is made about the hypotheses, two distinct errors are possible:

Type-I error: The type-I error is the action of rejecting H_0 when H_0 was actually true. The probability of such error is usually labeled by α and referred to as *significance level* of the test.

Type-II error: The type-II error is an action of failing to reject H_0 when H_1 was actually true. The probability of the type-II error is denoted by β. *Power* is defined as $1 - \beta$. In simple terms, the power is propensity of a test to reject wrong alternative hypothesis.

3.3.1 Test Properties

> I contend that the general acceptance of statistical hypothesis testing is one of the most unfortunate aspects of 20th century applied science
>
> Mark Nester (1996)

A test is *unbiased* if the power is always as high or higher in the region of H_1 than anywhere in H_0. A test is *consistent* if, over all of H_1, $\beta \to 0$ as the sample sizes go to infinity.

Suppose we have a hypothesis test of $H_0 : \theta = \theta_0$ versus $H_1 : \theta \neq \theta_0$. The *Wald* test of hypothesis is based on using a normal approximation for the test statistic. If we estimate the variance of the estimator θ_n by plugging in θ_n for θ in the variance term $\sigma_{\theta_n}^2$ (denote this $\hat{\sigma}_{\theta_n}^2$), we have the z-test statistic

$$z_0 = \frac{\theta_n - \theta_0}{\hat{\sigma}_{\theta_n}}.$$

The critical region (or rejection region) for the test is determined by the quantiles z_q of the normal distribution, where q is set to match the type-I error.

p-Values: The p-value is a popular but controversial statistic for describing the significance of a hypothesis given the observed data. Technically, it is the probability of observing a result as "rejectable" (according to H_0) as the observed statistic that actually occurred but from a new sample. So a p-value of 0.02 means that if H_0 is true, we would expect to see results more reflective of that hypothesis 98% of the time in repeated experiments. Note that if the p-value is less than the set α level of significance for the test, the null hypothesis should be rejected (and otherwise should not be rejected).

In the construct of classical hypothesis testing, the p-value has potential to be misleading with large samples. Consider an example in which $H_0 : \mu = 20.3$ versus $H_1 : \mu \neq 20.3$. As far as the experimenter is concerned, the null hypothesis might be conjectured only to three significant digits. However, if the sample is large enough, $\bar{x} = 20.30001$ will eventually be rejected as being too far away from H_0 (granted, the sample size will have to be *awfully* large, but you get our point?). This problem will be revisited when we learn about goodness-of-fit tests for distributions.

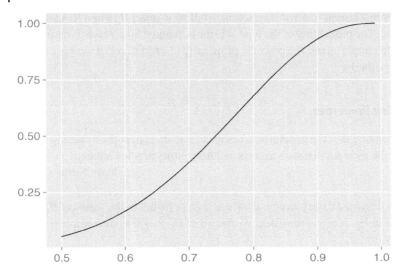

Figure 3.2 Graph of statistical test power for binomial test for specific alternative $H_1 : p = p_1$. Values of p_1 are given on the horizontal axis.

Binomial distribution: For binomial data, consider the test of hypothesis

$$H_0 : p \le p_0 \quad \text{versus} \quad H_1 : p > p_0.$$

If we fix the type-I error to α, we would have a critical region (or *rejection region*) of $\{x : x > x_0\}$, where x_0 is chosen so that $\alpha = P(X > x_0 \mid p = p_0)$. For instance, if $n = 10$, an $\alpha = 0.0547$ level test for $H_0 : p \le 0.5$ versus $H_1 : p > 0.5$ is to reject H_0 if $X \ge 8$. The test's power is plotted in Figure 3.2 based on the following R code. The figure illustrates how our chance at rejecting the null hypothesis in favor of specific alternative $H_1 : p = p_1$ increases as p_1 increases past 0.5:

```
> p1 <- seq(0,5,0.99,0.01)
> pow <- 1-pbinom(7,10,p1)
> ggplot()+geom_line(aes(x=p1,y=pow))
```

Example 3.2 A semiconductor manufacturer produces an unknown proportion p of defective integrative circuit (IC) chips, so that chip *yield* is defined as $1 - p$. The manufacturer's reliability target is 0.9. With a sample of 25 randomly selected microchips, the Wald test will reject $H_0 : p \le 0.10$ in favor of $H_1 : p > 0.10$ if

$$\frac{\hat{p} - 0.1}{\sqrt{(0.1)(0.9)/100}} > z_\alpha$$

or, for the case $\alpha = 0.05$, if the number of defective chips $X > 3.733$.

3.4   Confidence Intervals

A $1 - \alpha$ level *confidence interval* is a statistic, in the form of a region or interval, which contains an unknown parameter θ with probability $1 - \alpha$. For communicating uncertainty in layman's terms, confidence intervals are typically more suitable than tests of hypothesis, as the uncertainty is illustrated by the length of the interval constructed, along with the adjoining confidence statement.

A two-sided confidence interval has the form $(L(X), U(X))$, where X is the observed outcome, and $P(L(X) \leq \theta \leq U(X)) = 1 - \alpha$. These are the most commonly used intervals, but there are cases in which one-sided intervals are more appropriate. If one is concerned with how large a parameter might be, we would construct an *upper bound* $U(X)$ such that $P(\theta \leq U(X)) = 1 - \alpha$. If small values of the parameter are of concern to the experimenter, a *lower bound* $L(X)$ can be used where $P(L(X) \leq \theta) = 1 - \alpha$.

Example 3.3   t-distribution. Suppose that $X_1, \ldots, X_n \sim \mathcal{N}(\mu, \sigma^2)$ are independent and both μ and σ^2 are unknown. Since $T = \sqrt{n}(\overline{X} - \mu)/S$ has a t-distribution with $n - 1$ degrees of freedom, if $t_{n-1,\alpha/2}$ and $t_{n-1,1-\alpha/2}$ represent the $(\alpha/2, 1 - \alpha/2)$ quantiles of the t_{n-1} distribution, then $P(t_{n-1,\alpha/2} \leq T \leq t_{n-1,1-\alpha/2}) = 1 - \alpha$ leads to the $1 - \alpha$ confidence interval for μ, based on estimating σ with the sample standard deviation:

$$\overline{X} - t_{n-1,\alpha/2}\frac{S}{\sqrt{n}}, \overline{X} + t_{n-1,1-\alpha/2}\frac{S}{\sqrt{n}}.$$

Example 3.4   Binomial Distribution. To construct a two-sided $1 - \alpha$ confidence interval for p, we solve the equation

$$\sum_{k=0}^{x} \binom{n}{k} p^k (1 - p)^{n-k} = \alpha/2$$

for p to obtain the upper $1 - \alpha$ limit for p and solve

$$\sum_{k=x}^{n} \binom{n}{k} p^k (1 - p)^{n-k} = \alpha/2$$

to obtain the lower limit. One-sided $1 - \alpha$ intervals can be constructed by solving just one of the equations using α in place of $\alpha/2$. Use R function

```
binom.test(x,n,conf.level=1 - α).
```

This is named the Clopper–Pearson interval (Clopper and Pearson, 1934), where Pearson refers to Egon Pearson, Karl Pearson's son.

This exact interval is typically *conservative*, but not conservative like a GOP senator from Mississippi. In this case, conservative means the *coverage probability* of the confidence interval is at least as high as the *nominal* coverage probability $1 - \alpha$ and can be much higher. In general, "conservative" is synonymous with risk averse. The nominal and actual coverage probabilities disagree frequently with discrete data, where an interval with the exact coverage probability of $1 - \alpha$ may not exist. While the guaranteed confidence in a conservative interval is reassuring, it is potentially inefficient and misleading.

Example 3.5 If $n = 10$, $x = 3$, then $\hat{p} = 0.3$. Moreover, a 95% (two-sided) confidence interval for p is computed by finding the upper limit p_1 for which $F_X(3; p_1) = 0.025$ and lower limit p_2 for which $1 - F_X(2; p_2) = 0.025$, where F_X is the CDF for the binomial distribution with $n = 10$. The resulting interval $(0.06774, 0.65245)$ is not symmetric in p.

3.4.1 Intervals Based on Normal Approximation

The interval in Example 3.5 is "exact," in contrast to more commonly used intervals based on a normal approximation. Recall that $\bar{x} \pm z_{\alpha/2}\sigma/\sqrt{n}$ serves as a $1 - \alpha$ level confidence interval for μ with data generated from a normal distribution. Here z_α represents the α quantile of the standard normal distribution. With the normal approximation (see *Central Limit Theorem* in Chapter 2), $\hat{p}$ has an approximate normal distribution if n is large, so if we estimate the variance of $\hat{p}$, $\sigma_{\hat{p}}^2 = p(1 - p)/n$, with

$$\hat{\sigma}_{\hat{p}}^2 = \hat{p}(1 - \hat{p})/n,$$

an approximate $1 - \alpha$ interval for p is

$$\hat{p} \pm z_{\alpha/2}\sqrt{x(n - x)/n^3}.$$

This is called the Wald interval because it is based on inverting the (Wald) z-test statistic for $H_0 : p = p_0$ versus $H_1 : p \neq p_0$. Agresti (1998) points out that both the exact and Wald intervals perform poorly compared with the *score interval* that is based on the Wald z-test of hypothesis, but instead of using $\hat{p}$ in the error term, it uses the value p_0 for which $(\hat{p} - p_0)/\sqrt{p_0(1 - p_0)/n} = \pm z_{\alpha/2}$. The solution, first stated by Wilson (1927), is the interval

$$\frac{\hat{p} + \frac{z_{\alpha/2}^2}{2n} \pm z_{\alpha/2}\sqrt{\frac{\hat{p}(1-\hat{p})+z_{\alpha/2}^2/4n}{n}}}{1 + z_{\alpha/2}^2/n}.$$

This actually serves as an example of *shrinkage*, which is a statistical phenomenon where better estimators are sometimes produced by "shrinking" or adjusting treatment means toward an overall (sample) mean. In this case, one can show that the middle of the confidence interval shrinks a little from $\hat{p}$ toward 1/2, although the shrinking becomes negligible as n gets larger. Use R function

```
prop.test(x, n, conf.level=1-alpha, correct=FALSE)
```

to generate a two-sided Wilson's confidence interval. Alternatively, `binom.confint` function in `binom` package and `scoreci` function in `PropCIs` package provide the same result.

Example 3.6 In the previous example, with $n = 10$ and $x = 3$, the exact two-sided 95% confidence interval (0.06774, 0.65245) has the length of 0.5847, so the inference is rather vague (Figure 3.3). Using the normal approximation, the interval computes to (0.0160, 0.5840) and has the length of 0.5680. The shrinkage interval is (0.1078, 0.6032) and has the length of 0.4954.

Is the shrinkage interval accurate? In general, the exact interval will have coverage probability exceeding $1 - \alpha$, and the Wald interval sometimes has coverage probability below $1 - \alpha$. Overall, the shrinkage interval has coverage probability closer to $1 - \alpha$. In the case of the binomial, the word "exact" does not imply a confidence interval is better:

```
> x <- seq(0,10)
> y <- dbinom(x,10,0.3)
```

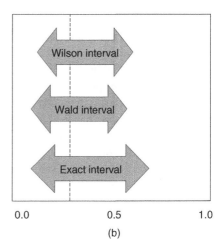

Figure 3.3   (a) The binomial $Bin(10, 0.3)$ PMF. (b) 95% confidence intervals based on exact, Wald, and Wilson methods.

```
> names(y) <- x
> ggplot()+geom_bar(aes(x=x,y=y),stat="identity")
```

Interval estimation is a helpful way to quantify and explain uncertainty with regard to statistical inference, and the confidence interval is the most practiced method of the bunch. In Chapter 5, we will learn about tolerance intervals, which are constructed to bound a proportion of the population. Prediction intervals, like confidence intervals, account for both the uncertainty in the estimation but also allow for uncertainty in the future observations, which is why confidence intervals are always narrower than prediction intervals (Chapter 12).

3.5 Likelihood

Sir Ronald Fisher, perhaps the greatest innovator of statistical methodology, developed the concepts of likelihood and sufficiency for statistical inference. With a set of random variables $X_1, \ldots, X_n$, suppose the joint distribution is a function of an unknown parameter θ: $f_n(x_1, \ldots, x_n; \theta)$. The *likelihood function* pertaining to the observed data $L(\theta) = f_n(x_1, \ldots, x_n; \theta)$ is associated with the probability of observing the data at each possible value θ of an unknown parameter. In case the sample consists of i.i.d. measurements with density function $f(x; \theta)$, the likelihood simplifies to

$$L(\theta) = \prod_{i=1}^{n} f(x_i; \theta).$$

The likelihood function has the same numerical value as the probability density function (PDF) of a random variable, but it is regarded as a function of the parameters θ and treats the data as fixed. The PDF, on the other hand, treats the parameters as fixed and is a function of the data points.

The *likelihood principle* states that, after x is observed, all relevant experimental information is contained in the likelihood function for the observed x and that θ_1 supports the data more than θ_2 if $L(\theta_1) \geq L(\theta_2)$. The *maximum likelihood estimate* (MLE) of θ is that value of θ in the parameter space maximizing $L(\theta)$. Although the MLE is based strongly on the parametric assumptions of the underlying density function $f(x; \theta)$, there is a sensible nonparametric version of the likelihood introduced in Chapter 10.

MLEs are known to have optimal performance if the sample size is sufficient and the densities are "regular"; for one, the support of $f(x; \theta)$ should not depend on θ. For example, if $\hat{\theta}$ is the MLE, then

$$\sqrt{n}(\hat{\theta} - \theta) \implies \mathcal{N}(0, i^{-1}(\theta)),$$

where $i(\theta) = \mathbb{E}([\partial \log f / \partial \theta]^2)$ is the *Fisher Information* of θ. The regularity conditions also demand that $i(\theta) \geq 0$ is bounded and $\int f(x; \theta) \, dx$ is thrice differentiable. In these cases, the Fisher information can be computed using the alternative formula

$$i(\theta) = -\mathbb{E}([\partial^2 \log f / \partial^2 \theta]).$$

For a comprehensive discussion about regularity conditions for maximum likelihood, sufficiency, and completeness, see Lehmann and Casella (1998). The optimality of the MLE is guaranteed by the following result:

Cramer–Rao lower bound: From an i.i.d. sample $X_1, \dots, X_n$ where X_i has density function $f_X(x)$, let $\hat{\theta}_n$ be an unbiased estimator for θ. Then

$$\mathbb{V}\mathrm{ar}(\hat{\theta}_n) \geq (i(\theta)n)^{-1}.$$

Example 3.7 If $X \sim \mathcal{N}(\mu, \sigma^2)$, then $\ln(f(x)) = -\ln(\sqrt{2\pi}\sigma) - (x - \mu)^2 / (2\sigma^2)$, and its derivative, with respect to μ, is $(x - \mu)/\sigma^2$. So, for a normal random variable, the Fisher information of X is computed as

$$i(\mu) = -\mathbb{E}([\partial^2 \log f / \partial^2 \mu]) = -\mathbb{E}(-1/\sigma^2) = 1/\sigma^2.$$

From the Cramer–Rao lower bound, we know that if $X_1, \dots, X_n \sim \mathcal{N}(\mu, \sigma^2)$, then $\mathbb{V}\mathrm{ar}(\hat{\mu}) \geq (i(\mu)n)^{-1} = \sigma^2/n$.

Delta method for MLE: The *invariance property* of MLEs states that if g is a one-to-one function of the parameter θ, then the MLE of $g(\theta)$ is $g(\hat{\theta})$. Assuming the first derivative of g (denoted g') exists, then

$$\sqrt{n}(g(\hat{\theta}) - g(\theta)) \implies \mathcal{N}(0, g'(\theta)^2 / i(\theta)).$$

Example 3.8 After waiting for the kth success in a repeated process with constant probabilities of success and failure, we recognize that the probability distribution of X = number of failures is *negative binomial*. To estimate the unknown success probability p, we can maximize

$$L(p) = p_X(x; p) \propto p^k (1 - p)^x, \quad 0 < p < 1.$$

Note the combinatoric part of p_X was left off the likelihood function because it plays no part in maximizing L. From $\log L(p) = k \log(p) + x \log(1 - p)$, $\partial L / \partial p = 0$ leads to $\hat{p} = k/(k + x)$, and $i(p) = k/(p^2(1 - p))$. Thus, for large n, $\hat{p}$ has an approximate normal distribution, i.e.

$$\hat{p} \overset{\mathrm{appr}}{\sim} \mathcal{N}(p, p^2(1 - p)/k).$$

Example 3.9 In Example 3.8, suppose that $k = 1$, so X has a geometric $\mathcal{G}(p)$ distribution. If we are interested in estimating $\theta =$ probability that m or more failures occur before a success occurs, then

$$\theta = g(p) = \sum_{j=m}^{\infty} p(1-p)^j = (1-p)^m,$$

and from the invariance principle, the MLE of θ is $\hat{\theta} = (1 - \hat{p})^m$. Furthermore,

$$\sqrt{n}(\hat{\theta} - \theta) \Longrightarrow \mathcal{N}(0, \sigma_\theta^2),$$

where $\sigma_\theta^2 = g'(p)^2 / i(p) = p^2(1-p)^{2m-1} m^2$.

3.5.1 Likelihood Ratio

The likelihood ratio function is defined for a parameter set θ as

$$R(\theta_0) = \frac{L(\theta_0)}{\sup_\theta L(\theta)}, \tag{3.2}$$

where $\sup_\theta L(\theta) = L(\hat{\theta})$ and $\hat{\theta}$ is the MLE of θ. Wilks (1938) showed that under the previously mentioned regularity conditions, $-2 \log R(\theta)$ is approximately distributed χ^2 with k degrees of freedom (when θ is a correctly specified vector of length k).

The likelihood ratio is useful in setting up tests and intervals via the parameter set defined by $C(\theta) = \{\theta : R(\theta) \geq r_0\}$ where r_0 is determined so that if $\theta = \theta_0$, $P(\hat{\theta} \in C) = 1 - \alpha$. Given the chi-square result above, we have the following $1 - \alpha$ confidence interval for θ based on the likelihood ratio:

$$\left\{ \theta : -2 \log R \leq \chi_p^2(1-\alpha) \right\}, \tag{3.3}$$

where $\chi_p^2(1 - \alpha)$ is the $1 - \alpha$ quantile of the χ_p^2 distribution. Along with the nonparametric MLE discussed in Chapter 10, there is also a nonparametric version of the likelihood ratio, called the *empirical likelihood* that we will introduce also in Chapter 10.

Example 3.10 If $X_1, \ldots, X_n \sim \mathcal{N}(\mu, 1)$, then

$$L(\mu) = \prod_{i=1}^{n} (2\pi)^{-n/2} e^{-\frac{1}{2}\sum_{i=1}^{n}(x_i - \mu)^2}.$$

Because $\hat{\mu} = \bar{x}$ is the MLE, $R(\mu) = L(\mu)/L(\bar{x})$, and the interval defined in (3.3) simplifies to

$$\left\{ \mu : \sum_{i=1}^{n} (x_i - \mu)^2 - \sum_{i=1}^{n} (x_i - \bar{x})^2 \leq \chi_1^2(1-\alpha) \right\}.$$

By expanding the sums of squares, one can show (see Exercise 3.5) that this interval is equivalent to the Fisher interval $\bar{x} \pm z_{\alpha/2}/\sqrt{n}$.

3.5.2 Efficiency

Let ϕ_1 and ϕ_2 be two different statistical tests (i.e. specified critical regions) based on the same underlying hypotheses. Let n_1 be the sample size for ϕ_1. Let n_2 be the sample size needed for ϕ_2 to make the type-I and type-II errors identical. The *relative efficiency* of ϕ_1 with respect to ϕ_2 is $RE = n_2/n_1$. The *asymptotic relative efficiency ARE* is the limiting value of RE as $n_1 \to \infty$. Nonparametric procedures are often compared with their parametric counterparts by computing the *ARE* for the two tests.

If a test or confidence interval is based on assumptions but tends to come up with valid answers even when some of the assumptions are not, the method is called *robust*. Most nonparametric procedures are more robust than their parametric counterparts but also less efficient. Robust methods are discussed in more detail in Chapter 12.

3.5.3 Exponential Family of Distributions

Let $f(y|\theta)$ be a member of the *exponential family* with natural parameter θ. Assume that θ is univariate. Then the log likelihood $\ell(\theta) = \sum_{i=1}^{n} \log(f(y_i|\theta)) = \sum_{i=1}^{n} \ell_i(\theta)$, where $\ell_i = \log f(y_i|\theta)$. The MLE for θ is solution of the equation

$$\frac{\partial \ell}{\partial \theta} = 0.$$

The following two properties (see Exercise 3.8) hold:

$$\text{(i) } \mathbb{E}\left(\frac{\partial \ell_i}{\partial \theta}\right) = 0 \quad \text{and} \quad \text{(ii) } \mathbb{E}\left(\frac{\partial^2 \ell_i}{\partial \theta^2}\right) + \mathbb{V}\text{ar}\left(\frac{\partial \ell}{\partial \theta}\right) = 0. \tag{3.4}$$

For the exponential family of distributions,

$$\ell_i = \ell(y_i, \theta, \phi) = \frac{y\theta - b(\theta)}{\phi} + c(y, \phi),$$

and $\frac{\partial \ell}{\partial \theta} = \frac{y - b'(\theta)}{\phi}$ and $\frac{\partial^2 \ell}{\partial \theta^2} = -\frac{b''(\theta)}{\phi}$. By properties (i) and (ii) from (3.4), if Y has PDF $f(y|\theta)$, then $\mathbb{E}(Y) = \mu = b'(\theta)$ and $\mathbb{V}\text{ar}(Y) = b''(\theta)\phi$. The function $b''(\theta)$ is called *variance function* and denoted by $V(\mu)$ (because θ depends on μ).

The *unit deviance* is defined as

$$d_i(y_i, \mu) = 2 \int_{\mu_i}^{y_i} \frac{y_i - u}{V(u)} \, du,$$

and the total deviance, a measure of the distance between y and μ, is defined as

$$D(y, \mu) = \sum_{i=1}^{n} w_i d(y_i, \mu),$$

where the summation is over the data and w_i are the prior weights. The quantity $D(y, \mu)/\phi$ is called the scaled deviance. For the normal distribution, the deviance is equivalent to the residual sum of squares, $\sum_{i=1}^{n} (y_i - \mu)^2$.

3.6 Exercises

3.1 With $n = 10$ observations and $x = 2$ observed successes in i.i.d. trials, construct 99% two-sided confidence intervals for the unknown binomial parameter p using the three methods discussed in this section (exact method, Wald method, and Wilson method). Compare your results.

3.2 From a manufacturing process, $n = 25$ items are manufactured. Let X be the number of defectives found in the lot. Construct a $\alpha = 0.01$ level test to see if the proportion of defectives is greater than 10%. What are your assumptions?

3.3 Let X be a Bernoulli random variable. The parameter p is of course estimable, since $\mathbb{E}X = p$, but it turns out the same is not true for p^2. Prove that p^2 is not estimable with a single Bernoulli observation.

3.4 Derive the MLE for μ with an i.i.d. sample of exponential random variables, and compare the confidence interval based on the Fisher information to an exact confidence interval based on the chi-square distribution.

3.5 Let $X \sim \mathcal{P}(\lambda)$. (i) Use moments of X to obtain an unbiased estimator of λ^2. (ii) Use the moment generating function for the Poisson distribution to find an unbiased estimator of $e^{2\lambda}$.

3.6 A single-parameter ("shape" parameter) Pareto distribution ($\mathcal{P}a(1, \alpha)$ on p. 22) has density function given by $f(x|\alpha) = \alpha/x^{\alpha+1}$, $x \geq 1$. For a given experiment, researchers believe that in Pareto model the shape parameter α exceeds 1 and that the first moment $\mathbb{E}X = \alpha/(\alpha - 1)$ is finite. Moment-matching estimators are solutions of equations in which theoretical moments are replaced empirical counterparts.
  (i) What is the moment-matching estimator of parameter α? In this case, the moment-matching equation is $\overline{X} = \alpha/(\alpha - 1)$.
  (ii) What is the MLE of α?
  (iii) Calculate the two estimators when $X_1 = 2, X_2 = 4$, and $X_3 = 3$ are observed.

3.7 Write an R simulation program to estimate the true coverage probability of a two-sided 90% Wald confidence interval for the case in which $n = 10$ and $p = 0.5$. Repeat the simulation at $p = 0.9$. Repeat the $p = 0.9$ case, but instead use the Wilson interval. To estimate, generate 1000 random binomial outcomes, and count the proportion of time the confidence interval contains the true value of p. Comment on your results.

3.8 Show that the confidence interval (for μ) derived from the likelihood ratio in Example 3.10 is equivalent to the Fisher interval.

3.9 Obtain the Fisher information and the Cramer–Rao lower bound on the variance of an unbiased estimator for p based on a sample of size n from the geometric $\mathcal{G}(p)$ model.

3.10 Let $X_1, \ldots, X_n$ be i.i.d. $\mathcal{P}(\lambda)$ and Y_k be the number of $X_1, \ldots, X_n$ equal to k. Derive the conditional distribution of Y_k given $T = \sum X_i = t$.

3.11 Consider the following i.i.d. sample generated from $F(x)$:

$$\{2.5, 5.0, 8.0, 8.5, 10.5, 11.5, 20\}.$$

Graph the empirical distribution, and estimate the probability $P(8 \leq X \leq 10)$, where X has distribution function $F(x)$.

3.12 Prove the equations in (3.4): (i) $\mathbb{E}(\frac{\partial \ell_i}{\partial \theta}) = 0$, (ii) $\mathbb{E}(\frac{\partial^2 \ell_i}{\partial \theta^2}) + \mathbb{V}\text{ar}(\frac{\partial \ell}{\partial \theta}) = 0$.

3.13 Write R code to determine a 95% two-sided confidence interval based on $n = 100$ Bernoulli observations with $X = 29$ successes. How does this exact interval compare with the interval based on a normal approximation?

3.14 Test the null hypothesis $H_0 : X \sim \mathcal{N}(20,9)$ versus the alternative hypothesis $H_1 : X \sim \mathcal{N}(28,16)$ using the critical region $\{x : x \geq 24\}$. What are the type-I and type-II error rates for this test?

3.15 Using Example 3.3, write R code to graph the length of a 90% t-interval as a function of sample size n when $X_1, \ldots, X_n \sim \mathcal{N}(\mu, \sigma^2)$, for $2 \leq n \leq 30$, using qt in R.

3.16 Let $X_1, \ldots, X_n$ be distributed Poisson with λ. Calculate the Cramer–Rao lower bound, and show that σ^2 (for $\overline{X}$) achieves this bound. Prove that $\overline{X}$ is also the MLE for λ.

3.17 Suppose $X_1, \ldots, X_n$ is distributed normal with mean 0 and variance θ. Find the MLE of θ, and derive its asymptotic distribution.

3.18 Consider two independent Bernoulli data sets: $X_1, \ldots, X_n \sim Ber(p_1)$ and $Y_1, \ldots, Y_m \sim Ber(p_2)$. Derive the likelihood ratio statistic for testing Ho: $p_1 = p_2$ versus H_1: $p_1 \neq p_2$. What is the asymptotic distribution of the likelihood ratio function R in this case?

3.19 Show that the binomial distribution is a member of the exponential family, and solve for the functions of b and c.

3.20 With an i.i.d. sample $(X_1, \ldots, X_n)$ from the Pareto distribution $(Pa(x, \alpha))$, show that the likelihood ratio test for $H_0 : \alpha = 1$ versus $H_1 : \alpha = 1$ is based on the test statistic

$$\frac{\prod_{i=1}^{n} x_i}{x_{1:n}^n},$$

where $x_{1:n} = \min(x_1, \ldots, x_n)$.

| RELEVANT R FUNCTIONS AND DATA SETS IN THIS CHAPTER |

R functions: `binom.test`, `prop.test`, `binom.confint`, `scoreci`
R package: `binom`, `PropCIs`

References

Agresti, A. (1998), "Approximate is Better than "Exact" for Interval Estimation of Binomial Proportions," *American Statistician*, 52, 119–126.

Clopper, C. J., and Pearson, E. S. (1934), "The Use of Confidence or Fiducial Limits Illustrated in the Case of the Binomial," *Biometrika*, 26, 404–413.

Lehmann, E. L., and Casella, G. (1998), *Theory of Point Estimation*, New York: Springer-Verlag.

Nester, M. (1996), "An Applied Statistician's Creed," *Journal of the Royal Statistical Society. Series C (Applied Statistics)*, 45, 401–410.

Wilks, S. S. (1938), "The Large-Sample Distribution of the Likelihood Ratio for Testing Composite Hypotheses," *Annals of Mathematical Statistics*, 9, 60–62.

Wilson, E. B. (1927), "Probability Inference, the Law of Succession, and Statistical Inference," *Journal of the American Statistical Association*, 22, 209–212.

4

Bayesian Statistics

To anyone sympathetic with the current neo-Bernoullian neo-Bayesian Ramseyesque Finettist Savageous movement in statistics, the subject of testing goodness of fit is something of an embarrassment.

F. J. Anscombe (1962)

4.1   The Bayesian Paradigm

There are several paradigms for approaching statistical inference, but the two dominant ones are *frequentist* (sometimes called classical or traditional) and *Bayesian*. The overview in the previous chapter covered mainly classical approaches. According to the Bayesian paradigm, the unobservable parameters in a statistical model are treated as random. When no data are available, a *prior distribution* is used to quantify our knowledge about the parameter. When data are available, we can update our prior knowledge using the conditional distribution of parameters, given the data. The transition from the prior to the posterior is possible via the Bayes theorem.

Suppose that before the experiment our prior distribution describing θ is $\pi(\theta)$. The data are coming from the assumed model (likelihood) that depends on the parameter and is denoted by $f(x|\theta)$. Bayes theorem updates the prior $\pi(\theta)$ to the posterior by accounting for the data x,

$$\pi(\theta|x) = \frac{f(x|\theta)\pi(\theta)}{m(x)}, \tag{4.1}$$

where $m(x)$ is a normalizing constant, $m(x) = \int_{\Theta} f(x|\theta)\pi(\theta)\, d\theta$.

Once the data x are available, θ is the only unknown Quantity, and the posterior distribution $\pi(\theta|x)$ completely describes the uncertainty. There are two key advantages of Bayesian paradigm: (i) once the uncertainty is expressed via the probability

Nonparametric Statistics with Applications to Science and Engineering with R, Second Edition. Paul Kvam, Brani Vidakovic, and Seong-joon Kim.

distribution and the statistical inference can be automated, it follows a conceptually simple recipe, and (ii) available prior information is coherently incorporated into the statistical model.

4.2 Ingredients for Bayesian Inference

The *model* for a typical observation X conditional on unknown parameter θ is the density function $f(x|\theta)$. As a function of θ, $f(x|\theta) = L(\theta)$ is called a *likelihood*. The functional form of f is fully specified up to a parameter θ. According to the *likelihood principle*, all experimental information about the data must be contained in this likelihood function.

The parameter θ, with values in the parameter space Θ, is considered a random variable. The random variable θ has a distribution $\pi(\theta)$ called the prior distribution. This prior describes uncertainty about the parameter before data are observed. If the prior for θ is specified up to a parameter τ, $\pi(\theta|\tau)$, τ is called a *hyperparameter*.

Our goal is to start with this prior information and update it using the data to make the best possible estimator of θ. We achieve this through the likelihood function to get $\pi(\theta|x)$, called the *posterior* distribution for θ, given $X = x$. Accompanying its role as the basis to Bayesian inference, the posterior distribution has been a source for an innumerable accumulation of tacky "butt" jokes by unimaginative statisticians with low-brow sense of humor, such as the authors of this book, for example.

To find $\pi(\theta|x)$, we use Bayes rule to divide *joint* distribution for X and θ ($h(x, \theta) = f(x|\theta)\pi(\theta)$) by the *marginal* distribution $m(x)$, which can be obtained by integrating out parameter θ from the joint distribution $h(x, \theta)$,

$$m(x) = \int_{\Theta} h(x, \theta) \, d\theta = \int_{\Theta} f(x|\theta)\pi(\theta) \, d\theta.$$

The marginal distribution is also called the *prior predictive* distribution. Finally we arrive at an expression for the posterior distribution $\pi(\theta|x)$:

$$\pi(\theta|x) = \frac{h(x, \theta)}{m(x)} = \frac{f(x|\theta)\pi(\theta)}{m(x)} = \frac{f(x|\theta)\pi(\theta)}{\int_{\Theta} f(x|\theta)\pi(\theta) \, d\theta}.$$

The following table summarizes the notation:

Likelihood	$f(x|\theta)$
Prior distribution	$\pi(\theta)$
Joint distribution	$h(x, \theta) = f(x|\theta)\pi(\theta)$
Marginal distribution	$m(x) = \int_{\Theta} f(x|\theta)\pi(\theta) \, d\theta$
Posterior distribution	$\pi(\theta|x) = f(x|\theta)\pi(\theta)/m(x)$

Example 4.1 *Normal Likelihood with Normal Prior.*
The normal likelihood and normal prior combination is important as it is often used in practice. Assume that an observation X is normally distributed with mean θ and known variance σ^2. The parameter of interest, θ, has a normal distribution as well with hyperparameters μ and τ^2. Starting with our Bayesian model of $X|\theta \sim \mathcal{N}(\theta, \sigma^2)$ and $\theta \sim \mathcal{N}(\mu, \tau^2)$, we will find the marginal and posterior distributions.
The exponent ζ in the joint distribution $h(x, \theta)$ is

$$\zeta = -\frac{1}{2\sigma^2}(x - \theta)^2 - \frac{1}{2\tau}(\theta - \mu)^2.$$

After straightforward but somewhat tedious algebra, ζ can be expressed as

$$\zeta = -\frac{1}{2\rho}\left(\theta - \rho\left(\frac{x}{\sigma^2} + \frac{\mu}{\tau^2}\right)\right)^2 - \frac{1}{2(\sigma^2 + \tau^2)}(x - \mu)^2,$$

where

$$\rho = \frac{\sigma^2\tau^2}{\sigma^2 + \tau^2}.$$

Recall that $h(x, \theta) = f(x|\theta)\pi(\theta) = \pi(\theta|x)m(x)$, so the marginal distribution simply resolves to $X \sim \mathcal{N}(\mu, \sigma^2 + \tau^2)$ and the posterior distribution comes out to be

$$\theta|X \sim \mathcal{N}\left(\frac{\tau^2}{\sigma^2 + \tau^2}X + \frac{\sigma^2}{\sigma^2 + \tau^2}\mu, \frac{\sigma^2\tau^2}{\sigma^2 + \tau^2}\right).$$

Below is the summary of MCMC output.

Node	Mean	sd	MC error	2.5%	Median	97.5%
θ	102.8	6.917	0.0214	89.17	102.8	116.3

If $X_1, X_2, \ldots, X_n$ are observed instead of a single observation X, then the sufficiency of $\overline{X}$ implies that the Bayesian model for θ is the same as for X with σ^2/n in place of σ^2. In other words, the Bayesian model is

$$\overline{X}|\theta \sim \mathcal{N}\left(\theta, \frac{\sigma^2}{n}\right) \quad \text{and} \quad \theta \sim \mathcal{N}(\mu, \tau^2),$$

producing

$$\theta|\overline{X} \sim \mathcal{N}\left(\frac{\tau^2}{\frac{\sigma^2}{n} + \tau^2}\overline{X} + \frac{\frac{\sigma^2}{n}}{\frac{\sigma^2}{n} + \tau^2}\mu, \frac{\frac{\sigma^2}{n}\tau^2}{\frac{\sigma^2}{n} + \tau^2}\right).$$

Notice that the posterior mean

$$\frac{\tau^2}{\frac{\sigma^2}{n} + \tau^2}\overline{X} + \frac{\frac{\sigma^2}{n}}{\frac{\sigma^2}{n} + \tau^2}\mu \tag{4.2}$$

is a weighted linear combination of the maximum likelihood estimate (MLE) $\overline{X}$ and the prior mean μ with weights

$$\lambda = \frac{n\tau^2}{\sigma^2 + n\tau^2}, \quad 1 - \lambda = \frac{\sigma^2}{\sigma^2 + n\tau^2}.$$

When the sample size n increases, $\lambda \to 1$, and the influence of the prior mean diminishes. On the other hand, when n is small and our prior opinion about μ is strong (i.e. τ^2 is small), the posterior mean is close to the prior mean μ. We will see later several more cases in which the posterior mean is a linear combination of a frequentist estimate and the prior mean.

For instance, suppose 10 observations are coming from $\mathcal{N}(\theta, 100)$. Assume that the prior on θ is $\mathcal{N}(20, 20)$. Using the numerical example in the R code below, the posterior is $\mathcal{N}(6.8352, 6.6667)$. These three densities are shown in Figure 4.1:

```
> source("BA.nornor2.r")
> dat <- c(2.9441,-13.3618,7.1432,16.2356,-6.9178,
+            8.5800,12.5400,-15.9373,-14.4096,5.7115)
> result <- BA.nornor2(dat,100,20,20)
> result
```

Figure 4.1 The normal $\mathcal{N}(\theta, 100)$ likelihood, $\mathcal{N}(20, 20)$ prior, and posterior for data $\{2.9441, -13.3618, \dots, 5.7115\}$.

4.2.1 Quantifying Expert Opinion

Errors using inadequate data are much less than those using no data at all.

Charles Babbage (1792–1871)

Bayesian statistics has become increasingly popular in engineering, and one reason for its increased application is that it allows researchers to input expert opinion as a catalyst in the analysis (through the prior distribution). Expert opinion might consist of subjective inputs from experienced engineers or perhaps a summary judgment of past research that yielded similar results.

There is always good reason to question authority, and that includes the authority that provides the expert opinion formulating a prior distribution. In most engineering problems, a weak prior is formed from previous data that cannot be mined directly. In most modern problems, data collections are usually large enough to make the specifics of the prior distribution inconsequential.

Example 4.2   Prior Elicitation for Reliability Tests. Suppose each of n independent reliability tests a machine revealing either a successful or unsuccessful outcome. If θ represents the reliability of the machine, let X be the number of successful missions the machine experienced in n independent trials. X is distributed binomial with parameters n (known) and θ (unknown). We will not probably expect an expert to quantify their uncertainty about θ directly into a prior distribution $\pi(\theta)$. Perhaps the researcher can elicit information such as the expected value and standard deviation of θ. If we suppose the prior distribution for θ is $Be(\alpha, \beta)$, where the hyperparameters α and β are known, then

$$\pi(\theta) = \frac{1}{B(\alpha, \beta)} \theta^{\alpha-1} (1 - \theta)^{\beta-1}, \quad 0 \le \theta \le 1.$$

With $X|\theta \sim Bin(n, \theta)$, the joint, marginal, and posterior distributions are

$$h(x, \theta) = \frac{\binom{n}{x}}{B(\alpha, \beta)} \theta^{\alpha+x-1} (1 - \theta)^{n-x+\beta-1}, \quad 0 \le \theta \le 1, x = 0, 1, \dots, n.$$

$$m(x) = \frac{\binom{n}{x} B(x + \alpha, n - x + \beta)}{B(\alpha, \beta)}, \quad x = 0, 1, \dots, n.$$

$$\pi(\theta|x) = \frac{1}{B(x + \alpha, n - x + \beta)} \theta^{\alpha+x-1} (1 - \theta)^{n-x+\beta-1}, \quad 0 \le \theta \le 1.$$

It is easy to see that the posterior distribution is $Be(\alpha + x, n - x + \beta)$. Suppose the experts suggest that the previous version of this machine was "reliable 93% of the time, plus or minus 2%." We might take $\mathbb{E}(\theta) = 0.93$ and insinuate that $\sigma_\theta = 0.04$

(or $\mathbb{V}\mathrm{ar}(\theta) = 0.0016$), using two-sigma rule as an argument. From the beta distribution,

$$\mathbb{E}\theta = \frac{\alpha}{\alpha + \beta} \quad \text{and} \quad \mathbb{V}\mathrm{ar}\theta = \frac{\alpha\beta}{(\alpha + \beta)^2(\alpha + \beta + 1)}.$$

We can actually solve for α and β as a function of the expected value μ and variance σ^2, as in (2.3),

$$\alpha = \mu(\mu - \mu^2 - \sigma^2)/\sigma^2, \quad \text{and} \quad \beta = (1 - \mu)(\mu - \mu^2 - \sigma^2)/\sigma^2.$$

In this example, $(\mu, \sigma^2) = (0.93, 0.0016)$ leads to $\alpha = 36.91$ and $\beta = 2.78$. To update the data X, we will use a $Be(36.91, 2.78)$ distribution for a prior on θ. Consider the weight given to the expert in this example. If we observe one test only and the machine happened to fail, our posterior distribution is then $Be(36.91, 3.78)$, which has a mean equal to 0.9071. The MLE for the average reliability is obviously zero, with such precise information elicited from the expert; the posterior is close to the prior. In some cases when you do not trust your expert, this might be Unsettling, and less informative priors may be a better choice.

4.3 Point Estimation

The posterior is the ultimate experimental summary for a Bayesian. The location measures (especially the mean) of the posterior are of great importance. The posterior mean represents the most frequently used Bayes estimator for the parameter. The posterior mode and median are less commonly used alternative Bayes estimators.

An objective way to choose an estimator from the posterior is through a penalty or loss function $L(\hat{\theta}, \theta)$ that describes how we penalize the discrepancy of the estimator $\hat{\theta}$ from the parameter θ. Because the parameter is viewed as a random variable, we seek to minimize *expected* loss, or *posterior risk*:

$$R(\hat{\theta}, x) = \int L(\hat{\theta}, \theta)\pi(\theta|x) \, d\theta.$$

For example, the estimator based on the common squared-error loss $L(\hat{\theta}, \theta) = (\hat{\theta} - \theta)^2$ minimizes $\mathbb{E}((\hat{\theta} - \theta)^2)$, where expectation is taken over the posterior distribution $\pi(\theta|X)$. It is easy to show that the estimator turns out to be the posterior expectation. Similar to squared-error loss, if we use absolute-error loss $L(\hat{\theta}, \theta) = |\hat{\theta} - \theta|$, the Bayes estimator is the posterior median.

The posterior mode maximizes the posterior density the same way MLE is maximizing the likelihood. The *generalized MLE* maximizes $\pi(\theta|X)$. Bayesians prefer the name MAP (maximum a posteriori) estimator or simply posterior mode.

The MAP estimator is popular in Bayesian analysis in part because it is often computationally less demanding than the posterior mean or median. The reason is simple; to find the maximum, the posterior need not to be fully specified because $\mathrm{argmax}_\theta \pi(\theta|x) = \mathrm{argmax}_\theta f(x|\theta)\pi(\theta)$, that is, one simply maximizes the product of likelihood and the prior.

In general, the posterior mean will fall between the MLE and the the prior mean. This was demonstrated in Example 4.1. As another example, suppose we flipped a coin four times and tails showed up on all four occasions. We are interested in estimating probability of heads, θ, in a Bayesian way. If the prior is $\mathcal{U}(0,1)$, the posterior is proportional to $\theta^0(1-\theta)^4$ that is beta $Be(1,5)$. The posterior mean *shrinks* the MLE toward the expected value of the prior $(1/2)$ to get $\hat{\theta}_B = 1/(1+5) = 1/6$, which is a more reasonable estimator of θ then the MLE.

Example 4.3   Binomial-Beta Conjugate Pair. Suppose $X|\theta \sim Bin(n,\theta)$. If the prior distribution for θ is $Be(\alpha,\beta)$, the posterior distribution is $Be(\alpha+x, n-x+\beta)$. Under squared-error loss $L(\hat{\theta},\theta) = (\hat{\theta}-\theta)^2$, the Bayes estimator of θ is the expected value of the posterior

$$\hat{\theta}_B = \frac{\alpha+x}{(\alpha+x)(\beta+n-x)} = \frac{\alpha+x}{\alpha+\beta+n}.$$

This is actually a weighted average of MLE, X/n, and the prior mean $\alpha/(\alpha+\beta)$. Notice that, as n becomes large, the posterior mean is getting close to MLE, because the weight $n/(n+\alpha+\beta)$ tends to 1. On the other hand, when α is large, the posterior mean is close to the prior mean. Large α indicates small prior variance (for fixed β, the variance of $Be(\alpha,\beta)$ behaves as $O(1/\alpha^2)$), and the prior is concentrated about its mean. Recall the Example 4.2; after one machine trial failure, the posterior distribution mean changed from 0.93 (the prior mean) to 0.9071, shrinking only slightly toward the MLE (which is zero).

Example 4.4   Jeremy's IQ. Jeremy, an enthusiastic Georgia Tech student, spoke in class and posed a statistical model for his scores on standard IQ tests. He thinks that, in general, his scores are normally distributed with unknown mean of θ and the variance of 80. Prior (and expert) opinion is that the IQ of Georgia Tech students, θ, is a normal random variable, with mean of 110 and the variance of 120. Jeremy took the test and scored 98. The traditional estimator of θ would be $\hat{\theta} = X = 98$. The posterior is $\mathcal{N}(102.8, 48)$, so the Bayes estimator of Jeremy's IQ score is $\hat{\theta}_B = 102.8$.

Example 4.5   Poisson-Gamma Conjugate Pair. Let $X_1,\dots,X_n$ given θ are Poisson $\mathcal{P}(\theta)$ with probability mass function

$$f(x_i|\theta) = \frac{\theta^{x_i}}{x_i!} e^{-\theta},$$

and $\theta \sim \mathcal{G}(\alpha, \beta)$ is given by $\pi(\theta) \propto \theta^{\alpha-1} e^{-\beta\theta}$. Then,

$$\pi(\theta|X_1, \dots, X_n) = \pi(\theta| \sum X_i) \propto \theta^{\sum X_i + \alpha - 1} e^{-(n+\beta)\theta},$$

which is $\mathcal{G}(\sum_i X_i + \alpha, n + \beta)$. The mean is $\mathbb{E}(\theta|X) = (\sum X_i + \alpha)/(n + \beta)$, and it can be represented as a weighted average of the MLE and the prior mean:

$$\mathbb{E}\theta|X = \frac{n}{n+\beta} \frac{\sum X_i}{n} + \frac{\beta}{n+\beta} \frac{\alpha}{\beta}.$$

4.3.1 Conjugate Priors

We have seen two convenient examples for which the posterior distribution remained in the same family as the prior distribution. In such a case, the effect of likelihood is only to "update" the prior parameters and not to change prior's functional form. We say that such priors are *conjugate* with the likelihood. Conjugacy is popular because of its mathematical convenience; once the conjugate pair likelihood/prior is found, the posterior is calculated with relative ease. In the years BC[1] and pre-MCMC (Markov chain Monte Carlo) era (see Chapter 18), conjugate priors have been extensively used (and overused and misused) precisely because of this computational convenience. Nowadays, the general agreement is that simple conjugate analysis is of limited practical value because, given the likelihood, the conjugate prior has limited modeling capability.

There are many univariate and multivariate instances of conjugacy. Table 4.1 provides several cases. For practice you may want to work out the posteriors in the table.

Table 4.1 Some conjugate pairs.

Likelihood	Prior	Posterior
$X|\theta \sim \mathcal{N}(\theta, \sigma^2)$	$\theta \sim \mathcal{N}(\mu, \tau^2)$	$\theta|X \sim \mathcal{N}\left(\frac{\tau^2}{\sigma^2+\tau^2}X + \frac{\sigma^2}{\sigma^2+\tau^2}\mu, \frac{\sigma^2\tau^2}{\sigma^2+\tau^2}\right)$
$X|\theta \sim B(n, \theta)$	$\theta \sim Be(\alpha, \beta)$	$\theta|X \sim Be(\alpha + x, n - x + \beta)$
$\mathbf{X}|\theta \sim P(\theta)$	$\theta \sim \mathcal{G}amma(\alpha, \beta)$	$\theta|\mathbf{X} \sim \mathcal{G}amma(\sum_i X_i + \alpha, n + \beta)$.
$\mathbf{X}|\theta \sim \mathcal{N}B(m, \theta)$	$\theta \sim Be(\alpha, \beta)$	$\theta|\mathbf{X} \sim Be(\alpha + mn, \beta + \sum_{i=1}^{n} x_i)$
$X \sim \mathcal{G}amma(n/2, 1/(2\theta))$	$\theta \sim \mathcal{I}\mathcal{G}(\alpha, \beta)$	$\theta|X \sim \mathcal{I}\mathcal{G}(n/2 + \alpha, x/2 + \beta)$
$\mathbf{X}|\theta \sim U(0, \theta)$	$\theta \sim Pa(\theta_0, \alpha)$	$\theta|\mathbf{X} \sim Pa(\max\{\theta_0, X_1, \dots, X_n\}, \alpha + n)$
$X|\theta \sim \mathcal{N}(\mu, \theta)$	$\theta \sim \mathcal{I}\mathcal{G}(\alpha, \beta)$	$\theta|X \sim \mathcal{I}\mathcal{G}(\alpha + 1/2, \beta + (\mu - X)^2/2)$
$X|\theta \sim \mathcal{G}amma(\nu, \theta)$	$\theta \sim \mathcal{G}a(\alpha, \beta)$	$\theta|X \sim \mathcal{G}amma(\alpha + \nu, \beta + x)$

Here $\mathbf{X}$ stands for a sample of size n, $X_1, \dots, X_n$.

---

1 For some, the BC era signifies *Before Christ*, rather than *Before Computers*.

4.4 Interval Estimation: Credible Sets

Bayesians call interval estimators of model parameters *credible sets*. Naturally, the measure used to assess the credibility of an interval estimator is the posterior distribution. Students learning concepts of classical confidence intervals (CIs) often err by stating that "the probability that the CI $[L, U]$ contains parameter θ is $1 - \alpha$." The correct statement seems more convoluted; one needs to generate data from the underlying model many times and for each generated data set to calculate the CI. The proportion of CIs covering the unknown parameter "tends to" $1 - \alpha$. The Bayesian interpretation of a credible set C is arguably more natural: the probability of a parameter belonging to the set C is $1 - \alpha$. A formal definition follows.

Assume the set C is a subset of Θ. Then, C is *credible set* with credibility $(1 - \alpha)100\%$ if

$$P(\theta \in C | X) = \mathbb{E}(I(\theta \in C) | X) = \int_C \pi(\theta | x)\, d\theta \geq 1 - \alpha.$$

If the posterior is discrete, then the integral is a sum (using the counting measure) and

$$P(\theta \in C | X) = \sum_{\theta_i \in C} \pi(\theta_i | x) \geq 1 - \alpha.$$

This is the definition of a $(1 - \alpha)100\%$ credible set, and for any given posterior distribution, such a set is not unique.

For a given credibility level $(1 - \alpha)100\%$, the shortest credible set has obvious appeal. To minimize size, the sets should correspond to highest posterior probability density areas (HPDs).

Definition 4.1 The $(1 - \alpha)100\%$ HPD credible set for parameter θ is a set C, subset of Θ of the form

$$C = \{\theta \in \Theta | \pi(\theta | x) \geq k(\alpha)\},$$

where $k(\alpha)$ is the largest constant for which

$$P(\theta \in C | X) \geq 1 - \alpha.$$

Geometrically, if the posterior density is cut by a horizontal line at the height $k(\alpha)$, the set C is projection on the θ axis of the part of line that lies below the density.

Example 4.6 Jeremy's IQ, Continued. Recall Jeremy, the enthusiastic Georgia Tech student from Example 4.4, who used Bayesian inference in modeling his IQ test scores. For a score $X | \theta$, he was using a $\mathcal{N}(\theta, 80)$ likelihood, and $\mathcal{N}(110, 120)$

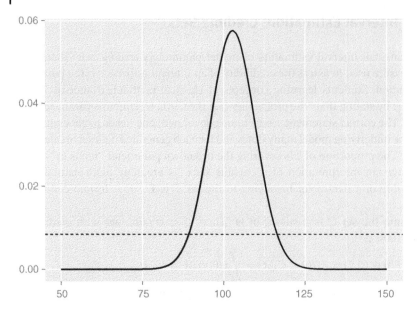

Figure 4.2 Bayesian credible set based on $\mathcal{N}(102.8, 48)$ density.

was chosen for the prior distribution. After the score of $X = 98$ was recorded, the resulting posterior was also normally distributed: $\mathcal{N}(102.8, 48)$.

Here, the MLE is $\hat{\theta} = 98$, and a 95% confidence interval is $[98 - 1.96\sqrt{80},\ 98 + 1.96\sqrt{80}] = [80.4692, 115.5308]$. The length of this interval is approximately 35.

The Bayesian counterparts are $\hat{\theta} = 102.8$, and $[102.8 - 1.96\sqrt{48},\ 102.8 + 1.96\sqrt{48}] = [89.2207, 116.3793]$. The length of 95% credible set is approximately 27. The Bayesian interval is shorter because the posterior variance is smaller than the likelihood variance; this is a consequence of the incorporation of information. The construction of the credible set is illustrated in Figure 4.2.

4.5 Bayesian Testing

Bayesian tests amount to comparison of posterior probabilities of the parameter regions defined by the two hypotheses.

Assume that Θ_0 and Θ_1 are two non-overlapping subsets of the parameter space Θ. We assume that Θ_0 and Θ_1 partition Θ, that is, $\Theta_1 = \Theta_0^c$, although cases in which $\Theta_1 \neq \Theta_0^c$ are easily formulated. Let $\theta \in \Theta_0$ signify the null hypothesis H_0, and let $\theta \in \Theta_1 = \Theta_0^c$ signify the alternative hypothesis H_1:

$$H_0 : \ \theta \in \Theta_0 \qquad H_1 : \ \theta \in \Theta_1.$$

Given the information from the posterior, the hypothesis with higher posterior probability is selected.

Example 4.7 We return again to Jeremy (Examples 4.4 and 4.6) and consider the posterior for the parameter θ, $\mathcal{N}(102.8, 48)$. Jeremy claims he had a bad day, and his genuine IQ is at least 105. After all, he is at Georgia Tech! The posterior probability of $\theta \geq 105$ is

$$p_0 = P^{\theta|X}(\theta \geq 105) = P\left(Z \geq \frac{105-102.8}{\sqrt{48}}\right) = 1 - \Phi(0.3175) = 0.3754,$$

less than 38%, so his claim is rejected. Posterior odds in favor of H_0 are $0.3754/(1-0.3754)=0.4652$, less than 50%.

We can represent the prior and posterior odds in favor of the hypothesis H_0, respectively, as

$$\frac{\pi_0}{\pi_1} = \frac{P^\theta(\theta \in \Theta_0)}{P^\theta(\theta \in \Theta_1)} \quad \text{and} \quad \frac{p_0}{p_1} = \frac{P^{\theta|X}(\theta \in \Theta_0)}{P^{\theta|X}(\theta \in \Theta_1)}.$$

The *Bayes factor* in favor of H_0 is the ratio of corresponding posterior to prior odds:

$$B_{01}^\pi(x) = \frac{\frac{P(\theta \in \Theta_0|X)}{P(\theta \in \Theta_1|X)}}{\frac{P(\theta \in \Theta_0)}{P(\theta \in \Theta_1)}} = \frac{p_0/p_1}{\pi_0/\pi_1}. \tag{4.3}$$

When the hypotheses are simple (i.e. $H_0 : \theta = \theta_0$ versus $H_1 : \theta = \theta_1$), and the prior is just the two-point distribution $\pi(\theta_0) = \pi_0$ and $\pi(\theta_1) = \pi_1 = 1 - \pi_0$, then the Bayes factor in favor of H_0 becomes the likelihood ratio:

$$B_{01}^\pi(x) = \frac{\frac{P^{\theta|X}(\theta \in \Theta_0)}{P^{\theta|X}(\theta \in \Theta_1)}}{\frac{P^\theta(\theta \in \Theta_0)}{P^\theta(\theta \in \Theta_1)}} = \frac{f(x|\theta_0)\pi_0}{f(x|\theta_1)\pi_1} \bigg/ \frac{\pi_0}{\pi_1} = \frac{f(x|\theta_0)}{f(x|\theta_1)}.$$

If the prior is a mixture of two priors, ξ_0 under H_0 and ξ_1 under H_1, then the Bayes factor is the ratio of two marginal (prior-predictive) distributions generated by ξ_0 and ξ_1. Thus, if $\pi(\theta) = \pi_0\xi_0(\theta) + \pi_1\xi_2(\theta)$, then

$$B_{01}^\pi(x) = \frac{\frac{\int_{\Theta_0} f(x|\theta)\pi_0\xi_0(\theta)\, d\theta}{\int_{\Theta_1} f(x|\theta)\pi_1\xi_1(\theta)\, d\theta}}{\frac{\pi_0}{\pi_1}} = \frac{m_0(x)}{m_1(x)}.$$

The Bayes factor measures relative change in prior odds once the evidence is collected. Table 4.2 offers practical guidelines for Bayesian testing of hypotheses depending on the value of log-Bayes factor. One could use $B_{01}^\pi(x)$ of course, but then $a \leq \log B_{10}(x) \leq b$ becomes $-b \leq \log B_{01}(x) \leq -a$. Negative values of the log-Bayes factor are handled by using symmetry and changed wording, in an obvious way.

Table 4.2 Treatment of H_0 according to the value of log-Bayes factor.

$0 \leq \log B_{10}(x) \leq 0.5$	Evidence against H_0 is **poor**
$0.5 \leq \log B_{10}(x) \leq 1$	Evidence against H_0 is **substantial**
$1 \leq \log B_{10}(x) \leq 2$	Evidence against H_0 is **strong**
$\log B_{10}(x) > 2$	Evidence against H_0 is **decisive**

4.5.1 Bayesian Testing of Precise Hypotheses

Testing precise hypotheses in Bayesian fashion has a considerable body of research. Berger (1985), pp. 148–157, has a comprehensive overview of the problem and provides a wealth of references. See also Berger and Selke (1987) and Berger and Delampady (1987).

If the priors are continuous, testing precise hypotheses in Bayesian fashion is impossible because with continuous priors and posteriors, the probability of a singleton is 0. Suppose $X|\theta \sim f(x|\theta)$ is observed and we are interested in testing

$$H_0 : \theta = \theta_0 \quad \text{versus} \quad H_1 : \theta \neq \theta_0.$$

The answer is to have a prior that has a point mass at the value θ_0 with prior weight π_0 and a spread distribution $\xi(\theta)$ that is the prior under H_1 having prior weight $\pi_1 = 1 - \pi_0$. Thus, the prior is the two-point mixture

$$\pi(\theta) = \pi_0 \delta_{\theta_0} + \pi_1 \xi(\theta),$$

where δ_{θ_0} is Dirac mass at θ_0.

The marginal density for X is

$$m(x) = \pi_0 f(x|\theta_0) + \pi_1 \int f(x|\theta)\xi(\theta) \, d\theta = \pi_0 f(x|\theta_0) + \pi_1 m_1(x).$$

The posterior probability of $\theta = \theta_0$ is

$$\pi(\theta_0|x) = f(x|\theta_0)\pi_0/m(x) = \frac{f(x|\theta_0)\pi_0}{\pi_0 f(x|\theta_0) + \pi_1 m_1(x)} = \left(1 + \frac{\pi_1}{\pi_0} \cdot \frac{m_1(x)}{f(x|\theta_0)}\right)^{-1}.$$

4.6 Bayesian Prediction

Statistical prediction fits naturally into the Bayesian framework. Suppose $Y \sim f(y|\theta)$ is to be observed. The posterior predictive distribution of Y given observed $X = x$ is

$$f(y|x) = \int_\Theta f(y|\theta)\pi(\theta|x) \, d\theta.$$

For example, in the normal distribution example, the predictive distribution of Y given $X_1, \ldots, X_n$ is

$$
Y | \overline{X} \sim \mathcal{N} \left(\frac{\tau^2}{\frac{\sigma^2}{n} + \tau^2} \overline{X} + \frac{\frac{\sigma^2}{n}}{\frac{\sigma^2}{n} + \tau^2} \mu, \ \sigma^2 + \frac{\frac{\sigma^2}{n} \tau^2}{\frac{\sigma^2}{n} + \tau^2} \right). \tag{4.4}
$$

Example 4.8   Martz and Waller (1985) suggest that Bayesian reliability inference is most helpful in applications where little system failure data exist, but past data from like systems are considered relevant to the present system. They use an example of heat exchanger reliability, where the lifetime X is the failure time for heat exchangers used in refining gasoline. From past research and modeling in this area, it is determined that X has a Weibull distribution with $\kappa = 3.5$. Furthermore, the scale parameter λ is considered to be in the interval $0.5 \le \lambda \le 1.5$ with no particular value of λ considered more likely than others.

From this argument, we have

$$
\pi(\lambda) = \begin{cases} 1, & 0.5 \le \lambda \le 1.5, \\ 0, & \text{otherwise}, \end{cases}
$$

$$
f(x|\lambda) = \kappa \lambda x^{\kappa - 1} e^{-(x\lambda)^\kappa},
$$

where $\kappa = 3.5$. With $n = 9$ observed failure times (measured in years of service) at $(0.41, 0.58, 0.75, 0.83, 1.00, 1.08, 1.17, 1.25, 1.35)$, the likelihood is

$$
f(x_1, \ldots, x_9 | \lambda) \propto \lambda^9 \left(\prod_{i=1}^{9} x_i^{2.5} \right) e^{-\lambda^{3.5} (\sum x_i^{3.5})},
$$

so the sufficient statistic is

$$
\sum_{i=1}^{n} x_i^{3.5} = 10.16.
$$

The resulting posterior distribution is not distributed Weibull (like the likelihood) or uniform (like the prior). It can be expressed as

$$
\pi(\lambda | x_1, \ldots, x_9) = \begin{cases} (1621.39)\lambda^9 \, e^{-10.16\lambda^{3.5}}, & 0.5 \le \lambda \le 1.5, \\ 0, & \text{otherwise}, \end{cases}
$$

and has expected value of $\lambda_B = 0.6896$. Figure 4.3a shows the posterior density from the prior distribution, $\mathbb{E}(\lambda) = 1$, so our estimate of λ has decreased in the process of updating the prior with the data.

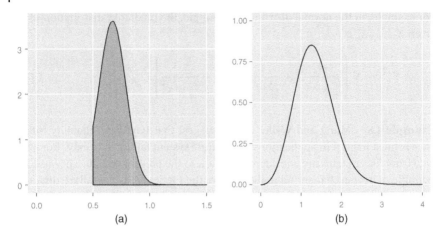

Figure 4.3 (a) Posterior density for λ. (b) Posterior predictive density for heat-exchanger lifetime.

Estimation of λ was not the focus of this study; the analysts were interested in predicting future lifetime of a generic (randomly picked) heat exchanger. Using the predictive density from (4.4),

$$f(y|x) = \int_{0.5}^{1.5} \left(3.5\lambda^{3.5}y^{2.5}\ e^{-(\lambda y)^{3.5}}\right) \left(1621.39\lambda^9\ e^{-10.16\lambda^{3.5}}\right) d\lambda.$$

The predictive density is a bit messy but straightforward to work with. The plot of the density in Figure 4.3b shows how uncertainty is gauged for the lifetime of a new heat exchanger. From $f(y|x)$, we might be interested in predicting early failure by the new item; for example, a 95% lower bound for heat-exchanger lifetime is found by computing the lower 0.05-quantile of $f(y|x)$, which is approximately 0.49.

4.7 Bayesian Computation and Use of WinBUGS

If the selection of an adequate prior was the major conceptual and modeling challenge of Bayesian analysis, the major implementational challenge is computation. When the model deviates from the conjugate structure, finding the posterior distribution and the Bayes rule is all but simple. A closed form solution is more an exception than the rule, and even for such exceptions, lucky mathematical coincidences, convenient mixtures, and other tricks are needed to uncover the explicit expression.

If the classical statistics relies on optimization, Bayesian statistics relies on integration. The marginal needed for the posterior is an integral

$$m(x) = \int_{\Theta} f(x|\theta)\pi(\theta) \, d\theta,$$

and the Bayes estimator of $h(\theta)$ with respect to the squared error loss is a ratio of integrals:

$$\delta_{\pi}(x) = \int_{\Theta} h(\theta)\pi(\theta|x) \, d\theta = \frac{\int_{\Theta} h(\theta)f(x|\theta)\pi(\theta) \, d\theta}{\int_{\Theta} f(x|\theta)\pi(\theta) \, d\theta}.$$

The difficulties in calculating the above Bayes rule come from the facts that (i) the posterior may not be representable in a finite form and (ii) the integral of $h(\theta)$ does not have a closed form even when the posterior distribution is explicit.

The last two decades of research in Bayesian statistics contributed to broadening the scope of Bayesian models. Models that could not be handled before by a computer are now routinely solved. This is done by MCMC methods and their introduction to the field of statistics revolutionized Bayesian statistics.

The MCMC methodology was first applied in statistical physics (Metropolis et al., 1953). Work by Gelfand and Smith (1990) focused on applications of MCMC to Bayesian models. The principle of MCMC is simple: to sample randomly from a target probability distribution one designs a Markov chain whose stationary distribution is the target distribution. By simulating long runs of such a Markov chain, the target distribution can be well approximated. Various strategies for constructing appropriate Markov chains that simulate form the desired distribution are possible: Metropolis–Hastings, Gibbs sampler, slice sampling, perfect sampling, and many specialized techniques. They are beyond the scope of this text, and the interested reader is directed to Robert (2001), Robert and Casella (2004), and Chen, Shao, and Ibrahim (2000) for an overview and a comprehensive treatment.

We will use WinBUGS for doing Bayesian inference on non-conjugate models. Appendix B offers a brief introduction to the front end of WinBUGS. Three volumes of examples are standard addition to the software; in the Examples menu of WinBUGS, see Spiegelhalter et al. (1996). It is recommended that you go over some of those in detail because they illustrate the functionality and real modeling power of WinBUGS. A wealth of examples on Bayesian modeling strategies using WinBUGS can be found in the monographs of Congdon (2001, 2003, 2005). The following example demonstrates the simulation power of WinBUGS, although it involves approximating probabilities of complex events and has nothing to do with Bayesian inference.

Example 4.9   Paradox DeMere in WinBUGS. In 1654, the Chevalier de Mere asked Blaise Pascal (1623–1662) the following question: *In playing a game with three dice, why the sum 11 is advantageous to sum 12 when both are results of six possible outcomes?* Indeed, there are six favorable triplets for each of the sums **11** and **12**:

11:	$(1, 4, 6), (1, 5, 5), (2, 3, 6), (2, 4, 5), (3, 3, 5), (3, 4, 4)$
12:	$(1, 5, 6), (2, 4, 6), (2, 5, 5), (3, 3, 6), (3, 4, 5), (4, 4, 4)$

The solution to this "paradox" DeMere is simple. By taking into account all possible permutations of the triples, the sum 11 has 27 favorable permutations while the sum 12 has 25 favorable permutation. But what if 300 fair dice are rolled and we are interested if the sum 1111 is advantageous to the sum 1112? Exact solution is unappealing, but the probabilities can be well approximated by WinBUGS model demere1:

```
model demere1;
{
for (i in 1:300) {
dice[i] ~ dcat(p.dice[]);
}
is1111 <- equals(sum(dice[]),1111)
is1112 <- equals(sum(dice[]),1112)
}
```

The data are

```
list(p.dice=c(0.1666666, 0.1666666,
0.1666667, 0.1666667, 0.1666667, 0.1666667))
```

and the initial values are generated. After five million rolls, WinBUGS outputs are is1111 = 0.0016 and is1112 = 0.0015, so the sum of 1111 is advantageous to the sum of 1112.

Example 4.10   Jeremy in WinBUGS. We will calculate a Bayes estimator for Jeremy's true IQ using BUGS. Recall the model in Example 4.4 was $X \sim \mathcal{N}(\theta, 80)$ and $\theta \sim \mathcal{N}(100, 120)$. In WinBUGS, we will use the precision parameters $1/120 = 0.00833$ and $1/80 = 0.0125$:

```
#Jeremy in WinBUGS
model{
x ~ dnorm( theta, tau)
theta ~ dnorm( 110, 0.008333333)
}
```

```
#data
list( tau=0.0125, x=98)
#inits
list(theta=100)
```

Below is the summary of MCMC output.

Node	Mean	sd	MC error	2.5%	Median	97.5%
θ	102.8	6.917	0.0214	89.17	102.8	116.3

Because this is a conjugate normal/normal model, the exact posterior distribution, $\mathcal{N}(102.8, 48)$, was easy to find (see Example 4.4). Note that in simulations, the MCMC approximation, when rounded, coincides with the exact posterior mean. The MCMC variance of θ is $6.917^2 = 47.84489$, close to the exact posterior variance of 48.

4.8 Exercises

4.1 A lifetime X (in years) of a particular machine is modeled by an exponential distribution with unknown failure rate parameter θ. The lifetimes of $X_1 = 5$, $X_2 = 6$, and $X_3 = 4$ are observed, and assume that an expert believes that θ should have exponential distribution as well and that on average θ should be $1/3$:
  (i) Write down the MLE of θ for those observations.
 (ii) Elicit a prior according to the expert's beliefs.
(iii) For the prior in (ii), find the posterior. Is the problem conjugate?
(iv) Find the Bayes estimator $\hat{\theta}_{Bayes}$, and compare it with the MLE estimator from (i). Discuss.

4.2 Suppose $X = (X_1, \ldots, X_n)$ is a sample from $\mathcal{U}(0, \theta)$. Let θ have Pareto $Pa(\theta_0, \alpha)$ distribution. Show that the posterior distribution is $Pa(\max\{\theta_0, x_1, \ldots, x_n\}, \alpha + n)$.

4.3 Let $X \sim \mathcal{G}(n/2, 2\theta)$, so that X/θ is χ_n^2. Let $\theta \sim \mathcal{IG}(\alpha, \beta)$. Show that the posterior is $\mathcal{IG}(n/2 + \alpha, (x/2 + \beta^{-1})^{-1})$.

4.4 If $X = (X_1, \ldots, X_n)$ is a sample from $\mathcal{NB}(m, \theta)$ and $\theta \sim Be(\alpha, \beta)$, show that the posterior for θ is beta $Be(\alpha + mn, \beta + \sum_{i=1}^n x_i)$.

4.5 In Example 4.5 on p. 57, show that the marginal distribution is negative binomial.

4.6 What is the Bayes factor B_{01}^{π} in Jeremy's case (Example 4.7)? Test H_0 using the Bayes factor and wording from the Table 4.2. Argue that the evidence against H_0 is poor.

4.7 Assume $X|\theta \sim \mathcal{N}(\theta, \sigma^2)$ and $\theta \sim \pi(\theta) = 1$. Consider testing $H_0 : \theta \leq \theta_0$ versus $H_1 : \theta > \theta_0$. Show that $p_0 = P^{\theta|X}(\theta \leq \theta_0)$ is equal to the classical p-value.

4.8 Show that the Bayes factor is $B_{01}^{\pi}(x) = f(x|\theta_0)/m_1(x)$.

4.9 Show that

$$p_0 = \pi(\theta_0|x) \geq \left[1 + \frac{\pi_1}{\pi_0} \cdot \frac{r(x)}{f(x|\theta_0)}\right]^{-1},$$

where $r(x) = \sup_{\theta \neq \theta_0} f(x|\theta)$. Usually, $r(x) = f(x|\hat{\theta}_{\text{MLE}})$, where $\hat{\theta}_{\text{MLE}}$ is MLE estimator of θ. The Bayes factor $B_{01}^{\pi}(x)$ is bounded from below:

$$B_{01}^{\pi}(x) \geq \frac{f(x|\theta_0)}{r(x)} = \frac{f(x|\theta_0)}{f(x|\hat{\theta}_{\text{MLE}})}.$$

4.10 Suppose $X = -2$ was observed from the population distributed as $N(0, 1/\theta)$, and one wishes to estimate the parameter θ. (Here θ is the reciprocal of variance σ^2 and is called the *precision parameter*. The precision parameter is used in WinBUGS to parameterize the normal distribution). A classical estimator of θ (e.g. the MLE) does exist, but one may be disturbed to estimate $1/\sigma^2$ based on a single observation. Suppose the analyst believes that the prior on θ is *Gamma*(1/2, 3) :
  (i) What is the MLE of θ?
  (ii) Find the posterior distribution and the Bayes estimator of θ. If the prior on θ is *Gamma*(α, β), represent the Bayes estimator as weighted average (sum of weights $= 1$) of the prior mean and the MLE.
  (iii) Find a 95% HPD credible set for θ.
  (iv) Test the hypothesis $H_0 : \theta \leq 1/4$ versus $H_1 : \theta > 1/4$.

4.11 *The Lindley (1957) paradox.* Suppose $\bar{y}|\theta \sim N(\theta, 1/n)$. We wish to test $H_0 : \theta = 0$ versus the two-sided alternative. Suppose a Bayesian puts the prior $P(\theta = 0) = P(\theta \neq 0) = 1/2$, and in the case of the alternative, the 1/2 is uniformly spread over the interval $[-M/2, M/2]$. Suppose $n = 40\,000$ and $\bar{y} = 0.01$ are observed, so $\sqrt{n}\,\bar{y} = 2$. The classical statistician rejects H_0 at level $\alpha = 0.05$. Show that posterior odds in favor of H_0 are 11 if $M = 1$, indicating that a Bayesian statistician strongly favors H_0, according to Table 4.2.

4.12 This exercise concerning Bayesian binary regression with a probit model using WinBUGS is borrowed from David Madigan's Bayesian Course Site. Finney (1947) describes a binary regression problem with data of size $n = 39$, two continuous predictors x_1 and x_2, and a binary response y. Here are the data in BUGS-ready format:

```
list(n=39,x1=c(3.7,3.5,1.25,0.75,0.8,0.7,0.6,1.1,
  0.9,0.9,0.8,0.55,0.6,1.4,0.75,2.3,3.2,0.85,1.7,1.8,
  0.4,0.95,1.35,1.5,1.6,0.6,1.8,0.95,1.9,1.6,2.7,
  2.35,1.1,1.1,1.2,0.8,0.95,0.75,1.3),
x2=c(0.825,1.09,2.5,1.5,3.2,3.5,0.75,1.7,0.75,0.45,
  0.57,2.75,3.0,2.33,3.75,1.64,1.6,1.415,1.06,1.8,
  2.0,1.36,1.35,1.36,1.78,1.5,1.5,1.9,0.95,0.4,0.75,
  0.03,1.83,2.2,2.0,3.33,1.9,1.9,1.625),
y=c(1,1,1,1,1,1,0,0,0,0,0,0,0,1,1,1,1,1,0,1,0,0,0,0,
  1,0,1,0,1,0,1,0,0,1,1,1,0,0,1))
```

The objective is to build a predictive model that predicts y from x_1 and x_2. Proposed approach is the probit model: $P(y = 1|x_1, x_2) = \Phi(\beta_0 + \beta_1 x_1 + \beta_2 x_2)$ where Φ is the standard normal CDF.
(i) Use WinBUGS to compute posterior distributions for β_0, β_1, and β_2 using diffuse normal priors for each.
(ii) Suppose instead of the diffuse normal prior for β_i, $i = 0, 1, 2$, you use a normal prior with mean zero and variance v_i, and assume the v_is are independently exponentially distributed with some hyperparameter γ. Fit this model using BUGS. How different are the two posterior distributions from this exercise?

4.13 The following WinBUGS code flips a coin; the outcome H is coded by 1 and tails by 0. Mimic the following code to simulate a rolling of a fair die:

```
#coin.bug:
model coin;
{
flip12 ~ dcat(p.coin[])
coin <- flip12 - 1
}
#coin.dat:
list(p.coin=c(0.5, 0.5))
# just generate initials
```

4.14 The highly publicized (recent TV reports) *in vitro fertilization* (IVF) success cases for women in their late 50s all involve donor's egg. If the egg is the woman's own, the story is quite different.

IVF, one of the assisted reproductive technology (ART) procedures, involves extracting a woman's eggs, fertilizing the eggs in the laboratory, and then transferring the resulting embryos into the woman's uterus through the cervix. Fertilization involves a specialized technique known as intracytoplasmic sperm injection (ICSI).

The table shows the live birth success rate per transfer rate from the recipients' eggs, stratified by age of recipient. The data are for year 1999, published by US Centers for Disease Control and Prevention (CDC):

(http://www.cdc.gov/reproductivehealth/ART99/
index99.htm)

Age (x)	24	25	26	27	28	29	30	31
Percentage (y)	38.7	38.6	38.9	41.4	39.7	41.1	38.7	37.6

Age (x)	32	33	34	35	36	37	38	39
Percentage(y)	36.3	36.9	35.7	33.8	33.2	30.1	27.8	22.7

Age (x)	40	41	42	43	44	45	46	
Percentage(y)	21.3	15.4	11.2	9.2	5.4	3.0	1.6	

Assume the change-point regression model

$$y_i = \beta_0 + \beta_1 x_i + \epsilon_i, \quad i = 1, \dots, \tau,$$
$$y_i = \gamma_0 + \gamma_1 x_i + \epsilon_i, \quad i = \tau + 1, \dots, n,$$
$$\epsilon_i \sim \mathcal{N}(0, \sigma^2).$$

(i) Propose priors (with possibly hyperpriors) on $\sigma^2, \beta_0, \beta_1, \gamma_0,$ and γ_1.

(ii) Take discrete uniform prior on τ, and write a WinBUGS program.

4.15 Is the cloning of humans moral? Recent Gallup Poll estimates that about 88% Americans opposed cloning humans. Results are based on telephone interviews with a randomly selected national sample of $n = 1000$ adults, aged 18 and older, conducted 2–4 May 2004. In these 1000 interviews, 882 adults opposed cloning humans.

(i) Write WinBUGS program to estimate the proportion p of people opposed to cloning humans. Use a non-informative prior for p.

(ii) Test the hypothesis that $p \leq 0.87$.

(iii) Pretend that the original poll had $n = 1062$ adults, i.e. results for 62 adults are missing. Estimate the number of people opposed to cloning among the 62 missing in the poll. *Hint*:

```
model {      anticlons ~ dbin(prob,npolled);
             lessthan87 <- step(prob-0.87)
             anticlons.missing ~ dbin(prob,nmissing)
             prob ~ dbeta(1,1)}
Data
list(anticlons=882,npolled= 1000, nmissing=62)
```

RELEVANT R FUNCTIONS AND DATA SETS IN THIS CHAPTER

R code: BA.nornor2.r

ex4.9.txt, ex4.10.txt, exer4.12.txt, exer4.13.txt, exer4.15.txt

---

References

Anscombe, F. J. (1962), "Tests of Goodness of Fit," *Journal of the Royal Statistical Society (B)*, 25, 81–94.

Bayes, T. (1763), "An Essay Towards Solving a Problem in the Doctrine of Chances," *Philosophical Transactions of the Royal Society, London*, 53, 370–418.

Berger, J. O. (1985), *Statistical Decision Theory and Bayesian Analysis*, Second Edition, New York: Springer-Verlag.

Berger, J. O., and Delampady, M. (1987), "Testing Precise Hypothesis," *Statistical Science*, 2, 317–352.

Berger, J. O., and Selke, T. (1987), "Testing a Point Null Hypothesis: The Irreconcilability of *p*-values and Evidence (with Discussion)," *Journal of American Statistical Association*, 82, 112–122.

Chen, M.-H., Shao, Q.-M., and Ibrahim, J. (2000), *Monte Carlo Methods in Bayesian Computation*, New York: Springer-Verlag.

Congdon, P. (2001), *Bayesian Statistical Modelling*, Hoboken, NJ: Wiley.

Congdon, P. (2003), *Applied Bayesian Models*, Hoboken, NJ: Wiley.

Congdon, P. (2005), *Bayesian Models for Categorical Data*, Hoboken, NJ: Wiley.

Finney, D. J. (1947), "The Estimation from Individual Records of the Relationship Between Dose and Quantal Response," *Biometrika*, 34, 320–334.

Gelfand, A. E., and Smith, A. F. M. (1990), "Sampling-based Approaches to Calculating Marginal Densities," *Journal of American Statistical Association*, 85, 398–409.

Lindley, D. V. (1957), "A Statistical Paradox," *Biometrika*, 44, 187–192.

Madigan, D. http://stat.rutgers.edu/ madigan/bayes02/. A Web Site for Course on Bayesian Statistics.

Martz, H., and Waller, R. (1985), *Bayesian Reliability Analysis*, New York: Wiley.

Metropolis, N., Rosenbluth, A., Rosenbluth, M., Teller, A., and Teller, E. (1953), "Equation of State Calculations by Fast Computing Machines," *The Journal of Chemical Physics*, 21, 1087–1092.

Robert, C. (2001), *The Bayesian Choice: From Decision-TheoreticMotivations to Computational Implementation*, Second Edition, New York: Springer-Verlag.

Robert, C., and Casella, G. (2004), *Monte Carlo Statistical Methods*, Second Edition, New York: Springer-Verlag.

Spiegelhalter, D. J., Thomas, A., Best, N. G., and Gilks, W. R. (1996), "BUGS Examples Volume 1," Version 0.5. Cambridge: Medical Research Council Biostatistics Unit (PDF).

5

Order Statistics

The early bird gets the worm, but the second mouse gets the cheese.

Steven Wright

Let $X_1, X_2, \ldots, X_n$ be an independent sample from a population with absolutely continuous cumulative distribution function F and density f. The continuity of F implies that $P(X_i = X_j) = 0$, when $i \neq j$ and the sample could be ordered with strict inequalities:

$$X_{1:n} < X_{2:n} < \cdots < X_{(n-1):n} < X_{n:n},\qquad(5.1)$$

where $X_{i:n}$ is called the *ith order statistic* (out of *n*). The *range* of the data is $X_{n:n} - X_{1:n}$, where $X_{n:n}$ and $X_{1:n}$ are, respectively, the sample maximum and minimum. The study of order statistics permeates through all areas of statistics, including nonparametric. There are some worthwhile books dedicated just to probability and statistics related to order statistics; the updated textbook by David and Nagaraja (2003) serves as the standard reference book, and the book by Arnold, Balakrishnan, and Nagaraja (1992) offers a lucid survey for anyone interested in learning about a wide range of applications for order statistics.

The marginal distribution of $X_{i:n}$ is not the same as X_i. Its distribution function $F_{i:n}(t) = P(X_{i:n} \leq t)$ is the probability that *at least i* out of *n* observations from the original sample are no greater than *t* or

$$F_{i:n}(t) = P(X_{i:n} \leq t) = \sum_{k=i}^{n} \binom{n}{k} F(t)^k (1 - F(t))^{n-k}.$$

Nonparametric Statistics with Applications to Science and Engineering with R, Second Edition.
Paul Kvam, Brani Vidakovic, and Seong-joon Kim.
© 2023 John Wiley & Sons, Inc. Published 2023 by John Wiley & Sons, Inc.

If F is differentiable, it is possible to show that the corresponding density function is

$$f_{i:n}(t) = i \binom{n}{i} F(t)^{i-1}(1 - F(t))^{n-i}f(t). \tag{5.2}$$

Example 5.1 Series Systems. In reliability, series and parallel systems are building blocks for system analysis and design. A *series system* is one that works only if all of its components are working. If the lifetimes of a n-component system $(X_1, \ldots, X_n)$ are independently and identically distributed (i.i.d.) with distribution function F, then, if the system is in series, the probability the system survives up to time t is the probability the failure time of all n components exceeds t, which is $P(X_1 > t, \ldots, X_n > t) = \prod_{i=1}^{n} P(X_i > t) = (1 - F(t))^n$. Thus, the lifetime of a series system is the minimum order statistic, $X_{1:n}$. If the individual components $X_i \sim \mathcal{Exp}(\lambda)$, then $X_{1:n} \sim \mathcal{Exp}(n\lambda)$.

Example 5.2 Parallel Systems. A *parallel system* is one that fails only if all of its components fail. If the components are i.i.d. with distribution function F, the distribution function for the parallel system can be computed $P(X_1 \le t, \ldots, X_n \le t) = F(t)^n$, so the lifetime of a parallel system is $X_{n:n}$. Diagrams for series and parallel systems are shown in Figure 5.1.

Example 5.3 Recall that for any continuous distribution F, the transformed sample $F(X_1), \ldots, F(X_n)$ is distributed $\mathcal{U}(0,1)$. Similarly, from (5.2) the distribution of $F(X_{i:n})$ is $Be(i, n - i + 1)$. Using the R code below, the densities are graphed in Figure 5.2:

```
> p<-ggplot()
> for(i in 1:5){
+ eval(parse(text=paste("p <- p + geom_line(aes(x=x,y=dbeta(x,",
+ i,",6-",i,")),lwd=0.8)")))
+ }
> p <- p+xlim(c(0,1))+ylim(c(0,5))+xlab("")+ylab("")
> print(p)
```

(a)   (b)

Figure 5.1 Diagram of simple system of three components in series (a) and parallel (b).

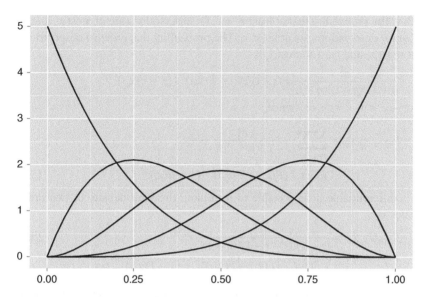

Figure 5.2 Distribution of order statistics from a sample of five $\mathcal{U}(0,1)$.

5.1 Joint Distributions of Order Statistics

Unlike the original sample $(X_1, X_2, \ldots, X_n)$, the set of order statistics is inevitably dependent. If the vector $(X_1, X_2, \ldots, X_n)$ has a joint density

$$f_{1,2,\ldots,n}(x_1, x_2, \ldots, x_n) = \prod_{i=1}^{n} f(x_i),$$

then the joint density for the order statistics $f_{1,2,\ldots,n:n}(x_1, \ldots, x_n)$ is

$$f_{1,2,\ldots,n:n}(x_1, \ldots, x_n) = \begin{cases} n! \prod_{i=1}^{n} f(x_i), & x_1 < x_2 < \cdots < x_n, \\ 0, & \text{otherwise.} \end{cases} \tag{5.3}$$

To understand why this is true, consider the conditional distribution of the order statistics $\mathbf{y} = (x_{1:n}, x_{2:n}, \ldots, x_{n:n})$ given $\mathbf{x} = (x_1, x_2, \ldots, x_n)$. Each one of the $n!$ permutations of $(X_1, X_2, \ldots, X_n)$ is equal in probability, so computing $f_{\mathbf{y}} = \int f_{\mathbf{y}|\mathbf{x}} \, dF_{\mathbf{x}}$ is incidental. The joint density can also be derived using a Jacobian transformation (see Exercise 5.3).

From (5.3) we can obtain the distribution of any subset of order statistics. The joint distribution of $X_{r:n}, X_{s:n}, 1 \leq r < s \leq n$ is defined as

$$F_{r,s:n}(x_r, x_s) = P(X_{r:n} \leq x_r, X_{s:n} \leq x_s),$$

which is the probability that at least r out of n observations are at most x_r *and* at least s of n observations are at most x_s. The probability that *exactly* i observations are at most x_r and j are at most x_s is

$$\frac{n!}{(i-1)!(j-i)!(n-j)!}F(x_r)^i\left(F(x_s)-F(x_r)\right)^{j-i}\left(1-F(x_s)\right)^{n-j},$$

where $-\infty < x_r < x_s < \infty$; hence,

$$F_{r,s:n}(x_r,x_s) = \sum_{j=s}^{n}\sum_{i=r}^{s}\frac{n!}{(i-1)!(j-i)!(n-j)!}$$
$$\times F(x_r)^i\left(F(x_s)-F(x_r)\right)^{j-i}\left(1-F(x_s)\right)^{n-j}. \tag{5.4}$$

If F is differentiable, it is possible to formulate the joint density of two order statistics as

$$f_{r,s:n}(x_r,x_s) = \frac{n!}{(r-1)!(s-r-1)!(n-s)!} \tag{5.5}$$
$$\times F(x_r)^{r-1}\left(F(x_s)-F(x_r)\right)^{s-r-1}\left(1-F(x_s)\right)^{n-s}f(x_r)f(x_s).$$

Example 5.4 Sample Range. The range of the sample, R, defined before as $X_{n:n}-X_{1:n}$, has density

$$f_R(u) = \int_{-\infty}^{\infty}n(n-1)[F(v)-F(v-u)]^{n-2}f(v-u)f(v)\,dv. \tag{5.6}$$

To find $f_R(u)$, start with the joint distribution of $(X_{1:n},X_{n:n})$ in (5.5):

$$f_{1,n:n}(y_1,y_n) = n(n-1)[F(y_n)-F(y_1)]^{n-2}f(y_1)f(y_n),$$

and make the transformation

$$u = y_n - y_1,$$
$$v = y_n.$$

The Jacobian of this transformation is 1, and $y_1 = v - u$, $y_n = v$. Plug y_1, y_n into the joint distribution $f_{1,n:n}(y_1,y_n)$, and integrate out v to arrive at (5.6). For the special case in which $F(t) = t$, the probability density function for the sample range simplifies to

$$f_R(u) = n(n-1)u^{n-2}(1-u), \quad 0 < u < 1.$$

5.2 Sample Quantiles

Recall that for a distribution F, the pth quantile (x_p) is the value x such that $F(x) = p$, if the distribution is continuous, and, more generally, such that $F(x) \geq p$ and $P(X \geq x) \geq 1 - p$, if the distribution is arbitrary. For example, if the distribution F is discrete, there may not be any value x for which $F(x) = p$.

Analogously, if $X_1, \ldots, X_n$ represents a sample from F, the *pth sample quantile* ($\hat{x}_p$) is a value of x such that $100p\%$ of the sample is smaller than x. This is also called the $100p\%$ *sample percentile*. With large samples, there is a number $1 \le r \le n$ such that $X_{r:n} \approx x_p$. Specifically, if n is large enough so that $p(n+1) = r$ for some $r \in \mathbb{Z}$, then $\hat{x}_p = X_{r:n}$ because there would be $r - 1$ values smaller than $\hat{x}_p$ in the sample and $n - r$ larger than it.

If $p(n+1)$ is not an integer, we can consider estimating the population quantile by an inner point between two order statistics, say, $X_{r:n}$ and $X_{(r+1):n}$, where $F(X_{r:n}) < p - \varepsilon$ and $F(X_{(r+1):n}) > p + \varepsilon$ for some small $\varepsilon > 0$. In this case, we can use a number that interpolates the value of $\hat{x}_p$ using the line between $(X_{r:n}, r/(n+1))$ and $(X_{(r+1):n}, (r+1)/(n+1))$:

$$\hat{x}_p = (-p(n+1) + r + 1)X_{r:n} + (p(n+1) - r)X_{(r+1):n}. \tag{5.7}$$

Note that if $p = 1/2$ and n is an even number, then $r = n/2$ and $r + 1 = n/2 + 1$ and $\hat{x}_p = (X_{\frac{n}{2}:n} + X_{(\frac{n}{2}+1):n})/2$. That is, the sample median is the average of the two middle sample order statistics.

We note that there are alternative definitions of sample quantile in the literature, but they all have the same large sample properties.

5.3 Tolerance Intervals

Unlike the confidence interval, which is constructed to contain an unknown parameter with some specified degree of uncertainty (say, $1 - \gamma$), a *tolerance interval* contains at least a proportion p of the population with probability γ. That is, a tolerance interval is a confidence interval for a distribution. Both p, the proportion of coverage, and $1 - \gamma$, the uncertainty associated with the confidence statement, are predefined probabilities. For instance, we may be 95% confident that 90% of the population will fall within the range specified by a tolerance interval.

Order statistics play an important role in the construction of tolerance intervals. From a sample $X_1, \ldots, X_n$ from (continuous) distribution F, two statistics $T_1 < T_2$ represent a $100\gamma\%$ tolerance interval for $100p\%$ of the distribution F if

$$P\left(F(T_2) - F(T_1) \ge p\right) \ge \gamma.$$

Obviously, the distribution $F(T_2) - F(T_1)$ should not depend on F. Recall that for an order statistic $X_{i:n}$, $U_{i:n} \equiv F(X_{i:n})$ is distributed $Be(i, n - i + 1)$. Choosing T_1 and T_2 from the set of order statistics satisfies the requirements of the tolerance interval, and the computations are not difficult.

One-sided tolerance intervals are related to confidence intervals for quantiles. For instance, a 90% upper tolerance bound for 95% of the population is identical to a 90% one-sided confidence interval for $x_{0.95}$, the 0.95 quantile of the distribution. With a sample of $x_1, \ldots, x_n$ from F, a γ interval for $100p\%$ of the population would be constructed as $(-\infty, x_{r:n})$ for some $r \in \{1, \ldots, n\}$.

Here are four simple steps to help determine r:

1. We seek r so that $P(-\infty < x_p < X_{r:n}) = \gamma = P(X_{r:n} > x_p)$.
2. At most $r - 1$ out of n observations are less than x_p.
3. Let Y = number of observations less than x_p so that $Y \sim Bin(n, p)$ if x_p is the pth quantile
4. Find r large enough so that $P(Y \leq r - 1) = \gamma$.

Example 5.5 A 90% upper confidence bound for the 75th percentile (or *upper quartile*) is found by assigning Y = number of observations less than $x_{0.75}$, where $Y \sim Bin(n, 0.75)$. Let $n = 20$. Note $P(Y \leq 16) = 0.7748$ and $P(Y \leq 17) = 0.9087$, so $r - 1 = 17$. The 90% upper bound for $x_{0.75}$, which is equivalent to a 90% upper tolerance bound for 75% of the population, is $x_{18:20}$ (the third largest observation out of 20).

For large samples, the normal approximation allows us to generate an upper bound more simply. For the upper bound $x_{r:n}$, r is approximated with

$$\tilde{r} = np + z_\gamma \sqrt{np(1-p)}.$$

In the example above, with $n = 20$ (of course, this is not exactly what we think of as "large"), $\tilde{r} = 20(0.75) + 1.28\sqrt{0.75(0.25)20} = 17.48$. According to this rule, $x_{17:20}$ is insufficient for the approximate interval, so $x_{18:20}$ is again the upper bound.

Example 5.6 Sample Range, Continued. From a sample of n, what is the probability that $100p\%$ of the population lies within the sample range $(X_{1:n}, X_{n:n})$?

$$P\left(F(X_{n:n}) - F(X_{1:n}) \geq p\right) = 1 - P\left(U_n < p\right),$$

where $U_n = U_{n:n} - U_{1:n}$. From (5.6) it was shown that $U_n \sim Be(n-1, 2)$. If we let $\gamma = P(U_n \geq p)$, then γ, the tolerance Coefficient, can be solved:

$$1 - \gamma = np^{n-1} - (n-1)p^n.$$

Example 5.7 The tolerance interval is especially useful in compliance monitoring at industrial sites. Suppose one is interested in maximum contaminant levels (MCLs). The tolerance interval already takes into account the fact that some values will be high. So if a few values exceed the MCL standard, a site may still not be in violation (because the calculated tolerance interval may still be lower than the MCL). However, if too many values are above the MCL, the calculated tolerance interval will extend beyond the acceptable standard. As few as three data points can be used to generate a tolerance interval, but the EPA recommends having at least eight points for the interval to have any usefulness (EPA/530-R-93-003).

Example 5.8 How large must a sample size n be so that at least 75% of the contamination levels are between $X_{2:n}$ and $X_{(n-1):n}$ with probability of at least 0.95? If we follow the approach above, the distribution of $V_n = U_{(n-1):n} - U_{2:n}$ is

$$Be((n-1)-2, n-(n-1)+2+1) = Be(n-3, 4).$$

We need n so that $P(V_n \geq 0.75) = \text{qbeta}(0.25, 4, n-3) \geq 0.95$ that occurs as long as $n \geq 29$.

5.4 Asymptotic Distributions of Order Statistics

Let $X_{r:n}$ be rth order statistic in a sample of size n from a population with an absolutely continuous distribution function F having a density f. Let $r/n \to p$, when $n \to \infty$. Then,

$$\sqrt{\frac{n}{p(1-p)}} f(x_p)(X_{r:n} - x_p) \Longrightarrow \mathcal{N}(0,1),$$

where x_p is pth quantile of F, i.e. $F(x_p) = p$.

Let $X_{r:n}$ and $X_{s:n}$ be rth and sth order statistics ($r < s$) in the sample of size n. Let $r/n \to p_1$ and $s/n \to p_2$, when $n \to \infty$. Then, for large n,

$$\begin{pmatrix} X_{r:n} \\ X_{s:n} \end{pmatrix} \overset{\text{appr}}{\underset{\sim}{}} \mathcal{N}\left(\begin{bmatrix} x_{p_1} \\ x_{p_2} \end{bmatrix}, \Sigma \right),$$

where

$$\Sigma = \begin{bmatrix} p_1(1-p_1)[f(x_{p_1})]^{-2}/n & p_1(1-p_2)/[nf(x_{p_1})f(x_{p_2})]^{-1} \\ p_1(1-p_2)/[nf(x_{p_1})f(x_{p_2})]^{-1} & p_2(1-p_2)[f(x_{p_2})]^{-2}/n \end{bmatrix},$$

and x_{p_i} is p_ith quantile of F.

Example 5.9 Suppose we are estimating the population median $x_{0.50}$. Then we use $r = n/2$, and our estimator is $\hat{x}_{0.50} = x_{(n/2):n}$. If X has an exponential distribution with density $f(x) = \theta \exp(-\theta x)$, for $x > 0$, then $x_{0.50} = \ln(2)/\theta$ and

$$\sqrt{n}\left(\hat{x}_{0.50} - x_{0.50}\right) \Longrightarrow \mathcal{N}\left(0, \theta^{-2}\right).$$

5.5 Extreme Value Theory

I don't have to outrun the bear, I just have to outrun you.

– Old proverb

Earlier we equated a series system lifetime (of n i.i.d. components) with the sample minimum $X_{1:n}$. The limiting distribution of the minima or maxima is not so interesting, e.g. if X has distribution function F, $X_{1:n} \rightarrow x_0$, where $x_0 = \inf_x \{x : F(x) > 0\}$. However, the *standardized limit* is more interesting. For an example involving sample maxima, with $X_1, \ldots, X_n$ from an exponential distribution with mean 1, consider the asymptotic distribution of $X_{n:n} - \log(n)$:

$$P(X_{n:n} - \log(n) \leq t)) = P(X_{n:n} \leq t + \log(n)) = [1 - \exp\{-t - \log(n)\}]^n$$
$$= [1 - e^{-t} n^{-1}]^n \rightarrow \exp\{-e^{-t}\}.$$

This is because $(1 + \alpha/n)^n \rightarrow e^\alpha$ as $n \rightarrow \infty$. This distribution, a special form of the Gumbel distribution, is also called the *extreme value distribution*.

Extreme value theory states that the standardized series system lifetime converges to one of the three following distribution types F^* (not including scale and location transformation) as the number of components increases to infinity:

Gumbel $\qquad F^*(x) = \exp(-\exp(-x)), -\infty < x < \infty.$

Fréchet $\qquad F^*(x) = \begin{cases} \exp(-x^{-a}), & x > 0, \, a > 0, \\ 0, & x \leq 0. \end{cases}$

Negative Weibull $\quad F^*(x) = \begin{cases} \exp(-(-x)^a), & x < 0, \, a > 0 \\ 0, & x \geq 0. \end{cases}$

For engineering applications, the class of extreme value distributions (the Weibull, in particular) is a popular choice for modeling in which the outcome represents the minimum of a large number of (non-i.i.d.) random factors. Extreme value distributions are commonly applied for the treatment of financial risks in which a possible investment's value will be more than three standard deviations away from its forecasted mean, according to the financial models (presumably based on assumptions of normality). A more comprehensive survey on extreme value theory is presented by Coles (2001).

5.6 Ranked Set Sampling

Suppose a researcher is sent out to Leech Lake, Minnesota, to ascertain the average weight of Walleye fish caught from that lake. She obtains her data by stopping the fishermen as they are returning to the dock after a day of fishing. In the time the researcher waited at the dock, three fishermen arrived, each with their daily limit of three Walleye. Because of limited time, she only has time to make one measurement with each fisherman, so at the end of her field study, she will get three measurements.

McIntyre (1952) discovered that with this forced limitation on measurements, there is an efficient way of getting information about the population mean. We might assume the researcher selected the fish to be measured randomly for each of the three fishermen that were returning to shore. McIntyre found that if she instead inspected the fish visually and selected them non-randomly, the data could beget a better estimator for the mean. Specifically, suppose the researcher examines the three Walleye from the first fisherman and selects the smallest one for measurement. She measures the second smallest from the next batch and the largest from the third batch.

Opposed to a simple random sample (SRS), this *ranked set sample* (RSS) consists of independent order statistics that we will denote by $X_{[1:3]}$, $X_{[2:3]}$, and $X_{[3:3]}$. If $\overline{X}$ is the sample mean from a SRS of size n, and $\overline{X}_{RSS}$ is the mean of a RSS $X_{[1:n]}, \ldots, X_{[n:n]}$, it is easy to show that like $\overline{X}$, $\overline{X}_{RSS}$ is an unbiased estimator of the population mean. Moreover, it has smaller variance. That is, $\mathbb{V}ar(\overline{X}_{RSS}) \leq \mathbb{V}ar(\overline{X})$.

This property is investigated further in the exercises. The key is that variances for order statistics are generally smaller than the variance of the i.i.d. measurements. If you think about the SRS estimator as a linear combination of order statistics, it differs from the linear combination of order statistics from an RSS by its covariance structure. It seems apparent, then, that the expected value of $\overline{X}_{RSS}$ must be the same as the expected value of a $\overline{X}_{RSS}$.

The sampling aspect of RSS has received the most attention. Estimators of other parameters can be constructed to be more efficient than SRS estimators, including nonparametric estimators of the cumulative distribution function (CDF) (Stokes and Sager, 1988). The book by Chen, Bai, and Sinha (2003) is a comprehensive guide about basic results and recent findings in RSS theory.

5.7 Exercises

5.1 In R, generate a sequence of 50 uniform random numbers, and find their range. Repeat this procedure $M = 1000$ times; you will obtain 1000 ranges for 1000 sequences of 50 uniforms. Next, simulate 1000 percentiles from a beta $Be(49,2)$ distribution for $p = (1 : 1000)/1001$. Use R function `qbeta(p, 49, 2)`. Produce a histogram for both sets of data, comparing the ordered ranges and percentiles of their theoretical distribution, $Be(49,2)$.

5.2 For a set of i.i.d. data from a continuous distribution $F(x)$, derive the probability density function of the order statistic $X_{i:n}$ in (5.2).

5.3 For a sample of $n = 3$ observations, use a Jacobian transformation to derive the joint density of the order statistics, $X_{1:3}, X_{2:3}, X_{3:3}$.

5.4 Consider a system that is composed of n identical components that have independent life distributions. In reliability, a *k-out-of-n system* is one for which at least k out of n components must work in order for the system to work. If the components have lifetime distribution F, find the distribution of the system lifetime, and relate it to the order statistics of the component lifetimes.

5.5 In 2003, the lab of Human Computer Interaction and Health Care Informatics at the Georgia Institute of Technology conducted empirical research on the performance of patients with diabetic retinopathy. The experiment included 29 participants placed either in the control group (without diabetic retinopathy) or the treatment group (with diabetic retinopathy). The visual acuity data of all participants are listed below. Normal visual acuity is 20/20, and 20/60 means a person sees at 20 ft what a normal person sees at 60 ft:

20/20	20/20	20/20	20/25	20/15	20/30	20/25	20/20
20/25	20/80	20/30	20/25	20/30	20/50	20/30	20/20
20/15	20/20	20/25	20/16	20/30	20/15	20/15	20/25

The data of five participants were excluded from the table due to their failure to meet the requirement of the experiment, so 24 participants are counted in all. To verify if the data can represent the visual acuity of the general population, a 90% upper tolerance bound for 80% of the population is calculated.

5.6 In R, repeat the following $M = 10\,000$ times:
- Generate a normal sample of size $n = 100$, $X_1, \ldots, X_{100}$.
- For a two-sided tolerance interval, fix the coverage probability as $p = 0.8$, and use the random interval $(X_{5:100}, X_{95:100})$. This interval will cover the proportion $F_X(X_{95:100}) - F_X(X_{5:100}) = U_{95:100} - U_{5:100}$ of the normal population.
- Count how many times in M runs $U_{95:100} - U_{5:100}$ exceeds the preassigned coverage p? Use this count to estimate γ.
- Compare the simulation estimator of γ with the theory, $\gamma = 1$ - pbeta (p, s-r, (n+1)-(s-r)).
  What if instead of normal sample you used an exponentially distributed sample?

5.7   With an i.i.d. sample of n generated from $\mathcal{U}(0,1)$, prove that the sample minima $X_{1:n}$ converges to zero and that $Y_n = nX_{1:n}$ converges in distribution to an exponential distribution.

5.8   Suppose that components of a system are i.i.d. $\mathcal{U}(0,1)$ lifetime. By standardizing with $1/n$ where n is the number of components in the system, find the limiting lifetime distribution of a parallel system as the number of components increases to infinity.

5.9   How large of a sample is needed in order for the sample range to serve as a 99% tolerance interval that contains 90% of the population?

5.10   How large must the sample be in order to have 95% confidence that at least 90% of the population is less than $X_{(n-1):n}$?

5.11   For a large sample of i.i.d. randomly generated $\mathcal{U}(0,1)$ variables, compare the asymptotic distribution of the sample mean with that of the sample median.

5.12   Prove that a RSS mean is unbiased for estimating the population mean by showing that $\sum_{i=1}^{n} \mathbb{E}(X_{[i:n]}) = n\mu$. In the case the underlying data are generated from $\mathcal{U}(0,1)$, prove that the sample variance for the RSS mean is strictly less than that of the sample mean from a SRS.

5.13   Find a 90% upper tolerance interval for the 99th percentile of a sample of size $n=1000$.

5.14   Suppose that N items, labeled by sequential integers as $\{1,2,\dots,N\}$, constitute the population. Let $X_1, X_2, \dots, X_n$ be a sample of size n (without repeating) from this population, and let $X_{1:n}, \dots, X_{n:n}$ be the order statistics. It is of interest to estimate the size of population, N.

This theoretical scenario is a basis for several interesting popular problems: tramcars in San Francisco, captured enemy tanks, maximal lottery number, etc. The most popular is the story about German tanks captured during the Second World War, featured in *The Guardian* (2006). The full story is quite interesting, but the bottom line is to estimate total size of production if five German tanks with "serial numbers" 12, 33, 37, 78, and 103 have been captured by Allied forces:

(i) Show that the distribution of $X_{i:n}$ is

$$P(X_{i:n} = k) = \frac{\binom{k-1}{i-1}\binom{N-k}{n-i}}{\binom{N}{n}}, \quad k = i, i+1, \ldots, N-n+i.$$

(ii) Using the identity $\sum_{k=i}^{N-n+i} \binom{k-1}{i-1}\binom{N-k}{n-i} = \binom{N}{n}$ and distribution from (i), show that $\mathbb{E}X_{i:n} = i(N+1)/(n+1)$.

(iii) Show that the estimator $Y_i = (n+1)/iX_{i:n} - 1$ is unbiased for estimating N for any $i = 1, 2, \ldots, n$. Estimate number of tanks N on basis of Y_5 from the observed sample $\{12, 33, 37, 78, 103\}$.

5.15 Let $X_1, \ldots, X_n$ be i.i.d. $\mathcal{U}(\theta - 1, \theta + 1)$. Find the joint density of the largest and smallest order statistics, and derive an expression for $P(X_{1:n} < \theta < X_{n:n})$.

5.16 Let $X_1, \ldots, X_n$ be i.i.d. $\mathcal{Exp}(\lambda)$. The *spacings* between order statistics are defined as $S_i = X_{i:n} - X_{(i-1):n}$, $i = 2, \ldots, n$, and $S_1 = X_{1:n}$. Show that the set of spacings $(S_1, \ldots, S_n)$ are independent variables and that, for $i \in \{1, \ldots, n\}$,

$$S_i \sim \mathcal{Gamma}(1, \lambda/(n-i+1)).$$

5.17 Suppose $X_1, \ldots, X_5$ are i.i.d. $\mathcal{Beta}(2,1)$. Find the joint density of $(X_{2:5}, X_{5:5})$, and compute the interval probability $P(X_{5:5} - X_{2:5} \geq 0.3)$.

5.18 Suppose the components of a series system have Weibull lifetimes, i.e. $X_1, \ldots, X_5$ are i.i.d. $\mathcal{W}(a, b)$. Find the distribution of the system lifetime, and compare its median lifetime with that of its components.

References

Arnold, B. C., Balakrishnan, N., and Nagaraja, H. N. (1992), *A First Course in Order Statistics*, New York: Wiley.

Chen, Z., Bai, Z., and Sinha, B. K. (2003), *Ranked Set Sampling: Theory and Applications*, New York: Springer-Verlag.

Coles, S. (2001), *An Introduction to Statistical Modeling of Extreme Values*, New York: Springer-Verlag.

David, H. A., and Nagaraj, H. N. (2003), *Order Statistics*, Third Edition, New York: Wiley.

McIntyre, G. A. (1952), "A Method for Unbiased Selective Sampling Using Ranked Sets," *Australian Journal of Agricultural Research*, 3, 385–390.

Stokes, S. L., and Sager, T. W. (1988), "Characterization of a Ranked-Set Sample with Application to Estimating Distribution Functions," *Journal of the American Statistical Association*, 83, 374–381.

The Guardian (2006), "Gavyn Davies Does the Maths: How a Statistical Formula Won the War," Thursday, July 20, 2006.

6

Goodness of Fit

Believe nothing just because a so-called wise person said it.
Believe nothing just because a belief is generally held.
Believe nothing just because it is said in ancient books.
Believe nothing just because it is said to be of divine origin.
Believe nothing just because someone else believes it.
Believe only what you yourself test and judge to be true.

paraphrased from the Buddha

Modern experiments are plagued by well-meaning assumptions postulating that the data are distributed according to some "textbook" cumulative distribution function (CDF). This chapter introduces methods to test the merits of a hypothesized distribution in fitting the data. The term *goodness of fit* was coined by Pearson in 1902 and refers to statistical tests that check the quality of a model or a distribution's fit to a set of data. The first measure of goodness of fit for general distributions was derived by Kolmogorov (1933). Andrei Nikolaevich Kolmogorov (1905–1987), perhaps the most accomplished and celebrated Soviet mathematician of all time, made fundamental contributions to probability theory, including test statistics for distribution functions – some of which bear his name. Nikolai Vasil'yevich Smirnov (1900–1966), another Soviet mathematician, extended Kolmogorov's results to two samples.

In this section we emphasize objective tests (with *p*-values) and later analyze *graphical* methods for testing goodness of fit. Recall the empirical distribution function (EDF) from p. 36. The *Kolmogorov statistic* (sometimes called the Kolmogorov–Smirnov [KS] test statistic)

$$D_n = \sup_t |F_n(t) - F(t)|$$

is a basis to many nonparametric goodness-of-fit tests for distributions, and this is where we will start.

Nonparametric Statistics with Applications to Science and Engineering with R, Second Edition.
Paul Kvam, Brani Vidakovic, and Seong-joon Kim.
© 2023 John Wiley & Sons, Inc. Published 2023 by John Wiley & Sons, Inc.

6.1 Kolmogorov–Smirnov Test Statistic

Let $X_1, X_2, \ldots, X_n$ be a sample from a population with continuous, but unknown CDF F. As in (3.1), let $F_n(x)$ be the empirical CDF based on the sample. To test the hypothesis

$$H_0 : F(x) = F_0(x), \quad (\forall x)$$

versus the alternative

$$H_1 : F(x) \neq F_0(x),$$

we use the modified statistics $\sqrt{n}D_n = sup_x \sqrt{n}|F_n(x) - F_0(x)|$ calculated from the sample as

$$\sqrt{n}D_n = \sqrt{n} \ \max\{\max_i |F_n(X_i) - F_0(X_i)|, \max_i |F_n(X_i^-) - F_0(X_i)|\}.$$

Figure 6.1 shows how the supremum is found based on an empirical distribution. In this example, we used a built in data set called `rivers` that lists the length (in miles) of 141 major rivers in North America. Although the KS test statistic is criticized for its reliance on sample extremes (we can visually pick out all the places in which the normal distribution does not fit the data), it still maintains optimal test qualities.

This is a simple discrete optimization problem because F_n is a step function and F_0 is nondecreasing so the maximum discrepancy between F_n and F_0 occurs at the observation points or at their left limits. When the hypothesis H_0 is true, the statistic $\sqrt{n}D_n$ is distributed free of F_0. In fact, Kolmogorov (1933) showed that under H_0,

$$P(\sqrt{n}D_n \leq d) \Longrightarrow H(d) = 1 - 2\sum_{j=1}^{\infty}(-1)^{j-1} \ e^{-2j^2 d^2}.$$

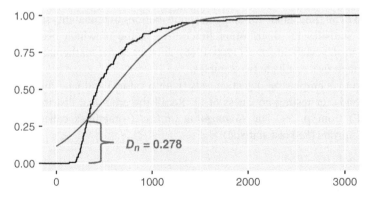

Figure 6.1 Comparing the EDF for river length data versus normal distribution.

In practice, most KS tests are two sided, testing whether the F is equal to F_0, the distribution postulated by H_0, or not. Alternatively, we might test to see if the distribution is larger or smaller than a hypothesized F_0. For example, to find out if X is stochastically smaller than Y ($F_X(x) \geq F_Y(x)$), two one-sided alternatives that can be tested are

$$H_{1,-} : F_X(x) \leq F_0(x) \quad \text{or} \quad H_{1,+} : F_X(x) \geq F_0(x).$$

Appropriate statistics for testing $H_{1,-}$ and $H_{1,+}$ are

$$\sqrt{n}D_n^- \equiv -\inf_x \sqrt{n}(F_n(x) - F_0(x)),$$

$$\sqrt{n}D_n^+ \equiv \sup_x \sqrt{n}(F_n(x) - F_0(x)),$$

which are calculated at the sample values as

$$\sqrt{n}D_n^- = \sqrt{n} \max \{\max_i (F_0(X_i) - F_n(X_i^-)), 0\} \text{and}$$

$$\sqrt{n}D_n^+ = \sqrt{n} \max \{\max_i (F_n(X_i) - F_0(X_i)), 0\}.$$

Obviously, $D_n = \max \{D_n^-, D_n^+\}$. In terms of order statistics,

$$D_n^+ = \max \{\max_i (F_n(X_i) - F_0(X_i)), 0\} = \max \{\max_i (i/n - F_0(X_{i:n}), 0\} \text{ and}$$

$$D_n^- = \max \{\max_i (F_0(X_{i:n}) - (i-1)/n), 0\}.$$

Under H_0, the distributions of D_n^+ and D_n^- coincide. Although conceptually straightforward, the derivation of the distribution for D_n^+ is quite involved. Under H_0, for $c \in (0,1)$, we have

$$P(D_n^+ < c) = P(i/n - U_{i:n} < c, \quad \text{for all } i = 1, 2, \ldots, n)$$

$$= P(U_{i:n} > i/n - c, \quad \text{for all } i = 1, 2, \ldots, n)$$

$$= \int_{1-c}^1 \int_{\frac{n-1}{n}-c}^1 \cdots \int_{\frac{2}{n}-c}^1 \int_{\frac{1}{n}-c}^1 f(u_1, \ldots, u_n) \, du_1 \cdots du_n,$$

where $f(u_1, \ldots, u_n) = n!\mathbf{1}(0 < u_1 < \cdots < u_n < 1)$ is the joint density of n order statistics from $\mathcal{U}(0,1)$.

Birnbaum and Tingey (1951) derived a more computationally friendly representation; if c is the observed value of D_n^+ (or D_n^-), then the p-value for testing H_0 against the corresponding one-sided alternative is

$$P(\sqrt{n}D_n^+ > c) = (1-c)^n + c \sum_{j=1}^{\lfloor n(1-c) \rfloor} \binom{n}{j} (1 - c - j/n)^{n-j}(c + j/n)^{j-1}.$$

This is an exact p-value. When the sample size n is large (enough so that the error of order $O(n^{-3/2})$ can be tolerated), an approximation can be used:

$$P\left[\frac{(6nD_n^+ + 1)^2}{18n} > x\right] = e^{-x}\left(1 - \frac{2x^2 - 4x - 1}{18n}\right) + O\left(n^{-3/2}\right).$$

To obtain the p-value approximation, take $x = (6nc + 1)^2/(18n)$, where c is the observed D_n^+ (or D_n^-), and plug in the right-hand side of the above equation.

Table 6.1, taken from Miller (1956), lists quantiles of D_n^+ for values of $n \leq 40$. The D_n^+ values refer to the one-sided test, so for the two-sided test, we would reject H_0 at level α if $D_n^+ > k_n(1 - \alpha/2)$, where $k_n(1 - \alpha)$ is the tabled quantile under α. If $n > 40$, we can approximate these quantiles $k_n(\alpha)$ as

k_n	$1.07/\sqrt{n}$	$1.22/\sqrt{n}$	$1.36/\sqrt{n}$	$1.52/\sqrt{n}$	$1.63/\sqrt{n}$
α	0.10	0.05	0.025	0.01	0.005

Later, we will discuss alternative tests for distribution goodness of fit. The KS test has advantages over exact tests based on the χ^2 goodness-of-fit statistic (see Chapter 9), which depend on an adequate sample size and proper interval assignments for the approximations to be valid. The KS test has important limitations, too. Technically, it only applies to continuous distributions. The KS statistic tends to be more sensitive near the center of the distribution than at the tails. Perhaps the most serious limitation is that the distribution must be fully specified. That is, if location, scale, and shape parameters are estimated from the data, the critical region of the KS test is no longer valid. It typically must be determined by simulation.

In R, ks.test can be used to perform the one-sample test. Along with the sample, the user specifies the null distribution using its CDF in R (e.g. "pnorm" and "pgamma").

Example 6.1 With five observations $\{0.1, 0.14, 0.2, 0.48, 0.58\}$, we wish to test H_0. Data are distributed $\mathcal{U}(0,1)$ versus H_1. Data are not distributed $\mathcal{U}(0,1)$. We check F_n and $F_0(x) = x$ at the five points of data along with their left-hand limits. $|F_n(x_i) - F_0(x_i)|$ equals $(0.1, 0.26, 0.4, 0.32, 0.42)$ at $i = 1, \ldots, 5$, and $|F_n(x_i^-) - F_0(x_i)|$ equals $(0.1, 0.06, 0.2, 0.12, 0.22)$, so that $D_n = 0.42$. According to the table, $k_5(0.10) = 0.44698$. This is a two-sided test, so the test statistic is not rejectable at $\alpha = 0.20$. This is due more to the lack of sample size than the evidence presented by the five observations.

Example 6.2 If we analyze the data from Example 6.1 using the R function ks.test, the p-value is computing using methods by Marsaglia, Tsang, and Wang (2003):

```
> x <- c(0.1, 0.14, 0.2, 0.48, 0.58)
> ks.test(x, "punif")

One-sample Kolmogorov-Smirnov test
```

Table 6.1 Upper quantiles for Kolmogorov–Smirnov test statistic.

n	α = 0.10	α = 0.05	α = 0.025	α = 0.01	α = 0.005
1	0.90000	0.95000	0.97500	0.99000	0.99500
2	0.68377	0.77639	0.84189	0.90000	0.92929
3	0.56481	0.63604	0.70760	0.78456	0.82900
4	0.49265	0.56522	0.62394	0.68887	0.73424
5	0.44698	0.50945	0.56328	0.62718	0.66853
6	0.41037	0.46799	0.51926	0.57741	0.61661
7	0.38148	0.43607	0.48342	0.53844	0.57581
8	0.35831	0.40962	0.45427	0.50654	0.54179
9	0.33910	0.38746	0.43001	0.47960	0.51332
10	0.32260	0.36866	0.40925	0.45662	0.48893
11	0.30829	0.35242	0.39122	0.43670	0.46770
12	0.29577	0.33815	0.37543	0.41918	0.44905
13	0.28470	0.32549	0.36143	0.40362	0.43247
14	0.27481	0.31417	0.34890	0.38970	0.41762
15	0.26588	0.30397	0.33760	0.37713	0.40420
16	0.25778	0.29472	0.32733	0.36571	0.39201
17	0.25039	0.28627	0.31796	0.35528	0.38086
18	0.24360	0.27851	0.30936	0.34569	0.37062
19	0.23735	0.27136	0.30143	0.33685	0.36117
20	0.23156	0.26473	0.29408	0.32866	0.35241
21	0.22617	0.25858	0.28724	0.32104	0.34427
22	0.22115	0.25283	0.28087	0.31394	0.33666
23	0.21645	0.24746	0.27490	0.30728	0.32954
24	0.21205	0.24242	0.26931	0.30104	0.32286
25	0.20790	0.23768	0.26404	0.29516	0.31657
26	0.20399	0.23320	0.25907	0.28962	0.31064
27	0.20030	0.22898	0.25438	0.28438	0.30502
28	0.19680	0.22497	0.24993	0.27942	0.29971
29	0.19348	0.22117	0.24571	0.27471	0.29466
30	0.19032	0.21756	0.24170	0.27023	0.28987

(Continued)

Table 6.1 (Continued)

n	α = 0.10	α = 0.05	α = 0.025	α = 0.01	α = 0.005
31	0.18732	0.21412	0.23788	0.26596	0.28530
32	0.18445	0.21085	0.23424	0.26189	0.28094
33	0.18171	0.20771	0.23076	0.25801	0.27677
34	0.17909	0.20472	0.22743	0.25429	0.27279
35	0.17659	0.20185	0.22425	0.25073	0.26897
36	0.17418	0.19910	0.22119	0.24732	0.26532
37	0.17188	0.19646	0.21826	0.24404	0.26180
38	0.16966	0.19392	0.21544	0.24089	0.25843
39	0.16753	0.19148	0.21273	0.23786	0.25518
40	0.16547	0.18913	0.21012	0.23494	0.25205

Source: Miller (1956).

```
data:  x
D = 0.42, p-value = 0.2589
alternative hypothesis: two-sided
```

Example 6.3 Galaxy velocity data, available on the book's website, was analyzed by Roeder (1990) and consists of the velocities of 82 distant galaxies, diverging from our own galaxy. A mixture model was applied to describe the underlying distribution. The first hypothesized fit is the normal distribution, specifically $\mathcal{N}(21, (\sqrt{21})^2)$, and the KS distance ($\sqrt{n}D_n = 1.6224$ with p-value of 0.0103). The following mixture of normal distributions with five components was also fit to the data:

$$\hat{F} = 0.1\Phi(9, 0.5^2) + 0.02\Phi(17, (\sqrt{0.8})^2) + 0.4\Phi(20, (\sqrt{5})^2)$$
$$+ 0.4\Phi(23, (\sqrt{8})^2) + 0.05\Phi(33, (\sqrt{2})^2),$$

where $\Phi(\mu, \sigma)$ is the CDF for the normal distribution. The KS statistics is $\sqrt{n}D_n = 1.1734$, and corresponding p-value is 0.1273. Figure 6.2 plots the CDF of the transformed variables $\hat{F}(X)$, so a good fit is indicated by a straight line. Recall, if $X \sim F$, then $F(X) \sim \mathcal{U}(0,1)$ and the straight line is, in fact, the CDF of $\mathcal{U}(0,1)$, $F(x) = x$, $0 \le x \le 1$. Panel (a) shows the fit for the $\mathcal{N}(21, (\sqrt{21})^2)$ model, while panel (b) shows the fit for the mixture model. Although not perfect itself, the mixture model shows significant improvement over the single normal model.

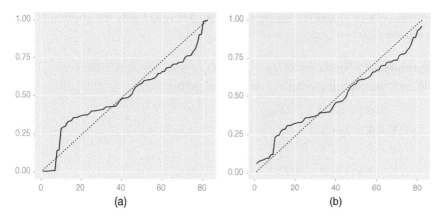

Figure 6.2 Fitted distributions: (a) $\mathcal{N}\left(21, \left(\sqrt{21}\right)^2\right)$ and (b) mixture of normals.

6.2 Smirnov Test to Compare Two Distributions

Smirnov (1939a, 1939b) extended the KS test to compare two distributions based on independent samples from each population. Let $X_1, X_2, \ldots, X_m$ and $Y_1, Y_2, \ldots, Y_n$ be two independent samples from populations with unknown CDFs F_X and G_Y. Let $F_m(x)$ and $G_n(x)$ be the corresponding EDFs.

We would like to test

$$H_0 : F_X(x) = G_Y(x) \ \forall x \ \text{versus} \ H_1 : F_X(x) \neq G_Y(x) \text{ for some } x.$$

We will use the analog of the KS statistic D_n:

$$D_{m,n} = \sup_x |F_m(x) - G_n(x)|, \tag{6.1}$$

where $D_{m,n}$ can be simplified (in terms of programming convenience) to

$$D_{m,n} = \max_i \{|F_m(Z_i) - G_n(Z_i)|\}$$

and $Z = Z_1, \ldots, Z_{m+n}$ is the *combined* sample $X_1, \ldots, X_m, Y_1, \ldots, Y_n$. $D_{m,n}$ will be large if there is a cluster of values from one sample after the samples are combined. The imbalance can be equivalently measured in how the *ranks* of one sample compare with those of the other after they are joined together. That is, values from the samples are not directly relevant except for how they are ordered when combined. This is the essential nature of rank tests that we will investigate later in Chapter 7.

The two-distribution test extends simply from two sided to one sided. The one-sided test statistics are $D_{m,n}^+ = \sup_x (F_m(x) - G_n(x))$ or $D_{m,n}^- = \sup_x (G_n(x) - F_m(x))$. Note that the ranks of the two groups of data determine the supremum difference in (6.1) and the values of the data determine only the position of the jumps for $G_n(x) - F_m(x)$.

Example 6.4 For the test of $H_1 : F_X(x) > G_Y(x)$ with $n = m = 2$, there are $\binom{4}{2} = 6$ different sample representations (with equal probability):

Sample order	$D^+_{m,n}$
$X < X < Y < Y$	1
$X < Y < X < Y$	1/2
$X < Y < Y < X$	1/2
$Y < X < X < Y$	1/2
$Y < X < Y < X$	0
$Y < Y < X < X$	0

The distribution of the test statistic is

$$P(D_{2,2} = d) = \begin{cases} 1/3, & \text{if } d = 0, \\ 1/2, & \text{if } d = 1/2, \\ 1/6, & \text{if } d = 1. \end{cases}$$

If we reject H_0 in the case $D_{2,2} = 1$ (for $H_1 : F_X(x) > G_Y(x)$), then our type-I error rate is $\alpha = 1/6$.

If $m = n$ in general, the null distribution of the test statistic simplifies to

$$P(D^+_{n,n} > d) = P(D^-_{n,n} > d) = \frac{\binom{2n}{\lfloor n(d+1) \rfloor}}{\binom{2n}{n}},$$

where $\lfloor a \rfloor$ denotes the greatest integer $\leq a$. For two-sided tests, this is doubled to obtain the p-value. If m and n are large ($m, n > 30$) and of comparable size, then an approximate distribution can be used:

$$P\left(\sqrt{\frac{mn}{m+n}} D_{m,n} \leq d\right) \approx 1 - 2\sum_{k=1}^{\infty} e^{-2k^2 d^2}.$$

A simpler large sample approximation, given in Table 6.2, works effectively if m and n are both larger than, say, 50.

Example 6.5 Suppose we have $n = m = 4$ with data $(x_1, x_2, x_3, x_4) = (16, 4, 7, 21)$ and $(y_1, y_2, y_3, y_4) = (56, 31, 15, 19)$. For the Smirnov test of $H_1 : F \neq G$, the only thing important about the data is how they are ranked within the group of eight combined observations:

$$x_{1:4} < x_{2:4} < y_{1:4} < x_{3:4} < y_{2:4} < x_{4:4} < y_{3:4} < y_{4:4}.$$

Table 6.2 Tail probabilities for Smirnov two-sample test.

One-sided test	$\alpha = 0.05$	$\alpha = 0.025$	$\alpha = 0.01$	$\alpha = 0.005$
Two-sided test	$\alpha = 0.10$	$\alpha = 0.05$	$\alpha = 0.02$	$\alpha = 0.01$
	$1.22\sqrt{\frac{m+n}{mn}}$	$1.36\sqrt{\frac{m+n}{mn}}$	$1.52\sqrt{\frac{m+n}{mn}}$	$1.63\sqrt{\frac{m+n}{mn}}$

$|F_n - G_m|$ is never larger than $1/2$, achieved in intervals $(7,15)$, $(16,19)$, and $(21, 31)$. The p-value for the two-sided test is

$$p\text{-value} = \frac{2\binom{2\times4}{[4\times1.5]}}{\binom{8}{4}} = \frac{2\binom{8}{6}}{\binom{8}{4}} = \frac{56}{70} = 0.80.$$

Example 6.6 Figure 6.3 shows the EDFs for two samples of size 100. One is generated from normal data, and the other from exponential data. They have identical mean ($\mu = 10$) and variance ($\sigma^2 = 100$). The R function `ks.test` can also be used for the two-sample test. The R code shows the p-value is 3.729e-05. If we compared the samples using a two-sample t-test, the significance value is 0.4777 because the

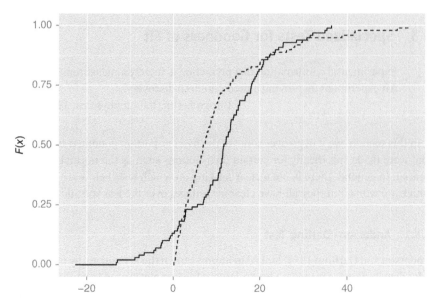

Figure 6.3 EDF for samples of $n = m = 100$ generated from normal and exponential with $\mu = 10$ and $\sigma^2 = 100$.

t-test is testing only the means and not the distribution (which is assumed to be normal). Note that sup $_x|F_m(x) - G_n(x)| = 0.33$, and according to Table 6.2, the 0.99 quantile for the two-sided test is 0.2305:

```
> xn<-rnorm(100,10,10)
> xe<-rexp(100,0.1)
> y<-1:100/100
>
> p <- ggplot()+geom_step(aes(x=sort(xn),y=y))
> p <- p + geom_step(aes(x=sort(xe),y=y),lty=2) + xlab("") + ylab("F(x)")
> print(p)
>
> ks.test(xn,xe)
        Two-sample Kolmogorov-Smirnov test
data:  xn and xe
D = 0.33, p-value = 3.729e-05
alternative hypothesis: two-sided

> t.test(xn,xe)
        Welch Two Sample t-test
data:  xn and xe
t = 0.7114, df = 197.823, p-value = 0.4777
alternative hypothesis: true difference in means is not equal to 0
95 percent confidence interval:
 -1.949861  4.150360
sample estimates:
mean of x mean of y
 11.44810  10.34785
```

6.3 Specialized Tests for Goodness of Fit

> Experimental confirmation of a prediction is merely a measurement.
> An experiment disproving a prediction is a discovery.
>
> Enrico Fermi, Italian physicist, 1901–1954

In this section, we will go over some of the most important goodness-of-fit tests that were made specifically for certain distributions such as the normal or exponential. In general, there is not a clear ranking on which tests below are best and which are worst, but they all have clear advantages over the less specific KS test.

6.3.1 Anderson–Darling Test

Anderson and Darling (1954) looked to improve upon the KS statistic by modifying it for distributions of interest. The Anderson–Darling test is used to verify if a sample of data came from a population with a specific distribution. It is a modification of the KS test that accounts for the distribution and test and gives more attention to the tails. As mentioned before, the KS test is distribution free, in the sense that the critical values do not depend on the specific distribution being tested. The Anderson–Darling test makes use of the specific distribution in calculating the critical values. The advantage is that this sharpens the test, but the disadvantage is that critical values must be calculated for each hypothesized distribution.

The statistics for testing $H_0 : F(x) = F_0(x)$ versus the two-sided alternative is $A^2 = -n - S$, where

$$S = \sum_{i=1}^{n} \frac{2i-1}{n} \left[\log F_0(X_{i:n}) + \log(1 - F_0(X_{n+1-i:n}))\right].$$

Tabulated values and formulas have been published Stephens (1974, 1976) for the normal, lognormal, and exponential distributions. The hypothesis that the distribution is of a specific form is rejected if the test statistic A^2 (or modified A^*, A^*) is greater than the critical value given in Table 6.3. Cases 0, 1, and 2 do not need modification, i.e. observed A^2 is directly compared with those in the table. Case 3 and (c) compare a modified A^2 (A^* or A^{**}) to the critical values in Table 6.3. In (b), $A^* = A^2(1 + \frac{0.75}{n} + \frac{2.25}{n^2})$, and in (c), $A^{**} = A^2(1 + \frac{0.3}{n})$.

Example 6.7 The following example has been used extensively in testing for normality. The weights of 11 men (in pounds) are given: 148, 154, 158, 160, 161, 162, 166, 170, 182, 195, and 236. The sample mean is 172, and sample standard deviation is 24.952. Because mean and variance are estimated, this refers to Case 3 in Table 6.3. The standardized observations are $w_1 = (148 - 172)/24.952 = -0.9618, \ldots, w_{11} = 2.5649$ and $z_1 = \Phi(w_1) = 0.1681, \ldots, z_{11} = 0.9948$. Next we calculate $A^2 = 0.9468$ and modify it as $A^* = A^2(1 + 0.75/11 + 0.25/121) = 1.029$. From the table, we see that this is significant at all levels except for $\alpha = 0.01$, e.g. the null hypothesis of normality is rejected at level $\alpha = 0.05$. In R, ad.test(x) function provides a statistic A^2 and p-value of the Anderson–Darling test. Note that it requires to load the nortest package. Here is the corresponding R code:

Table 6.3 Null distribution of Anderson–Darling test statistic: modifications of A^2 and upper tail percentage points.

Modification A^*, A^{**}	Upper tail probability α			
	0.10	0.05	0.025	0.01
(a) Case 0: fully specified $\mathcal{N}(\mu, \sigma^2)$	1.933	2.492	3.070	3.857
(b) Case 1: $\mathcal{N}(\mu, \sigma^2)$, only σ^2 known	0.894	1.087	1.285	1.551
Case 2: σ^2 estimated by s^2, μ known	1.743	2.308	2.898	3.702
Case 3: μ and σ^2 estimated, A^*	0.631	0.752	0.873	1.035
(c) Case 4: $\mathcal{E}\mathrm{xp}(\theta)$, A^{**}	1.062	1.321	1.591	1.959

```
> library(nortest)
> weights <- c(148, 154, 158, 160, 161, 162, 166, 170, 182, 195, 236)
> ad.test(weights)
        Anderson-Darling normality test
data:  weights
A = 0.9468, p-value = 0.01045
> # Here is the manual code for the A-D test
> n <- length(weights)
> ws <- (weights-mean(weights))/sd(weights)
> zs <- pnorm(ws)
> # transformation to uniform o.s.
> # calculation of anderson-darling
> s <- 0
> for(i in 1:n){
+     s<-s+(2*i-1)/n*(log(zs[i])+log(1-zs[n+1-i]))
+ }
> a2 <- -n-s
> a2
[1] 0.9467719
> astar <- a2*(1+0.75/n+2.25/n^2)
[1] 1.02893
```

6.3.2 Cramér–von Mises Test

The Cramér–von Mises test measures the weighted distance between the empirical CDF F_n and postulated CDF F_0. Based on a squared-error function, the test statistic is

$$\omega_n^2(\psi(F_0)) = \int_{-\infty}^{\infty} \left(F_n(x) - F_0(x)\right)^2 \psi(F_0(x)) \, dF_0(x). \tag{6.2}$$

There are several popular choices for the (weight) functional ψ. When $\psi(x) = 1$, this is the "standard" Cramér–von Mises statistic $\omega_n^2(1) = \omega_n^2$, in which case the test statistic becomes

$$n\omega_n^2 = \frac{1}{12n} + \sum_{i=1}^{n} \left(F_0(X_{i:n}) - \frac{2i-1}{2n}\right)^2.$$

When $\psi(x) = x^{-1}(1-x)^{-1}$, $\omega_n^2(1/(F_0(1-F_0))) = A^2/n$, and A^2 is the Anderson–Darling statistic. Under the hypothesis $H_0 : F = F_0$, the asymptotic distribution of $\omega_n^2(\psi(F))$ is

$$\lim_{n\to\infty} P(n\omega_n^2 < x) = \frac{1}{\sqrt{2x}} \sum_{j=0}^{\infty} \frac{\Gamma(j+1/2)}{\Gamma(1/2)\Gamma(j+1)} \sqrt{4j+1}$$

$$\times \exp\left\{-\frac{(4j+1)^2}{16x}\right\} \cdot \left[J_{-1/4}\left(\frac{(4j+1)^2}{16x}\right) - J_{1/4}\left(\frac{(4j+1)^2}{16x}\right)\right],$$

where $J_k(z)$ is the modified Bessel function (in R, `besselJ(z,k)`).

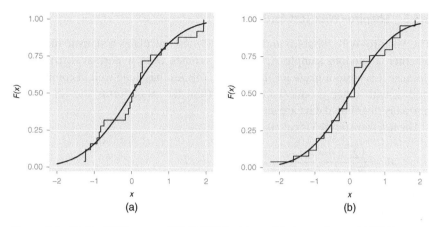

Figure 6.4 Plots of EDF versus $\mathcal{N}(0,1)$ CDF for (a) $n = 25$ observations of $\mathcal{N}(0,1)$ data and (b) standardized $Bin(100, 0.5)$ data.

In R, the particular Cramér–von Mises test for *normality* can be applied to a sample x with the function in *nortest* package

```
cvm.test(x),
```

where the weight function is one. The R code below shows how it works and plots the data and theoretical distribution (Figure 6.4):

```
> library(nortest)
> xx <- seq(-2,2,by=0.1)
> x <- rnorm(25,0,1)
> cvm.test(x)
        Cramer-von Mises normality test
data:  x
W = 0.0693, p-value = 0.2743

> y <- rbinom(25,100,0.5)
> y2 <- (y-mean(y))/sd(y)
> cvm.test(y)
        Cramer-von Mises normality test
data:  y
W = 0.0454, p-value = 0.5678

> p <- ggplot() + geom_line(aes(x=xx,y=pnorm(xx)),lwd=0.8)
> p <- p + geom_step(aes(x=sort(x),y=1:length(x)/length(x)))
> p <- p + xlab("x") + ylab("F(x)")
> print(p)
>
> p2 <- ggplot() + geom_line(aes(x=xx,y=pnorm(xx)),lwd=0.8)
> p2 <- p2 + geom_step(aes(x=sort(y2),y=1:length(y2)/length(y2)))
> p2 <- p2 + xlab("x") + ylab("F(x)")
> print(p2)
```

6.3.3 Shapiro–Wilk Test for Normality

The Shapiro–Wilk (Shapiro and Wilk, 1965) test calculates a statistic that tests whether a random sample $X_1, X_2, \ldots, X_n$ comes from a normal distribution. Because it is custom made for the normal, this test has done well in comparison studies with other goodness of fit tests (and far outperforms the KS test) if normally distributed data are involved.

The test statistic (W) is calculated as

$$W = \frac{\left(\sum_{i=1}^{n} a_i X_{i:n}\right)^2}{\sum_{i=1}^{n} (X_i - \bar{X})^2},$$

where the $X_{1:n} < X_{2:n} < \cdots < X_{n:n}$ are the ordered sample values and the a_i are constants generated from the means, variances, and covariances of the order statistics of a sample of size n from a normal distribution (see Table 6.5). If H_0 is true, W is close to one; otherwise, $W < 1$, and we reject H_0 for small values of W. Table 6.4 lists Shapiro–Wilk test statistic quantiles for sample sizes up to $n = 39$.

The weights a_i are defined as the components of the vector

$$a = M'V^{-1}\left((M'V^{-1})(V^{-1}M)\right)^{-1/2},$$

where M denotes the expected values of standard normal order statistic for a sample of size n and V is the corresponding covariance matrix. While some of these values are tabled here, most likely you will see the test statistic (and critical value) listed in computer output.

Example 6.8 For $n = 5$, the coefficients a_i given in Table 6.5 lead to

$$W = \frac{(0.6646(x_{5:5} - x_{1:5}) + 0.2413(x_{4:5} - x_{2:5}))^2}{\sum (x_i - \bar{x})^2}.$$

If the data resemble a normally distributed set, then the numerator will be approximately to $\sum (x_i - \bar{x})^2$, and $W \approx 1$. Suppose $(x_1, \ldots, x_5) = (-2, -1, 0, 1, 2)$, so that $\sum (x_i - \bar{x})^2 = 10$ and $W = 0.1(0.6646[2 - (-2)] + 0.2413[1 - (-1)])^2 = 0.987$. From Table 6.4, $w_{0.10} = 0.806$, so our test statistic is clearly not significant. In fact, $W \approx w_{0.95} = 0.986$, so the critical value (p-value) for this goodness-of-fit test is nearly 0.95. Undoubtedly the perfect symmetry of the invented sample is a cause for this.

6.3.4 Choosing a Goodness-of-Fit Test

At this point, several potential goodness-of-fit tests have been introduced with nary a word that recommends one over another. There are several other specialized tests we have not mentioned, such as the Lilliefors tests (for exponentiality and normality), the D'Agostino–Pearson test, and the Bowman–Shenton test. These

Table 6.4 Quantiles for Shapiro–Wilk test statistic.

					α				
n	0.01	0.02	0.05	0.10	0.50	0.90	0.95	0.98	0.99
3	0.753	0.756	0.767	0.789	0.959	0.998	0.999	1.000	1.000
4	0.687	0.707	0.748	0.792	0.935	0.987	0.992	0.996	0.997
5	0.686	0.715	0.762	0.806	0.927	0.979	0.986	0.991	0.993
6	0.713	0.743	0.788	0.826	0.927	0.974	0.981	0.986	0.989
7	0.730	0.760	0.803	0.838	0.928	0.972	0.979	0.985	0.988
8	0.749	0.778	0.818	0.851	0.932	0.972	0.978	0.984	0.987
9	0.764	0.791	0.829	0.859	0.935	0.972	0.978	0.984	0.986
10	0.781	0.806	0.842	0.869	0.938	0.972	0.978	0.983	0.986
11	0.792	0.817	0.850	0.876	0.940	0.973	0.979	0.984	0.986
12	0.805	0.828	0.859	0.883	0.943	0.973	0.979	0.984	0.986
13	0.814	0.837	0.866	0.889	0.945	0.974	0.979	0.984	0.986
14	0.825	0.846	0.874	0.895	0.947	0.975	0.980	0.984	0.986
15	0.835	0.855	0.881	0.901	0.950	0.975	0.980	0.984	0.987
16	0.844	0.863	0.887	0.906	0.952	0.976	0.981	0.985	0.987
17	0.851	0.869	0.892	0.910	0.954	0.977	0.981	0.985	0.987
18	0.858	0.874	0.897	0.914	0.956	0.978	0.982	0.986	0.988
19	0.863	0.879	0.901	0.917	0.957	0.978	0.982	0.986	0.988
20	0.868	0.884	0.905	0.920	0.959	0.979	0.983	0.986	0.988
21	0.873	0.888	0.908	0.923	0.960	0.980	0.983	0.987	0.989
22	0.878	0.892	0.911	0.926	0.961	0.980	0.984	0.987	0.989
23	0.881	0.895	0.914	0.928	0.962	0.981	0.984	0.987	0.989
24	0.884	0.898	0.916	0.930	0.963	0.981	0.984	0.987	0.989
25	0.888	0.901	0.918	0.931	0.964	0.981	0.985	0.988	0.989
26	0.891	0.904	0.920	0.933	0.965	0.982	0.985	0.988	0.989
27	0.894	0.906	0.923	0.935	0.965	0.982	0.985	0.988	0.990
28	0.896	0.908	0.924	0.936	0.966	0.982	0.985	0.988	0.990
29	0.898	0.910	0.926	0.937	0.966	0.982	0.985	0.988	0.990
30	0.900	0.912	0.927	0.939	0.967	0.983	0.985	0.988	0.900

(Continued)

Table 6.4 (Continued)

n	0.01	0.02	0.05	0.10	0.50	0.90	0.95	0.98	0.99
					α				
31	0.902	0.914	0.929	0.940	0.967	0.983	0.986	0.988	0.990
32	0.904	0.915	0.930	0.941	0.968	0.983	0.986	0.988	0.990
33	0.906	0.917	0.931	0.942	0.968	0.983	0.986	0.989	0.990
34	0.908	0.919	0.933	0.943	0.969	0.983	0.986	0.989	0.990
35	0.910	0.920	0.934	0.944	0.969	0.984	0.986	0.989	0.990
36	0.912	0.922	0.935	0.945	0.970	0.984	0.986	0.989	0.990
37	0.914	0.924	0.936	0.946	0.970	0.984	0.987	0.989	0.990
38	0.916	0.925	0.938	0.947	0.971	0.984	0.987	0.989	0.990
39	0.917	0.927	0.939	0.948	0.971	0.984	0.987	0.989	0.991

Table 6.5 Coefficients for the Shapiro–Wilk test.

n	i = 1	i = 2	i = 3	i = 4	i = 5	i = 6	i = 7	i = 8
2	0.7071							
3	0.7071	0.0000						
4	0.6872	0.1677						
5	0.6646	0.2413	0.0000					
6	0.6431	0.2806	0.0875					
7	0.6233	0.3031	0.1401	0.0000				
8	0.6052	0.3164	0.1743	0.0561				
9	0.5888	0.3244	0.1976	0.0947	0.0000			
10	0.5739	0.3291	0.2141	0.2141	0.1224	0.0399		
11	0.5601	0.3315	0.2260	0.1429	0.0695	0.0000		
12	0.5475	0.3325	0.2347	0.1586	0.0922	0.0303		
13	0.5359	0.3325	0.2412	0.1707	0.1099	0.0539	0.0000	
14	0.5251	0.3318	0.2460	0.1802	0.1240	0.0727	0.0240	
15	0.5150	0.3306	0.2495	0.1878	0.1353	0.0880	0.0433	0.0000
16	0.5056	0.3290	0.2521	0.1939	0.1447	0.1005	0.0593	0.0196

last two tests are extensions of the Shapiro–Wilk test. Obviously, the specialized tests will be more powerful than an omnibus test such as the KS test. D'Agostino and Stephens (1986) warn

> …for testing for normality, the Kolmogorov-Smirnov test is only a historical curiosity. It should never be used. It has poor power in comparison to [specialized tests such as Shapiro-Wilk, D'Agostino-Pearson, Bowman-Shenton, and Anderson-Darling tests].

These top-performing tests fail to distinguish themselves across a broad range of distributions and parameter values. Statistical software programs often list two or more test results, allowing the analyst to choose the one that will best support their research grants.

There is another way, altogether different, for testing the fit of a distribution to the data. This is detailed in the upcoming section on probability plotting. One problem with all of the analytical tests discussed thus far involves the large sample behavior. As the sample size gets large, the test can afford to be pickier about what is considered a departure from the hypothesized null distribution F_0. In short, your data might look normally distributed to you, for all practical purposes, but if it is not *exactly* normal, the goodness-of-fit test will eventually find this out. Probability plotting is one way to avoid this problem.

6.4   Probability Plotting

A probability plot is a graphical way to show goodness of fit. Although it is more subjective than the analytical tests (e.g. KS, Anderson–Darling, and Shapiro–Wilk), it has important advantages over them. First, it allows the practitioner to see what observations of the data are in agreement (or disagreement) with the hypothesized distribution. Second, while no significance level is attached to the plotted points, the analytical tests can be misleading with large samples (this will be illustrated below). There is no such problem with large samples in probability plotting – the bigger the sample, the better.

The plot is based on transforming the data with the hypothesized distribution. After all, if $X_1, \dots, X_n$ have distribution F, we know $F(X_i), \dots, F(X_n)$ are $\mathcal{U}(0,1)$. Specifically, if we find a transformation with F that linearizes the data, we can find a linear relationship to plot.

Example 6.9   Normal Distribution. If Φ represents the CDF of the standard normal distribution function, then the quantile for a normal distribution with parameters (μ, σ^2) can be written as

$$x_p = \mu + \Phi^{-1}(p)\sigma.$$

The plot of x_p versus $\Phi^{-1}(p)$ is a straight line. If the line shows curvature, we know Φ^{-1} was not the right inverse distribution that transformed the percentile to the normal quantile.

A vector consisting of 1000 generated variables from $\mathcal{N}(0,1)$ and 100 from $\mathcal{N}(0.1, 1)$ is tested for normality. For this case, we used the Cramér–von Mises test using the R function `cvm.test(z)`. We input a vector z of data to test, and α represents the test level. The plot in Figure 6.5 shows the EDF of the 1100 observations versus the best fitting normal distribution. In this case, the Cramér–von Mises test rejects the hypothesis that the data are normally distributed at level $\alpha = 0.001$. However, the data are not discernably non-normal for all practical purposes. The probability plot in Figure 6.5 is constructed with the R function

```
qqnorm
```

and confirms this conjecture.

As the sample size increases, the goodness-of-fit tests grow increasingly sensitive to slight perturbations in the normality assumption. In fact, the Cramér–von Mises test has correctly found the non-normality in the data that was generated by a normal mixture:

```
> library(nortest)
> x <- rnorm(1000,0,1); xx <- seq(-2,2,by=0.1)
> y <- rnorm(100,0.1,1)
> z <- c(x,y)
> cvm.test(z)
        Cramer-von Mises normality test
data:  z
```

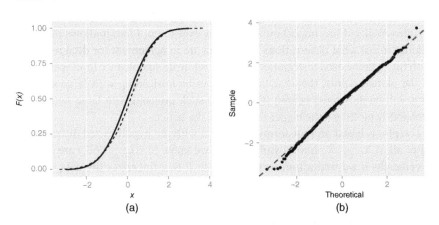

Figure 6.5 (a) Plot of EDF versus normal CDF and (b) normal probability plot.

```
W = 0.292, p-value = 0.0004112
>
> p <- ggplot()+geom_line(aes(x=sort(z),y=1:length(z)/length(z)),lty=2)
> p <- p + geom_line(aes(x=xx,y=pnorm(xx)),lwd=0.8) + xlab("x") + ylab("F(x)")
> print(p)
>
> qqnorm(z,xaxs="i",yaxs="i",xlim=c(-3.2,3.2),ylim=c(-3.2,3.2),main="")
> abline(c(0,1))
> # The following code can be used for better visualization
> ggplot()+stat_qq(aes(sample=z))+geom_abline(intercept=0,
+ slope=1,lwd=0.8,lty=2)
```

Example 6.10   Thirty observations were generated from a normal distribution. The Weibull probability plot in Figure 6.6 shows a slight curvature that suggests the model is misfit. To linearize the Weibull CDF, if the CDF is expressed as $F(x) = 1 - \exp(-(x/\gamma)^\beta)$, then

$$\ln(x_p) = \frac{1}{\beta}\ln(-\ln(1-p)) + \ln(\gamma).$$

The plot of $\ln(x_p)$ versus $\ln(-\ln(1-p))$ is a straight line determined by the two parameters β^{-1} and $\ln(\gamma)$. By using the R function $\mathtt{lm}$, the scale parameter *scale* and the shape parameter *shape* can be estimated by the method of least

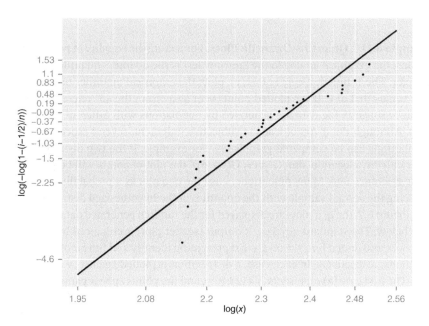

Figure 6.6   Weibull probability plot of 30 observations generated from a normal distribution.

squares. The R code below estimates the Weibull distribution and plots the fitted line:

```
> x <- rnorm(30,10,1)
> y <- (seq(1,length(x))-0.5)/length(x)
> fit <- lm(log(sort(x))~log(-log(1-y)))
> shape <- 1/coef(fit)[2]
> scale <- exp(coef(fit)[1])
>
> # log-log coordinate transformation function
> loglog_trans <- function(){
+    trans<-function(y){log(-log(1-y))}; inv<-function(y){exp(-exp(1-y))}
+    trans_new("loglog", trans, inv)
+ }

>
> p <- ggplot()+geom_point(aes(x=sort(x),y=y))+coord_trans("log","loglog")
> p <- p + geom_line(aes(x=seq(7,13,by=0.1),y=pweibull(seq(7,13,by=0.1),
+ shape=shape,scale=scale)),lwd=0.8)
> p <- p + scale_x_continuous(breaks=5:17,labels=round(log(5:17),2))
> p <- p + scale_y_continuous(breaks=c(0.01,seq(0.1,0.9,by=0.1),0.95,0.99),
+ labels=round(log(-log(1-c(0.01,seq(0.1,0.9,by=0.1),0.95,0.99))),2))
> p <- p + xlab("log(x)") + ylab("log(-log(1-(i-1/2)/n))")
> print(p)
> shape
    12.13703
> scale
    10.62437
```

Example 6.11   Quantile–Quantile Plots. For testing the equality of two distributions, the graphical analog to the Smirnov test is the quantile–quantile plot, or q–q plot. The R function qqplot (x,y) plots the empirical quantiles of the vector x versus that of y. If the plotted points veer away from the 45° reference line, evidence suggests the data are generated by populations with different distributions. Although the q–q plot leads to subjective judgment, several aspects of the distributions can be compared graphically. For example, if the two distributions differ only by a location shift $(F(x) = G(x + \delta))$, the plot of points will be parallel to the reference line. Many practitioners use the q–q plot as a probability plot by replacing the second sample with the quantiles of the hypothesized distribution.

In Figure 6.7, the q–q plots are displayed for the random generated data in the R code below. The standard qqplot R output (scatter plot) is enhanced by dashed line $y = x$ and dotted line $y = a \cdot x + b$ representing identity and fitness of two distributions, respectively. In each case, a distribution is plotted against $\mathcal{N}(100,10^2)$ data. The first case (a) represents $\mathcal{N}(120,10^2)$, and the points appear parallel to the reference line because the only difference between the two distributions is a shift in the mean. In (b) the second distribution is distributed $\mathcal{N}(100,40^2)$. The only difference is in variance, and this is reflected in the slope change in the plot. In the cases (c) and (d), the discrepancy is due to the lack of distribution fit; the data

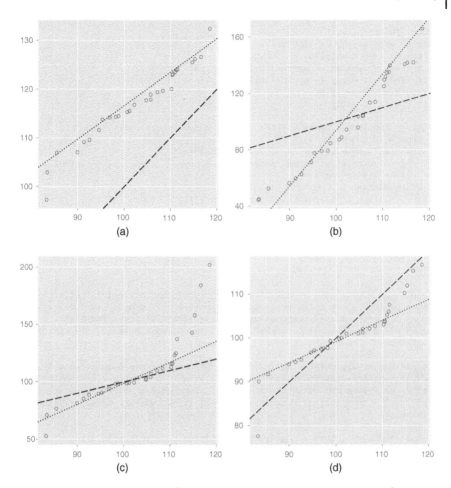

Figure 6.7   Data from $\mathcal{N}(100,10^2)$ are plotted against data from (a) $\mathcal{N}(120,10^2)$, (b) $\mathcal{N}(100,40^2)$, (c) t_1, and (d) $Gamma(200,2)$. The standard qqplot R output (scatter plot) is enhanced by dashed and dotted lines representing identity and fitness of two distributions, respectively.

in (c) are generated from the t-distribution with one degree of freedom, so the tail behavior is much different than that of the normal distribution. This is evident in the left and right end of the q–q plot. In (d), the data are distributed gamma, and the illustrated difference between the two samples is more clear:

```
> x <- rnorm(30,100,10)
> y1 <- rnorm(30,120,10)
> y2 <- rnorm(30,100,40)
> y3 <- 100+ 10*rt(30,1)
> y4 <- rgamma(30,200,2)
```

```
>
> qqp<-function(x,y){
+ sx <- sort(x); sy <- sort(y);
+ lenx <- length(sx);leny <- length(sy);
+ if (leny < lenx)sx <- approx(1L:lenx, sx, n = leny)$y;
+ if (leny > lenx)sy <- approx(1L:leny, sy, n = lenx)$y;
+ return(cbind(sx,sy))
+ }
>
> qql <- function(y,datax = FALSE, dist = qnorm, probs = c(0.25,0.75),
+ qtype = 7, ...){
+ stopifnot(length(probs) == 2, is.function(dist));
+     y <- quantile(y, probs, names = FALSE, type = qtype, na.rm = TRUE);
+     x <- dist(probs);
+     if (datax)  }
+         slope <- diff(x)/diff(y); int <- x[1L] - slope * y[1L];
+     }
+     else {
+         slope <- diff(y)/diff(x); int <- y[1L] - slope * x[1L];
+     }
+     return(c(int,slope))
+ }
>
> qqp1<-qqp(x,y1); qql1 <- qql(y1,dist=function(p)qnorm(p,mean(x),sd(x)));
> p <- ggplot()+geom_point(aes(x=qqp1[,1],y=qqp1[,2]),pch=1,size=3,col=2)
> p <- p + geom_abline(intercept=qql1[1],slope=qql1[2],lty=3,lwd=0.8)
> p <- p + geom_abline(intercept=0,slope=1,lty=2,lwd=0.8)+xlab("")+ylab("")
> print(p)
>
> qqp2<-qqp(x,y2); qql2 <- qql(y2,dist=function(p)qnorm(p,mean(x),sd(x)));
> p2 <- ggplot()+geom_point(aes(x=qqp2[,1],y=qqp2[,2]),pch=1,size=3,col=2)
> p2 <- p2 + geom_abline(intercept=qql2[1],slope=qql2[2],lty=3,lwd=0.8)
> p2 <- p2 + geom_abline(intercept=0,slope=1,lty=2,lwd=0.8)+xlab("")+ylab("")
> print(p2)
>
> qqp3<-qqp(x,y3); qql3 <- qql(y3,dist=function(p)qnorm(p,mean(x),sd(x)));
> p3 <- ggplot()+geom_point(aes(x=qqp3[,1],y=qqp3[,2]),pch=1,size=3,col=2)
> p3 <- p3 + geom_abline(intercept=qql3[1],slope=qql3[2],lty=3,lwd=0.8)
> p3 <- p3 + geom_abline(intercept=0,slope=1,lty=2,lwd=0.8)+xlab("")+ylab("")
> print(p3)
>
> qqp4<-qqp(x,y4); qql4 <- qql(y4,dist=function(p)qnorm(p,mean(x),sd(x)));
> p4 <- ggplot()+geom_point(aes(x=qqp4[,1],y=qqp4[,2]),pch=1,size=3,col=2)
> p4 <- p4 + geom_abline(intercept=qql4[1],slope=qql4[2],lty=3,lwd=0.8)
> p4 <- p4 + geom_abline(intercept=0,slope=1,lty=2,lwd=0.8)+xlab("")+ylab("")
> print(p4)
```

6.5   Runs Test

A chief concern in the application of statistics is to find and understand patterns in data apart from the randomness (noise) that obscures them. While humans are good at deciphering and interpreting patterns, we are much less able to detect randomness. For example, if you ask any large group of people to randomly choose an integer from 1 to 10, the numbers 7 and 4 are chosen nearly half the time, while the endpoints (1, 10) are rarely chosen. Someone trying to think of a random number in that range imagines something toward the middle, but not exactly in the middle. Anything else just does not look "random" to us.

In this section, we use statistics to look for randomness in a simple string of dichotomous data. In many examples, the runs test will not be the most efficient statistical tool available, but the runs test is intuitive and easier to interpret than more computational tests. Suppose items from the sample $X_1, X_2, \dots, X_n$ could be classified as type 1 or type 2. If the sample is random, the 1's and 2's are well mixed, and any clustering or pattern in 1's and 2's is violating the hypothesis of randomness. To decide whether or not the pattern is random, we consider the statistic R, defined as the number of homogenous runs in a sequence of ones and twos. In other words R represents the number of times the symbols change in the sequence (including the first one). For example, $R = 5$ in this sequence of $n = 11$:

$$1 \quad 2 \quad 2 \quad 2 \quad 1 \quad 1 \quad 2 \quad 2 \quad 1 \quad 1 \quad 1.$$

Obviously if there were only two runs in that sequence, we could see the pattern where the symbols are separated right and left. On the other hand, if $R = 11$, the symbols are intermingling in a non-random way. If R is too large, the sequence is showing anti-correlation, a repulsion of same symbols, and zigzag behavior. If R is too small, the sample is suggesting trends, clustering, and groupings in the order of the dichotomous symbols. If the null hypothesis claims that the pattern of randomness exists, then if R is either too big or too small, the alternative hypothesis of an existing trend is supported.

Assume that a dichotomous sequence has n_1 ones and n_2 twos, $n_1 + n_2 = n$. If R is the number of subsequent runs, then if the hypothesis of randomness is true (*sequence is made by random selection of 1's and 2's from the set containing n_1 1's and n_2 2's*), then

$$f_R(r) = \begin{cases} \dfrac{2\binom{n_1-1}{r/2-1}\cdot\binom{n_2-1}{r/2-1}}{\binom{n}{n_1}}, & \text{if } r \text{ is even,} \\[3ex] \dfrac{\binom{n_1-1}{(r-1)/2}\binom{n_2-1}{(r-3)/2}+\binom{n_1-1}{(r-3)/2}\binom{n_2-1}{(r-1)/2}}{\binom{n}{n_1}}, & \text{if } r \text{ is odd,} \end{cases}$$

for $r = 2, 3, \dots, n$. Here is a hint for solving this: first, note that the number of ways to put n objects into r groups *with no cell being empty* is $\binom{n-1}{r-1}$.

The null hypothesis is that the sequence is random, and alternatives could be one sided and two sided. Also, under the hypotheses of randomness, the symbols 1 and 2 are interchangeable, and without loss of generality we assume that $n_1 \leq n_2$. The first three central moments for R (under the hypothesis of randomness) are

$$\mu_R = 1 + \frac{2n_1 n_2}{n},$$

$$\sigma_R{}^2 = \frac{2n_1 n_2 (2n_1 n_2 - n)}{n^2(n-1)}, \text{ and}$$

$$E(R - \mu_R)^3 = -\frac{2n_1 n_2 (n_2 - n_1)^2 (4n_1 n_2 - 3n)}{n^3(n-1)(n-2)},$$

and whenever $n_1 > 15$ and $n_2 > 15$ the normal distribution can be used to approximate lower and upper quantiles. Asymptotically, when $n_1 \to \infty$ and $\varepsilon \le n_1/(n_1 + n_2) \le 1 - \varepsilon$ (for some $0 < \varepsilon < 1$),

$$P(R \le r) = \Phi\left(\frac{r + 0.5 - \mu_R}{\sigma_R}\right) + O(n_1^{-1/2}).$$

The hypothesis of randomness is rejected at level α if the number of runs is either too small (smaller than some $g(\alpha, n_1, n_2)$) or too large (larger than some $G(\alpha, n_1, n_2)$). Thus there is no statistical evidence to reject H_0 if

$$g(\alpha, n_1, n_2) < R < G(\alpha, n_1, n_2).$$

Based on the normal approximation, critical values are

$$g(\alpha, n_1, n_2) \approx \lfloor \mu_R - z_\alpha \sigma_R - 0.5 \rfloor,$$
$$G(\alpha, n_1, n_2) \approx \lfloor \mu_R + z_\alpha \sigma_R + 0.5 \rfloor.$$

For the two-sided rejection region, one should calculate critical values with $z_{\alpha/2}$ instead of z_α. One-sided critical regions, again based on the normal approximation, are values of R for which

$$\frac{R - \mu_R + 0.5}{\sigma_R} \le -z_\alpha,$$
$$\frac{R - \mu_R - 0.5}{\sigma_R} \ge z_\alpha,$$

while the two-sided critical region can be expressed as

$$\frac{(R - \mathbb{E}R)^2}{\sigma_R{}^2} \ge \left(z_{\alpha/2} + \frac{1}{2\sigma_R}\right)^2.$$

When the ratio n_1/n_2 is small, the normal approximation becomes unreliable. If the exact test is still too cumbersome for calculation, a better approximation is given by

$$P(R \le r) \approx I_{1-x}(N - r + 2, r - 1) = I_x(r - 1, N - r + 2),$$

where $I_x(a, b)$ is the incomplete beta function (see Chapter 2) and

$$x = 1 - \frac{n_1 n_2}{n(n-1)} \quad \text{and} \quad N = \frac{(n-1)(2n_1 n_2 - n)}{n_1(n_1 - 1) + n_2(n_2 - 1)}.$$

Critical values are then approximated by $g(\alpha, n_1, n_2) \approx \lfloor g^* \rfloor$ and $G(\alpha, n_1, n_2) \approx 1 + \lfloor G^* \rfloor$, where g^* and G^* are solutions to

$$I_{1-x}(N - g^* + 2, g^* - 1) = \alpha,$$
$$I_x(G^* - 1, N - G^* + 3) = \alpha.$$

Example 6.12 The tourism officials, in Santa Cruz worried about global warming and El Niño effect, compared daily temperatures (1–21 July 2003) with averages of corresponding daily temperatures in 1993–2002. If the temperature in year 2003 is above the same day average in 1993–2002, then symbol *A* is recorded; if it is below, the symbol *B* is recorded. The following sequence of 21 letters was obtained:

AAABBAA|AABAABA|AAABBBB.

We wish to test the hypothesis of random direction of deviation from the average temperature against the alternative of non-randomness at level $\alpha = 5\%$. The R function for computing the test is `runs.test`:

```
> source("runs.test.codes.r")
> cruz <- c(1, 1, 1, 2, 2, 1, 1, 1, 1, 2, 1, 1, 2, 1,
+ 1, 1, 1, 2, 2, 2, 2)
> result <- runs.test(cruz)
> result
      problow         probup        nrun expectedruns
   0.12779498     0.04200206   8.00000000  10.90476190
> count.runs(cruz)
runs1 runs2 truns    n1    n2     n
    4     4     8    13     8    21
>
> n1 <- 13; n2 <- 8
> dfrvec<-rep(0,16);names(dfrvec)<-2:17
> for( i in 2:17){
+   dfrvec[i-1]<-dfr(i,n1,n2)
+ }
>
> p <- ggplot()+geom_bar(aes(x=2:17,y=dfrvec),
+ fill="white",col="black",stat="identity")
> p <- p + geom_bar(aes(x=2:8,y=dfrvec[1:7]),
+ fill="gray",col="black",stat="identity")
> p <- p + scale_x_continuous(breaks=1:17,labels=1:17)
> p <- p + xlab("R") + ylab("P(R)")
> print(p)
```

If observed number of runs is less than expected, `problow` is

$$P(R = 2) + \cdots + P(R = n\text{runs}),$$

and `probup` is

$$P(R = n - n\text{runs} + 2) + \cdots + P(R = n).$$

Alternatively, if $n$runs is larger than expected, then `problow` is

$$P(R = 2) + \cdots + P(R = n - n\text{runs} + 2),$$

and `probup` is

$$P(R = n\text{runs}) + \cdots + P(R = n).$$

In this case, the number of runs (8) was less than expected (10.9048), and the probability of seeing 8 or fewer runs in a random scattering is 0.1278 See Figure 6.8. However, this is a two-sided test. This R test implies we should use $P(R \geq n - n_2 + 2) = P(R \geq 15) = 0.0420$ as the "other tail" to include in the critical region (which would make the p-value equal to 0.1698). Yet, using $P(R \geq 15)$ is slightly misleading, because there is no symmetry in the null distribution of R; instead, we suggest using $2 * \text{problow} = 0.2556$ as the critical value for a two-sided test.

Example 6.13 The following are 30 time lapses, measured in minutes, between eruptions of Old Faithful geyser in Yellowstone National Park. In the R code below, `forruns` stores 2 if the temperature is below average; otherwise stores 1.

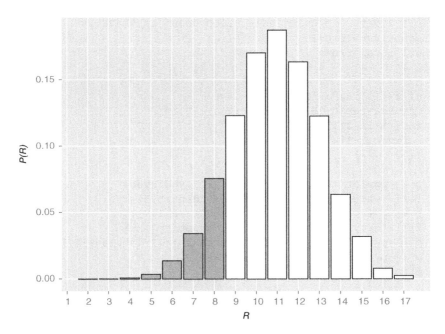

Figure 6.8 Probability distribution of runs under H_0.

The expected number of runs (15.9333) is larger than what was observed (13), and the *p*-value for the two-sided runs test is 2*0.1678 = 0.3356:

```
> source("runs.test.codes.r")
> oldfaithful <- c(68, 63, 66, 63, 61, 44, 60, 62, 71, 62, 62,
+ 55, 62, 67, 73, 72, 55, 67, 68, 65, 60, 61, 71, 60, 68,
+ 67, 72, 69, 65, 66)
> mean(oldfaithful)
[1] 64.16667
> forruns <- as.numeric( (oldfaithful - 64.1667) > 0 ) + 1
> runs.test(forruns)
      problow        probup        nrun  expectedruns
    0.1804463     0.1677907  13.0000000   15.9333333
> count.runs(forruns)
runs1 runs2 truns    n1    n2     n
    6     7    13    14    16    30
```

Example 6.14 Kvam and Chen (2017) analyzed winning streaks for every major league baseball team between 1962 and 2016. There are various theories about how streaky baseball teams can be, whether due to home field advantage, varying quality of starting pitcher, or other factors. Teams play 162 total games, and it was found that across 1455 team seasons (20–30 teams across 55 years), 94.5% of the observations produced run statistics within two standard deviations of the expected value. In fact, only one team (the *least* streakiest team of all time) produced a run statistic outside of 3 standard deviations: the 2005 St. Louis Cardinals.

The Cardinals produced 99 streaks in 2005 and stopped losing streaks at one game 39 times (30% more frequent than expected for a team with a 0.617 winning

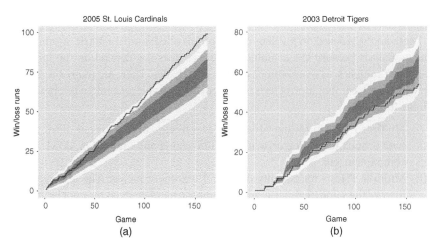

Figure 6.9 Runs versus games for (a) 2005 St. Louis Cardinals and (b) 2003 Detroit Tigers. The darkened parts represent areas within one, two, and three standard deviations of the expected value.

percentage). Figure 6.9 shows how the number of runs increases as a function of games played. Also shown is the most streaky team of all time: the 2003 Detroit Tigers. Coincidentally, this team finished 43–119 and is also known as one of the worst teams of the modern baseball era. The Tigers endured 10 different losing streaks of 6 or more games in 2003.

Before we finish with the runs test, we are compelled to make note of its limitations. After its inception by Mood (1940), the runs test was used as a cure-all nonparametric procedure for a variety of problems, including two-sample comparisons. However, it is inferior to more modern tests we will discuss in Chapter 7. More recently, Mogull (1994) showed an anomaly of the one-sample runs test; it is unable to reject the null hypothesis for series of data with run length of two.

6.6 Meta Analysis

A good compromise is one where everybody makes a contribution.

Angela Merkel, Chancellor of Germany

Meta-analysis is concerned with combining the inference from several studies performed under similar conditions and experimental design. From each study, an "effect size" is derived before the effects are combined and their variability assessed. However, for optimal meta-analysis, the analyst needs substantial information about the experiment such as sample sizes, values of the test statistics, the sampling scheme, and the test design. Such information is often not provided in the published work. In many cases, only the p-values of particular studies are available to be combined.

Meta-analysis based on p-values only is often called nonparametric or omnibus meta-analysis because the combined inference does not depend on the form of data, test statistics, or distributions of the test statistics. There are many situations in which such combination of tests is needed. For example, one might be interested in the following:

(i) Multiple t tests in testing equality of two treatments versus one-sided alternative. Such tests often arise in function testing and estimation, fMRI, DNA comparison, etc.

(ii) Multiple F tests for equality of several treatment means. The test may not involve the same treatments, and parametric meta-analysis may not be appropriate.

(iii) Multiple χ^2 tests for testing the independence in contingency tables (see Chapter 9). The table counts may not be given, or the tables could be of different size (the same factor of interest could be given at different levels).

Most of the methods for combining the tests on basis of their p-values use the facts that, (i) under H_0 and assuming the test statistics have a continuous distribution, the p-values are uniform and, (ii) if G is a monotone CDF and $U \sim \mathcal{U}(0,1)$, then $G^{-1}(U)$ has distribution G. A nice overview can be found in Folks (1984) and the monograph by Hedges and Olkin (1985).

Tippett–Wilkinson method: If the p-values from n studies $p_1, p_2, \ldots, p_n$ are ordered in increasing order $p_{1:n}, p_{2:n}, \ldots, p_{n:n}$, then, for a given k, $1 \leq k \leq n$, the kth smallest p-value, $p_{k:n}$, is distributed $Be(k, n - k + 1)$ and

$$p = P(X \leq p_{k:n}), \quad X \sim Be(k, n - k + 1).$$

Beta random variables are related to the F distribution via

$$P\left(V \leq \frac{\alpha}{\alpha + \beta w}\right) = P(W \geq w),$$

for $V \sim Be(\alpha, \beta)$ and $W \sim \mathcal{F}(2\beta, 2\alpha)$. Thus, the combined significance level p is

$$p = P\left(X \geq \frac{k}{n - k + 1} \frac{1 - p_{k:n}}{p_{k:n}}\right),$$

where $X \sim \mathcal{F}(2(n - k + 1), 2k)$. This single p represents a measure of the uniformity of $p_1, \ldots, p_n$ and can be thought as a combined p-value of all n tests. The nonparametric nature of this procedure is unmistakable. This method was proposed by Tippett (1931) with $k = 1$ and $k = n$ and later generalized by Wilkinson (1951) for arbitrary k between 1 and n. For $k = 1$, the test of level α rejects H_0 if $p_{1:n} \leq 1 - (1 - \alpha)^{1/n}$.

Fisher's inverse χ^2 method: Maybe the most popular method of combining the p-values is Fisher's inverse χ^2 method (Fisher, 1932). Under H_0, the random variable $-2 \log p_i$ has χ^2 distribution with two degrees of freedom, so that $\sum_i \chi^2_{k_i}$ is distributed as χ^2 with $\sum_i k_i$ degrees of freedom. The combined p-value is

$$p = P\left(\chi^2_{2k} \geq -2 \sum_{i=1}^{n} \log p_i\right).$$

This test is, in fact, based on the product of all p-values due to the fact that

$$-2 \sum_i \log p_i = -2 \log \prod_i p_i.$$

Averaging p-values by inverse normals: The following method for combining p-values is based on the fact that if $Z_1, Z_2, \ldots, Z_n$ are i.i.d. $\mathcal{N}(0,1)$, then $(Z_1 + Z_2 + \cdots + Z_n)/\sqrt{n}$ is distributed $\mathcal{N}(0,1)$, as well. Let Φ^{-1} denote the inverse function to the standard normal CDF Φ, and let $p_1, p_2, \ldots, p_n$ be the p-values to be averaged. Then the averaged p-value is

$$p = P\left(Z > \frac{\Phi^{-1}(1 - p_1) + \cdots + \Phi^{-1}(1 - p_n)}{\sqrt{n}}\right),$$

where $Z \sim \mathcal{N}(0,1)$. This procedure can be extended by using weighted sums:

$$p = P\left(Z > \frac{\lambda_1 \Phi^{-1}(1 - p_1) + \cdots + \lambda_n \Phi^{-1}(1 - p_n)}{\sqrt{\lambda_1^2 + \cdots + \lambda_n^2}}\right).$$

There are several more approaches in combining the p-values. Good (1955) suggested use of weighted product

$$-2\sum_i \log p_i = -2\log \prod_i p_i^{\lambda_i},$$

but the distributional theory behind this statistic is complex. Mudholkar and George (1979) suggest transforming the p-values into logits, that is, logit(p) = log($p/(1 - p)$). The combined p-value is

$$p \approx P\left(t_{5n+4} > \frac{-\sum_{i=1}^n \text{logit}(p_i)}{\sqrt{\pi^2 n/3}}\right).$$

As an alternative, Lancaster (1961) proposes a method based on inverse gamma distributions.

Example 6.15 This example is adapted from a presentation by Jessica Utts from University of California, Irvine. Two scientists, Professors A and B, each have a theory they would like to demonstrate. Each plans to run a fixed number of Bernoulli trials and then test $H_0 : p = 0.25$ versus $H_1 : p > 0.25$.

Professor A has access to large numbers of students each semester to use as subjects. He runs the first experiment with 100 subjects, and there are 33 successes ($p = 0.04$). Knowing the importance of replication, Professor A then runs an additional experiment with 100 subjects. He finds 36 successes ($p = 0.009$).

Professor B only teaches small classes. Each quarter, she runs an experiment on her students to test her theory. Results of her 10 studies are given in the table below.

At first glance Professor A's theory has much stronger support. After all, the p-values are 0.04 and 0.009. None of the 10 experiments of professor B was found significant. However, if the results of the experiment for each professor are aggregated, Professor B actually demonstrated a higher level of success than Professor

A, with 71 out of 200 as opposed to 69 out of 200 successful trials. The p-values for the combined trials are 0.0017 for Professor A and 0.0006 for Professor B:

n	No. of successes	p-Value
10	4	0.22
15	6	0.15
17	6	0.23
25	8	0.17
30	10	0.20
40	13	0.18
18	7	0.14
10	5	0.08
15	5	0.31
20	7	0.21

Now suppose that reports of the studies have been incomplete and only p-values are supplied. Nonparametric meta-analysis performed on 10 studies of Professor B reveals an overall omnibus test significant. The R code for Fisher's and inverse-normal methods are below; the combined p-values for Professor B are 0.0235 and 0.021:

```
> pvals <- c(0.22, 0.15, 0.23, 0.17, 0.20, 0.18, 0.14, 0.08, 0.31, 0.21)
> fisherstat <- -2*sum(log(pvals))
> fisherstat
[1] 34.40158
> 1-pchisq(fisherstat,2*10)
[1] 0.0235331
> 1-pnorm(sum(qnorm(1-pvals))/sqrt(length(pvals)))
[1] 0.002113619
```

6.7 Exercises

6.1 Derive the exact distribution of the Kolmogorov test statistic D_n for the case $n = 1$.

6.2 Go the NIST link below to download 31 measurements of polished window strength data for a glass airplane window. In reliability tests such as this one, researchers rely on parametric distributions to characterize the observed lifetimes, but the normal distribution is not commonly used. Does this data follow any well-known distribution? Use probability plotting to make your point.

```
http://www.itl.nist.gov/div898/handbook/eda/
section4/eda4291.htm
```

6.3 Go to the NIST link below to download 100 measurements of the speed of light in air. This classic experiment was carried out by a US Naval Academy teacher Albert Michelson is 1879. Do the data appear to be normally distributed? Use three tests (Kolmogorov, Anderson–Darling, Shapiro–Wilk) and compare answers.

```
http://www.itl.nist.gov/div898/strd/univ/data/
Michelso.dat
```

6.4 Do those little peanut bags handed out during airline flights actually contain as many peanuts as they claim? From a box of peanut bags that have 14 g label weights, 15 bags are sampled and weighed: 16.4, 14.4, 15.5, 14.7, 15.6, 15.2, 15.2, 15.2, 15.3, 15.4, 14.6, 15.6, 14.7, 15.9, and 13.9. Are the data approximately normal so that a t-test has validity?

6.5 Generate a sample S_0 of size $m = 47$ from the population with normal $\mathcal{N}(3,1)$ distribution. Test the hypothesis that the sample is standard normal $H_0 : F = F_0 = \mathcal{N}(0,1)$ (not at $\mu = 3$) versus the alternative $H_1 : F < F_0$. You will need to use D_n^- in the test. Repeat this testing procedure (with new samples, of course) 1000 times. What proportion of p-values exceeded 5%?

6.6 Generate two samples of sizes $m = 30$ and $m = 40$ from $\mathcal{U}(0,1)$. Square the observations in the second sample. What is the theoretical distribution of the squared uniforms? Next, "forget" that you squared the second sample and test by Smirnov test equality of the distributions. Repeat this testing procedure (with new samples, of course) 1000 times. What proportion of p-values exceeded 5%?

6.7 In R, generate two data sets of size $n = 10\,000$: the first from $\mathcal{N}(0,1)$ and the second from the t distribution with five degrees of freedom. These are your two samples to be tested for normality. Recall the asymptotic properties of order statistics from Chapter 5, and find the approximate distribution of $X_{[3000]}$. Standardize it appropriately (here $p = 0.3$, and $\mu = $ qnorm(0.3) $= -0.5244$, and find the two-sided p-values for the goodness-of-fit test of the normal distribution. If the testing is repeated 10 times, how many times will you reject the hypothesis of normality for the second t distributed sequence? What if the degrees of freedom in the t sequence increase from 5 to 10 and to 40? Comment.

6.8 For two samples of size $m = 2$ and $n = 4$, find the exact distribution of the Smirnov test statistics for the test of $H_0 : F(x) \le G(x)$ versus $H_1 : F(x) > G(x)$.

6.9 Let $X_1, X_2, \ldots, X_{n_1}$ be a sample from a population with distribution F_X and $Y_1, Y_2, \ldots, Y_{n_2}$ be a sample from distribution F_Y. If we are interested in testing $H_0 : F_X = F_Y$, one possibility is to use the runs test in the following way. Combine the two samples, and let $Z_1, Z_2, \ldots, Z_{n_1+n_2}$ denote the respective order statistics. Let dichotomous variables 1 and 2 signify if Z is from the first or the second sample. Generate 50 $\mathcal{U}(0,1)$ numbers and 50 $\mathcal{N}(0,1)$ numbers. Concatenate and sort them. Keep track of each number's source by assigning 1 if the number came from the uniform distribution and 2 otherwise. Test the hypothesis that the distributions are the same.

6.10 Combine the p-values for Professor B from the meta-analysis example using the Tippett–Wilkinson method with the smallest p-value and Lancaster's method.

6.11 Derive the exact distribution of the number of runs for $n = 4$ when there are $n_1 = n_2 = 2$ observations of ones and twos. Base your derivation on the exhausting all $\binom{4}{2}$ possible outcomes.

6.12 The link below connects you to the Dow-Jones Industrial Average (DJIA) closing values from 1900 to 1993. First column contains the date (yym-mdd); second column contains the value. Use the runs test to see if there is a non-random pattern in the increases and decreases in the sequence of closing values. Consult

    `http://lib.stat.cmu.edu/datasets/djdc0093`

6.13 Recall Exercise 5.1. Repeat the simulation and make a comparison between the two populations using `qqplot`. Because the sample range has a beta $Be(49,2)$. distribution, this should be verified with a straight line in the plot.

6.14 Consider the Cramér–von Mises test statistic with $\psi(x) = 1$. With a sample of $n = 1$, derive the test statistic distribution, and show that it is minimized at $X = 1/2$.

6.15 Generate two samples S_1 and S_2 of sizes $m = 30$ and $m = 40$ from the uniform distribution. Square the observations in the second sample. What is the theoretical distribution of the squared uniforms? Next, "forget" that you squared the second sample, and test equality of the distributions. Repeat this testing procedure (with new samples, of course) 1000 times. What proportion of p-values exceeded 5%?

6.16 Recall the Gumbel distribution (or *extreme value distribution*) from Chapter 5. Linearize the CDF of the Gumbel distribution to show how a probability plot could be constructed.

6.17 The table below displays the accuracy of meteorological forecasts for the city of Marietta, Georgia. Results are supplied for the month of February 2005. If the forecast differed for the real temperature for more than 3 °F, the symbol 1 was assigned. If the forecast was in error limits < 3 °F, the symbol 2 was assigned. Is it possible to claim that correct and wrong forecasts group at random?

2	2	2	2	2	2	2	2	2	2	2	1	1	1
1	1	2	2	1	1	2	2	2	2	2	1	2	2

6.18 Previous records have indicated that the total points of Olympic dives are normally distributed. Here are the records for *Men 10-meter Platform Preliminary* in 2004. Test the normality of the point distribution. For a computational exercise, generate 1000 sets of 33 normal observations with the same mean and variance as the diving point data. Use the Smirnov test to see how often the p-value corresponding to the test of equal distributions exceeds 0.05. Comment on your results:

Rank	Name	Country	Points	Lag
1	HELM, Mathew	AUS	513.06	
2	DESPATIE, Alexandre	CAN	500.55	12.51
3	TIAN, Liang	CHN	481.47	31.59
4	WATERFIELD, Peter	GBR	474.03	39.03
5	PACHECO, Rommel	MEX	463.47	49.59
6	HU, Jia	CHN	463.44	49.62
7	NEWBERY, Robert	AUS	461.91	51.15
8	DOBROSKOK, Dmitry	RUS	445.68	67.38

Rank	Name	Country	Points	Lag
9	MEYER, Heiko	GER	440.85	72.21
10	URAN-SALAZAR, Juan G.	COL	439.77	73.29
11	TAYLOR, Leon	GBR	433.38	79.68
12	KALEC, Christopher	CAN	429.72	83.34
13	GALPERIN, Gleb	RUS	427.68	85.38
14	DELL'UOMO, Francesco	ITA	426.12	86.94
15	ZAKHAROV, Anton	UKR	420.3	92.76
16	CHOE, Hyong Gil	PRK	419.58	93.48
17	PAK, Yong Ryong	PRK	414.33	98.73
18	ADAM, Tony	GER	411.3	101.76
19	BRYAN, Nickson	MAS	407.13	105.93
20	MAZZUCCHI, Massimiliano	ITA	405.18	107.88
21	VOLODKOV, Roman	UKR	403.59	109.47
22	GAVRIILIDIS, Ioannis	GRE	395.34	117.72
23	GARCIA, Caesar	USA	388.77	124.29
24	DURAN, Cassius	BRA	387.75	125.31
25	GUERRA-OLIVA, Jose Antonio	CUB	375.87	137.19
26	TRAKAS, Sotirios	GRE	361.56	151.5
27	VARLAMOV, Aliaksandr	BLR	361.41	151.65
28	FORNARIS, ALVAREZ Erick	CUB	351.75	161.31
29	PRANDI, Kyle	USA	346.53	166.53
30	MAMONTOV, Andrei	BLR	338.55	174.51
31	DELALOYE, Jean Romain	SUI	326.82	186.24
32	PARISI, Hugo	BRA	325.08	187.98
33	HAJNAL, Andras	HUN	305.79	207.27

RELEVANT R FUNCTIONS AND DATA SETS IN THIS CHAPTER

R code: `runs.test.codes.r`

R functions: `ad.test`, `cvm.test`, `ks.test`, `qqnorm`, `qqplot`, `runs.test`, `t.test`

R package: `nortest`

References

Anderson, T. W., and Darling, D. A. (1954), "A Test of Goodness of Fit," *Journal of the American Statistical Association*, 49, 765–769.

Birnbaum, Z. W., and Tingey, F. (1951), "One-sided Confidence Contours for Probability Distribution Functions," *Annals of Mathematical Statistics*, 22, 592–596.

D'Agostino, R. B., and Stephens, M. A. (1986), *Goodness-of-Fit Techniques*, New York: Marcel Dekker.

Fisher, R. A. (1932), *Statistical Methods for Research Workers*, 4th Edition, Edinburgh, UK: Oliver and Boyd.

Folks, J. L. (1984), "Combination of Independent Tests," in *Handbook of Statistics* 4, Nonparametric Methods, Eds. P. R. Krishnaiah and P. K. Sen, Amsterdam, North-Holland: Elsevier Science, pp. 113–121.

Good, I. J. (1955), "On the Weighted Combination of Significance Tests," *Journal of the Royal Statistical Society (B)*, 17, 264–265.

Hedges, L. V., and Olkin, I. (1985), *Statistical Methods for Meta-Analysis*, New York: Academic Press.

Kolmogorov, A. N. (1933), "Sulla Determinazione Empirica di Una Legge di Distribuzione," *Giornio Instituto Italia Attuari*, 4, 83–91.

Kvam, P. H., and Chen, Z. (2017), "A Comprehensive Analysis of Team Streakiness in Major League Baseball: 1962–2016," *Baseball Research Journal*, 46 (2), 112–115.

Lancaster, H. O. (1961), "The Combination of Probabilities: An Application of Orthonormal Functions," *Australian Journal of Statistics*, 3, 20–33.

Marsaglia, G., Tsang, W.-W., and Wang, J. (2003), "Evaluating Kolmogorov's Distribution," *Journal of Statistical Software*, 8/18. doi: 10.18637/jss.v008.i18.

Miller, L. H. (1956), "Table of Percentage Points of Kolmogorov Statistics," *Journal of the American Statistical Association*, 51, 111–121.

Mogull, R. G. (1994). "The One-Sample Runs Test: A Category of Exception," *Journal of Educational and Behavioral Statistics*, 19, 296–303.

Mood, A. (1940), "The Distribution Theory of Runs," *Annals of Mathematical Statistics*, 11, 367–392.

Mudholkar, G. S., and George, E. O. (1979), "The Logit Method for Combining Probabilities," in *Symposium on Optimizing Methods in Statistics*, Ed. J. Rustagi, New York: Academic Press, pp. 345–366.

Roeder, K. (1990), "Density Estimation with Confidence Sets Exemplified by Superclusters and Voids in the Galaxies," *Journal of the American Statistical Association*, 85, 617–624.

Shapiro, S. S., and Wilk, M. B. (1965), "An Analysis of Variance Test for Normality (Complete Samples)," *Biometrika*, 52, 591–611.

Smirnov, N. V. (1939a), "On the Derivations of the Empirical Distribution Curve," *Matematicheskii Sbornik*, 6, 2–26.

Smirnov, N. V. (1939b), "On the Estimation of the Discrepancy Between Empirical Curves of Distribution for Two Independent Samples," *Bulletin Moscow University*, 2, 3–16.

Stephens, M. A. (1974), "EDF Statistics for Goodness of Fit and Some Comparisons," *Journal of the American Statistical Association*, 69, 730–737.

Stephens, M. A. (1976), "Asymptotic Results for Goodness-of-Fit Statistics with Unknown Parameters," *Annals of Statistics*, 4, 357–369.

Tippett, L. H. C. (1931), *The Method of Statistics*, First Edition, London: Williams and Norgate.

Wilkinson, B. (1951), "A Statistical Consideration in Psychological Research," *Psychological Bulletin*, 48, 156–158.

7

Rank Tests

Each of us has been doing statistics all his life, in the sense that each of us has been busily reaching conclusions based on empirical observations ever since birth.

William Kruskal

All those old basic statistical procedures – the *t*-test, the correlation coefficient, and the analysis of variance (ANOVA) – depend strongly on the assumption that the sampled data (or the sufficient statistics) are distributed according to a well-known distribution, hardly the fodder for a nonparametrics textbook. However, for every classical test, there is a nonparametric alternative that does the same job with fewer assumptions made of the data. Even if the assumptions from a parametric model are modest and relatively non-constraining, they will undoubtedly be false in the most pure sense. Life, along with your experimental data, is too complicated to fit perfectly into a framework of independently and identically distributed (i.i.d.) errors and exact normal distributions.

Mathematicians have been researching ranks and order statistics since ages ago, but it was not until the 1940s that the idea of rank tests gained prominence in the statistics literature. Hotelling and Pabst (1936) Pabst, M. wrote one of the first papers on the subject, focusing on rank correlations.

There are nonparametric procedures for one sample, for comparing two or more samples, matched samples, bivariate correlation, and more. The key to evaluating data in a nonparametric framework is to compare observations based on their *ranks* within the sample rather than entrusting the actual data measurements to your analytical verdicts. The following table shows nonparametric counterparts to

Nonparametric Statistics with Applications to Science and Engineering with R, Second Edition.
Paul Kvam, Brani Vidakovic, and Seong-joon Kim.
© 2023 John Wiley & Sons, Inc. Published 2023 by John Wiley & Sons, Inc.

the well-known parametric procedures (WSiRT stands for Wilcoxon signed rank test and WSuRT stands for Wilcoxon sum rank test):

Parametric	Nonparametric
Pearson coefficient of correlation	Spearman coefficient of correlation
One-sample *t*-test for the location	Sign test, WSiRT
Paired test *t*-test	Sign test, WSiRT
Two-sample *t* test	WSuRT, Mann–Whitney
ANOVA	Kruskal–Wallis test
Block design ANOVA	Friedman test

To be fair, it should be said that many of these nonparametric procedures come with their own set of assumptions. We will see, in fact, that some of them are rather obtrusive on an experimental design. Others are much less so. Keep this in mind when a nonparametric test is touted as "assumption free." Nothing in life is free.

In addition to properties of ranks and basic sign test, in this chapter, we will present the following nonparametric procedures:

- *Spearman coefficient*: two-sample correlation statistic.
- *Wilcoxon test*: one-sample median test (also see *sign test*).
- *Wilcoxon sum rank test*: two-sample test of distributions.
- *Mann–Whitney test*: two-sample test of medians.

7.1 Properties of Ranks

Let $X_1, X_2, \ldots, X_n$ be a sample from a population with continuous cumulative distribution function (CDF) F_X. The nonparametric procedures are based on how observations within the sample are *ranked*, whether in terms of a parameter μ or another sample. The ranks connected with the sample $X_1, X_2, \ldots, X_n$ denoted as

$$r(X_1), r(X_2), \ldots, r(X_n),$$

are defined as

$$r(X_i) = \#\{X_j | X_j \leq X_i, \ j = 1, \ldots, n\}.$$

Equivalently, ranks can be defined via the *order statistics* of the sample, $r(X_{i:n}) = i$, or

$$r(X_i) = \sum_{j=1}^{n} I(X_i \geq X_j).$$

Since $X_1, \ldots, X_n$ is a random sample, it is true that $X_1, \ldots, X_n \overset{d}{=} X_{\pi_1}, \ldots, X_{\pi_n}$ where $\pi_1, \ldots, \pi_n$ is a permutation of $1, 2, \ldots, n$ and $\overset{d}{=}$ denotes equality in distribution. Consequently, $P(r(X_i) = j) = 1/n$, $1 \leq j \leq n$, i.e. *ranks in an i.i.d. sample are distributed as discrete uniform random variables*. Corresponding to the data r_i, let $R_i = r(X_i)$, the rank of the random variable X_i.

From Chapter 2, the properties of integer sums lead to the following properties for ranks:

(i) $\mathbb{E}(R_i) = \sum_{j=1}^{n} \frac{j}{n} = \frac{n+1}{2}$.

(ii) $\mathbb{E}(R_i^2) = \sum_{j=1}^{n} \frac{j^2}{n} = \frac{n(n+1)(2n+1)}{6n} = \frac{(n+1)(2n+1)}{6}$.

(iii) $\mathbb{V}\mathrm{ar}(R_i) = \frac{n^2-1}{12}$.

(iv) $\mathbb{E}(X_i R_i) = \frac{1}{n} \sum_{i=1}^{n} i\mathbb{E}(X_{i:n})$,

where

$$\mathbb{E}(X_{r:n}) = F_X^{-1}\left(\frac{r}{n+1}\right) \quad \text{and}$$

$$\mathbb{E}(X_i R_i) = \mathbb{E}(\mathbb{E}(R_i X_i)|R_i = k) = \mathbb{E}(\mathbb{E}(kX_{k:n})) = \frac{1}{n}\sum_{i=1}^{n} i\mathbb{E}(X_{i:n}).$$

In the case of ties, it is customary to average the tied rank values. The R function `rank` does just that:

```
> rank(c(3, 1, 4, 1, 5, 9, 2, 6, 5, 3, 5, 8, 9))
 [1]  4.5  1.5  6.0  1.5  8.0 12.5  3.0 10.0  8.0  4.5  8.0 11.0 12.5
```

Property (iv) can be used to find the correlation between observations and their ranks. Such correlation depends on the sample size and the underlying distribution. For example, for $X \sim \mathcal{U}(0,1)$, $\mathbb{E}(X_i R_i) = (2n+1)/6$, which gives $\mathbb{C}\mathrm{ov}(X_i, R_i) = (n-1)/12$ and $\mathbb{C}\mathrm{orr}(X_i, R_i) = \sqrt{(n-1)/(n+1)}$.

With two samples, comparisons between populations can be made in a nonparametric way by comparing ranks for the combined ordered samples. Rank statistics that are made up of sums of indicator variables comparing items from one sample with those of the other are called *linear rank statistics*.

7.2 Sign Test

Suppose we are interested in testing the hypothesis H_0 that a population with continuous CDF has a median m_0 against one of the alternatives $H_1 : m > m_0$, $H_1 : m < m_0$, or $H_1 : m \neq m_0$. Designate the sign $+$ when $X_i > m_0$ (i.e. when the difference $X_i - m_0$ is positive) and the sign $-$ otherwise. For continuous distributions, the case $X_i = m$ (a tie) is theoretically impossible, although in practice ties are often possible, and this feature can be accommodated. For now, we assume the ideal situation in which the ties are not present.

Assumptions: Actually, no assumptions are necessary for the sign test other than the data being at least ordinal.

If m_0 is the median, i.e. if H_0 is true, then by definition of the median, $P(X_i > m_0)$ $= P(X_i < m_0) = 1/2$. If we let T be the total number of $+$ signs, that is,

$$T = \sum_{i=1}^{n} I(X_i > m_0),$$

then $T \sim Bin(n, 1/2)$.

Let the level of test, α, be specified. When the alternative is $H_1 : m > m_0$, the critical values of T are integers larger than or equal to t_α, which is defined as the smallest integer for which

$$P(T \geq t_\alpha | H_0) = \sum_{t=t_\alpha}^{n} \binom{n}{t} \left(\frac{1}{2}\right)^n < \alpha.$$

Likewise, if the alternative is $H_1 : m < m_0$, the critical values of T are integers smaller than or equal to t'_α, which is defined as the largest integer for which

$$P(T \leq t'_\alpha | H_0) = \sum_{t=0}^{t'_\alpha} \binom{n}{t} \left(\frac{1}{2}\right)^n < \alpha.$$

If the alternative hypothesis is two sided ($H_1 : m \neq m_0$), the critical values of T are integers smaller than or equal to $t'_{\alpha/2}$ and integers larger than or equal to $t_{\alpha/2}$, which are defined via

$$\sum_{t=0}^{t'_{\alpha/2}} \binom{n}{t} \left(\frac{1}{2}\right)^n < \alpha/2, \quad \text{and} \quad \sum_{t=t_{\alpha/2}}^{n} \binom{n}{t} \left(\frac{1}{2}\right)^n < \alpha/2.$$

If the value T is observed, then in testing against alternative $H_1 : m > m_0$, large values of T serve as evidence against H_0, and the p-value is

$$p = \sum_{i=T}^{n} \binom{n}{i} 2^{-n} = \sum_{i=0}^{n-T} \binom{n}{i} 2^{-n}.$$

When testing against the alternative $H_1 : m < m_0$, small values of T are critical, and the p-value is

$$p = \sum_{i=0}^{T} \binom{n}{i} 2^{-n}.$$

When the hypothesis is the two-sided, take $T' = \min\{T, n - T\}$ and calculate p-value as

$$p = 2 \sum_{i=0}^{T'} \binom{n}{i} 2^{-n}.$$

Example 7.1 In Chapter 3 of his textbook, Conover (1999) recognizes the analysis of baptism data by Arbuthnot (1710) as the first application of a sign test. Arbuthnot compared the number of male with female baptisms across London from 1629 to 1710, noting there were more male baptisms in each of the 82 recorded years. He argued that nature

> *"brings forth more males than females", and so if one argues that birth rates were the same for a given year and one "undertakes to do the same thing 82 times running, his lot will be $\left(\frac{1}{2}\right)^{82}$."*

With this, Arbuthnot computed a p-value for his one-sided test where he lays out his alternative hypothesis that "there shall be born more males than females."[1]

7.2.1 Paired Samples

Consider now the case in which two samples are paired:

$$\{(X_1, Y_1), \dots, (X_n, Y_n)\}.$$

Suppose we are interested in finding out whether the median of the population differences is 0. In this case we let $T = \sum_{i=1}^{n} I(X_i > Y_i)$, which is the total number of strictly positive differences.

For two population means, it is true that the hypothesis of equality of means is equivalent to the hypothesis that the mean of the population differences is equal to zero. This is not always true for the test of medians. That is, if $D = X - Y$, then it is quite possible that $m_D \neq m_X - m_Y$. With the sign test we are not testing the *equality* of two medians, but whether the *median of the difference* is 0.

Under H_0, *equal population medians*,

$$\mathbb{E}(T) = \sum P(X_i > Y_i) = n/2$$

---

1 Arbuthnot overlooked potential biases, including the effect of child mortality prior to baptism, and his conclusion from the study seems dramatically abrupt: "From hence it follows, that Polygamy is contrary to the Law of Nature and Justice, and to the Propagation of Human Races.?"

and $\mathbb{V}\mathrm{ar}(T) = n \cdot \mathbb{V}\mathrm{ar}(I(X > Y)) = n/4$. With large enough n, T is approximately normal, so for the statistical test of H_1, *the medians are not equal*. We would reject H_0 if T is far enough away from $n/2$; that is,

$$z_0 = \frac{T - n/2}{\sqrt{n/2}} : \quad \text{reject } H_0 \text{ if } |z_0| > z_{\alpha/2}.$$

Example 7.2 According to The Rothstein Catalog on Disaster Recovery, the median number of violent crimes per state dropped from the year 1999 to 2000. Of 50 states, if X_i is number of violent crimes in state i in 1999 and Y_i is the number for 2000, the median of sample differences is $X_i - Y_i$. This number decreased in 38 out of 50 states in one year. With $T = 38$ and $n = 50$, we find $z_0 = 3.67$, which has a p-value of 0.00012 for the one-sided test (medians decreased over the year) or 0.00024 for the two-sided test.

Example 7.3 Let X_1 and X_2 be independent random variables distributed as Poisson with parameters λ_1 and λ_2. We would like to test the hypothesis $H_0 : \lambda_1 = \lambda_2 \ (= \lambda)$. If H_0 is true (Figure 7.1),

$$P(X_1 = k, X_2 = l) = \frac{(2\lambda)^{k+l}}{(k+l)!} \, e^{-2\lambda} \binom{k+l}{k} \left(\frac{1}{2}\right)^{k+l}.$$

Figure 7.1 Nineteenth-century country carolers singing "Hogmanay, Trollolay, Give us your white bread, and none of your grey!"

If we observe X_1 and X_2 and if $X_1 + X_2 = n$, then testing H_0 is exactly the sign test, with $T = X_1$. Indeed,

$$P(X_1 = k \mid X_1 + X_2 = n) = \binom{n}{k} \left(\frac{1}{2}\right)^n.$$

For instance, if $X_1 = 10$ and $X_2 = 20$ are observed, then the p-value for the two-sided alternative $H_1 : \lambda_1 \neq \lambda_2$ is $2 \sum_{i=0}^{10} \binom{30}{i} \left(\frac{1}{2}\right)^{30} = 2 \cdot 0.0494 = 0.0987$.

Example 7.4 Hogmanay Celebration.[2] Roger van Gompel and Shona Falconer at the University of Dundee conducted an experiment to examine the drinking patterns of Members of the Scottish Parliament over the festive holiday season.

Being elected to the Scottish Parliament is likely to have created in members a sense of stereotypical conformity so that they appear to fit in with the traditional ways of Scotland, pleasing the tabloid newspapers and ensuring popular support. One stereotype of the Scottish people is that they drink a lot of whisky and that they enjoy celebrating both Christmas and Hogmanay (Hervey 1888). However, it is possible that members of parliament tend to drink more whisky at one of these times than the other, and an investigation into this was carried out.

The measure used to investigate any such bias was the number of units of single malt scotch whisky ("drams") consumed over two 48-hour periods: Christmas Eve/Christmas Day and Hogmanay/New Year's Day. The hypothesis is that Members of the Scottish Parliament drink a significantly different amount of whisky over Christmas than over Hogmanay (either consistently more or consistently less). The following data were collected:

MSP	1	2	3	4	5	6	7	8	9
Drams at Christmas	2	3	3	2	4	0	3	6	2
Drams at Hogmanay	5	1	5	6	4	7	5	9	0
MSP	10	11	12	13	14	15	16	17	18
Drams at Christmas	2	5	4	3	6	0	3	3	0
Drams at Hogmanay	4	15	6	8	9	0	6	5	12

The R function `sign.test(x,y,tietreat)` lists five summary statistics from the data for the sign test. The first is a p-value based on randomly assigning a "+" or "−" to tied values (see next subsection), and the second is the p-value based

---

2 Hogmanay is the Scottish New Year, celebrated on 31st December every year. The night involves a celebratory drink or two, fireworks, and kissing complete strangers (not necessarily in that order).

on the normal approximation, where ties are counted as half. n is the number of non-tied observations, *plus* are the number of plusses in $y - x$, and *tie* is the number of tied observations:

```
> source("sign.test.r")
> x <- c(2,3,3,2,4,0,3,6,2,2,5,4,3,6,0,3,3,0)
> y <- c(5,1,5,6,4,7,5,9,0,4,15,6,8,9,0,6,5,12)
> result <- sign.test(x,y,"I")
> result
         pvae          pvaa            n        pluses          ties
  0.002090454   0.002979763  16.000000000   2.000000000   2.000000000
```

7.2.2 Treatments of Ties

Tied data present numerous problems in derivations of nonparametric methods and are frequently encountered in real-world data. Even when observations are generated from a continuous distribution, due to limited precision on measurement and application, ties may appear. To deal with ties, R function `sign.test` does one of three things via the third input argument `tietreat`:

R Randomly assigns "+" or "-" to tied values.
C Uses least favorable assignment in terms of H_0.
I Ignores tied values in test statistic computation.

The preferable way to deal with ties is the first option (to randomize). Another equivalent way to deal with ties is to add a slight bit of "noise" to the data. That is, complete the sign test after modifying D by adding a small enough random variable that will not affect the ranking of the differences; i.e. $\tilde{D}_i = D_i + \varepsilon_i$, where $\varepsilon_i \sim \mathcal{N}(0, 0.0001)$. Using the second or third options in `sign.test` will lead to biased or misleading results, in general.

7.3 Spearman Coefficient of Rank Correlation

Charles Edward Spearman (1863–1945) was a late bloomer, academically. He received his PhD at the age of 48, after serving as an officer in the British army for 15 years. He is most famous in the field of psychology, where he theorized that "general intelligence" was a function of a comprehensive mental competence rather than a collection of multi-faceted mental abilities. His theories eventually led to the development of factor analysis.

Spearman (1904) proposed the rank correlation coefficient long before statistics became a scientific discipline. For bivariate data, an observation has two coupled components (X, Y) that may or may not be related to each other. Let $\rho = \mathrm{Corr}(X, Y)$

represent the unknown correlation between the two components. In a sample of n, let $R_1, \ldots, R_n$ denote the ranks for the first component X and $S_1, \ldots, S_n$ denote the ranks for Y. For example, if $x_1 = x_{n:n}$ is the largest value from $x_1, \ldots, x_n$ and $y_1 = y_{1:n}$ is the smallest value from $y_1, \ldots, y_n$, then $(r_1, s_1) = (n, 1)$. Corresponding to Pearson's (parametric) coefficient of correlation, the Spearman coefficient of correlation is defined as

$$\hat{\rho} = \frac{\sum_{i=1}^{n}(R_i - \bar{R})(S_i - \bar{S})}{\sqrt{\sum_{i=1}^{n}(R_i - \bar{R})^2 \cdot \sum_{i=1}^{n}(S_i - \bar{S})^2}}. \tag{7.1}$$

This expression can be simplified. From (7.1), $\bar{R} = \bar{S} = (n+1)/2$, and $\sum(R_i - \bar{R})^2 = \sum(S_i - \bar{S})^2 = n\mathbb{V}\text{ar}(R_i) = n(n^2 - 1)/12$. Define D as the difference between ranks, i.e. $D_i = R_i - S_i$. With $\bar{R} = \bar{S}$, we can see that

$$D_i = (R_i - \bar{R}) - (S_i - \bar{S}),$$

and

$$\sum_{i=1}^{n} D_i^2 = \sum_{i=1}^{n}(R_i - \bar{R})^2 + \sum_{i=1}^{n}(S_i - \bar{S})^2 - 2\sum_{i=1}^{n}(R_i - \bar{R})(S_i - \bar{S}),$$

that is,

$$\sum_{i=1}^{n}(R_i - \bar{R})(S_i - \bar{S}) = \frac{n(n^2 - 1)}{12} - \frac{1}{2}\sum_{i=1}^{n} D_i^2.$$

By dividing both sides of the equation with $\sqrt{\sum_{i=1}^{n}(R_i - \bar{R})^2 \cdot \sum_{i=1}^{n}(S_i - \bar{S})^2} = \sum_{i=1}^{n}(R_i - \bar{R})^2 = n(n^2 - 1)/12$, we obtain

$$\hat{\rho} = 1 - \frac{6\sum_{i=1}^{n} D_i^2}{n(n^2 - 1)}. \tag{7.2}$$

Consistent with Pearson's coefficient of correlation (the standard parametric measure of covariance), the Spearman coefficient of correlation ranges between -1 and 1. If there is perfect agreement, that is, all the differences are 0, then $\hat{\rho} = 1$. The scenario that maximizes $\sum D_i^2$ occurs when ranks are perfectly opposite: $r_i = n - s_i + 1$.

If the sample is large enough, the Spearman statistic can be approximated using the normal distribution. It was shown that if $n > 10$,

$$Z = (\hat{\rho} - \rho)\sqrt{n - 1} \sim \mathcal{N}(0,1).$$

Assumptions: Actually, no assumptions are necessary for testing ρ other than the data being at least ordinal.

Example 7.5 Stichler, Richey, and Mandel (1953) list tread wear for tires (see table below), each tire measured by two methods based on (i) weight loss and (ii) groove wear. In R, the function

```
cor(x,y,method="spearman")
```

computes the Spearman coefficient. For this example, $\hat{\rho} = 0.9265$. Note that if we opt for the parametric measure of correlation, the Pearson coefficient is 0.948:

Weight	Groove	Weight	Groove
45.9	35.7	41.9	39.2
37.5	31.1	33.4	28.1
31.0	24.0	30.5	28.7
30.9	25.9	31.9	23.3
30.4	23.1	27.3	23.7
20.4	20.9	24.5	16.1
20.9	19.9	18.9	15.2
13.7	11.5	11.4	11.2

```
> weight <- c(45.9, 37.5, 31.0, 30.9, 30.4, 20.4, 20.9, 13.7,
+             41.9, 33.4, 30.5, 31.9, 27.3, 24.5, 18.9, 11.4)
> groove <- c(35.7, 31.1, 24.0, 25.9, 23.1, 20.9, 19.9,
+             11.5, 39.2, 28.1, 28.7, 23.3, 23.7, 16.1, 15.2,
+             11.2)
>
> cor(weight, groove, method="spearman")
[1] 0.9264706
```

7.3.1 Ties in the Data

The statistics in (7.1) and (7.2) are not designed for paired data that include tied measurements. If ties exist in the data, a simple adjustment should be made. Define $u' = \sum u(u^2 - 1)/12$ and $v' = \sum v(v^2 - 1)/12$ where the u's and v's are the ranks for X and Y adjusted (e.g. averaged) for ties. Then,

$$\hat{\rho}' = \frac{n(n^2 - 1) - 6\sum_{i=1}^{n} D_i^2 - 6(u' + v')}{\{[n(n^2 - 1) - 12u'][n(n^2 - 1) - 12v']\}^{1/2}}$$

and it holds that, for large n,

$$Z = (\hat{\rho}' - \rho)\sqrt{n - 1} \sim \mathcal{N}(0,1).$$

7.3.2 Kendall's Tau

Kendall (1938) derived an alternative measure of bivariate dependence by finding out how many pairs in the sample are "concordant," which means the signs between X and Y agree in the pairs. That is, out of $\binom{n}{2}$ pairs such as (X_i, Y_i) and (X_j, Y_j), we compare the sign of $(X_i - X_j)$ to that of $(Y_i - Y_j)$. Pairs for which one sign is plus and the other is minus are "discordant."

The Kendall's τ statistic is defined as

$$\tau = \frac{2S_\tau}{n(n-1)}, \quad S_\tau = \sum_{i=1}^{n} \sum_{j=i+1}^{n} \text{sign}\{r_i - r_j\},$$

where r_is are defined via ranks of the second sample corresponding to the ordered ranks of the first sample, $\{1, 2, \dots, n\}$, that is,

$$\begin{pmatrix} 1 & 2 & \dots & n \\ r_1 & r_2 & \dots & r_n \end{pmatrix}.$$

In this notation, $\sum_{i=1}^{n} D_i^2$ from the Spearman's coefficient of correlation becomes $\sum_{i=1}^{n} (r_i - i)^2$. In terms of the number of concordant (n_c) and discordant $\left(n_D = \binom{n}{2} - n_c \right)$ pairs,

$$\tau = \frac{n_c - n_D}{\binom{n}{2}},$$

and in the case of ties, use

$$\tau = \frac{n_c - n_D}{n_c + n_D},$$

or

$$\tau = \frac{n_c - n_D}{\sqrt{(n(n-1)/2 - n_1)(n(n-1)/2 - n_2)}},$$

where n_1 and n_2 are the numbers of pairs with a tie in X and Y, respectively.

Example 7.6 Trends in Indiana's water use from 1986 to 1996 were reported by Arvin and Spaeth (1997) for Indiana Department of Natural Resources. About 95% of the surface water taken annually is accounted for by two categories: surface water withdrawal and groundwater withdrawal. Kendall's tau statistic showed no apparent trend in total surface water withdrawal over time (p-value ≈ 0.59), but groundwater withdrawal increased slightly over the 10-year span (p-value ≈ 0.13):

```
> x <- 1986:1996
> y1 <- c(2.96,3.00,3.12,3.22,3.21,2.96,2.89,3.04,2.99,3.08,3.12)
> y2 <- c(0.175,0.173,0.197,0.182,0.176,0.205,0.188,0.186,0.202,
+         0.208,0.213)
>
```

```
> y1.rank <- rank(y1)
> y2.rank <- rank(y2)
> n <- length(x); s1 <- 0; s2 <- 0
>
> for(i in 1:(n-1)){
+ for(j in (i+1):n){
+ s1 <- s1 + sign(y1.rank[j]-y1.rank[i])
+ s2 <- s2 + sign(y2.rank[j]-y2.rank[i])
+ }}
>
> u <- sapply(unique(x),function(val){length(which(x==val))})
> v <- sapply(unique(y1),function(val){length(which(y1==val))})
> n0<-n*(n-1)/2; n1<-sum(u*(u-1)/2); n2<-sum(v*(v-1)/2)
>
> ktau1 <- s1/sqrt((n0-n1)*(n0-n2)) # tied values are observed in y1.
[1] 0.09260847
>
> ktau2 <- 2*s2/(n*(n-1)) # no ties present.
[1] 0.6363636
>
> # same values can be obtained from 'cor' function.
> cor(x,y1,method="kendall")
[1] 0.09260847
> cor(x,y2,method="kendall")
[1] 0.6363636
```

With large sample size n, we can use the following z-statistic as a normal approximation:

$$z_\tau = \frac{3\tau\sqrt{n(n-1)}}{\sqrt{2(2n+5)}}.$$

This can be used to test the null hypothesis of zero correlation between the populations. Kendall's tau is natural measure of the relationship between X and Y. We can describe it as an odds ratio by noting that

$$\frac{1+\tau}{1-\tau} = \frac{P(C)}{P(D)},$$

where C is the event that any pair in the population is concordant and D is the event any pair is discordant. Spearman's coefficient, on the other hand, cannot be explained this way. For example, in a population with $\tau = 1/3$, any two sets of observations are twice as likely to be concordant than discordant. On the other hand, computations for τ grow as $O(n^2)$, compared with the Spearman coefficient, which grows as $O(n \ln n)$.

7.4 Wilcoxon Signed Rank Test

Recall that the sign test can be used to test differences in medians for two independent samples. A major shortcoming of the sign test is that only the sign of

$D_i = X_i - m_0$ or $D_i = X_i - Y_i$ (depending if we have a one- or two-sample problem) contributes to the test statistics. Frank Wilcoxon suggested that, in addition to the sign, the absolute value of the discrepancy between the pairs should matter as well, and it could increase the efficiency of the sign test.

Suppose that, as in the sign test, we are interested in testing the hypothesis that a median of the unknown distribution is m_0. We make an important assumption of the data.

> **Assumption:** The differences D_i, $i = 1, \ldots, n$ are symmetrically distributed about zero.

This implies that positive and negative differences are equally likely. For this test, the absolute values of the differences $(|D_1|, |D_2|, \ldots, |D_n|)$ are ranked. The idea is to use $(|D_1|, |D_2|, \ldots, |D_n|)$ as a set of weights for comparing the differences between $(S_1, \ldots, S_n)$.

Under H_0 (the median of distribution is m_0), the expectation of the sum of positive differences should be equal to the expectation of the sum of the negative differences. Define

$$T^+ = \sum_{i=1}^n S_i \, r(|D_i|),$$

and

$$T^- = \sum_{i=1}^n (1 - S_i) \, r(|D_i|),$$

where $S_i \equiv S(D_i) = I(D_i > 0)$. Thus $T^+ + T^- = \sum_{i=1}^n i = n(n+1)/2$ and

$$T = T^+ - T^- = 2\sum_{i=1}^n r(|D_i|)S_i - n(n+1)/2. \tag{7.3}$$

Under H_0, $(S_1, \ldots, S_n)$ are i.i.d. Bernoulli random variables with $p = 1/2$, independent of the corresponding magnitudes. Thus, when H_0 is true, $\mathbb{E}(T^+) = n(n+1)/4$ and $\mathbb{V}\mathrm{ar}(T^+) = n(n+1)(2n+1)/24$. Quantiles for T^+ are listed in Table 7.1. In R, the signed rank test based on T^+ is

```
wilcoxon.signed2
```

Large sample tests are typically based on a normal approximation of the test statistic, which is even more effective if there are ties in the data.

> **Rule:** For the Wilcoxon signed rank test, it is suggested to use T from (7.3) instead of T^+ in the case of large sample approximation.

Table 7.1 Quantiles of T^+ for the Wilcoxon signed rank test.

n	0.01	0.025	0.05	n	0.01	0.025	0.05
8	2	4	6	24	70	82	92
9	4	6	9	25	77	90	101
10	6	9	11	26	85	99	111
11	8	11	14	27	94	108	120
12	10	14	18	28	102	117	131
13	13	18	22	29	111	127	141
14	16	22	26	30	121	138	152
15	16	20	26	31	131	148	164
16	24	30	36	32	141	160	176
17	28	35	42	33	152	171	188
18	33	41	48	34	163	183	201
19	38	47	54	35	175	196	214
20	44	53	61	36	187	209	228
21	50	59	68	37	199	222	242
22	56	67	76	38	212	236	257
23	63	74	84	39	225	250	272

In this case, $\mathbb{E}(T) = 0$ and $\mathbb{V}\mathrm{ar}(T) = \sum_i (R(|D_i|)^2) = n(n+1)(2n+1)/6$ under H_0. Normal quantiles

$$P\left(\frac{T}{\sqrt{\mathbb{V}\mathrm{ar}(T)}} \le t \right) = \Phi(t),$$

can be used to evaluate p-values of the observed statistics T with respect to a particular alternative (see the r-file `wilcoxon.signed.r`).

Example 7.7 Twelve sets of identical twins underwent psychological tests to measure the amount of aggressiveness in each person's personality. We are interested in comparing the twins with each other to see if the firstborn twin tends to be more aggressive than the other. The results are as follows; the higher score indicates more aggressiveness:

First twin X_i:	86	71	77	68	91	72	77	91	70	71	88	87
Second twin Y_i:	88	77	76	64	96	72	65	90	65	80	81	72

The hypotheses are as follows: H_0 where the first twin does not tend to be more aggressive than the other, that is, $\mathbb{E}(X_i) \leq \mathbb{E}(Y_i)$, and H_1 where the first twin tends to be more aggressive than the other, that is, $\mathbb{E}(X_i) > \mathbb{E}(Y_i)$. The Wilcoxon signed rank test is appropriate if we assume that $D_i = X_i - Y_i$ are independent and symmetric and have the same mean. Below is the output of `wilcoxon.signed`, where T statistics have been used:

```
> fb <- c(86,71,77,68,91,72,77,91,70,71,88,87)
> sb <- c(88,77,76,64,96,72,65,90,65,80,81,72)
>
> source("wilcoxon.signed.r")
> result1 <- wilcoxon.signed(fb, sb, 1)
> result1
   Tp1          Tp          p
41.5000000  0.7564901  0.2382353
>
```

The following is the output of `wilcoxon.signed2` where T^+ statistics have been used. The p-values are identical, and there is insufficient evidence to conclude the first twin is more aggressive than the next:

```
> source("wilcoxon.signed2.r")
> result2 <- wilcoxon.signed2(fb, sb, 1)
> result2
      R           T          p
17.0000000  0.7564901  0.2382353
```

7.5 Wilcoxon (Two-Sample) Sum Rank Test

The WSuRT is often used in place of a two sample t-test when the populations being compared are not normally distributed. It requires independent random samples of sizes n_1 and n_2.

> **Assumption:** Actually, no additional assumptions are needed for the Wilcoxon two-sample test.

An example of the sort of data for which this test could be used is responses on a Likert scale (e.g. 1 = much worse, 2 = worse, 3 = no change, 4 = better, 5 = much better). It would be inappropriate to use the t-test for such data because it is only of an ordinal nature. The Wilcoxon rank sum test tells us more generally whether the groups are homogeneous or one group is "better" than the other. More generally, the basic null hypothesis of the Wilcoxon sum rank test is that the two populations are equal. That is, $H_0 : F_X(x) = F_Y(x)$. This test assumes that the shapes of the distributions are similar.

Let $\mathbf{X} = X_1, \ldots, X_{n_1}$ and $\mathbf{Y} = Y_1, \ldots, Y_{n_2}$ be two samples from populations that we want to compare. The $n = n_1 + n_2$ ranks are assigned as they were in the sign test. The test statistic W_n is the sum of ranks (1 to n) for $\mathbf{X}$. For example, if $X_1 = 1$, $X_2 = 13$, $X_3 = 7$, $X_4 = 9$, and $Y_1 = 2$, $Y_2 = 0$, $Y_3 = 18$, then the value of W_n is $2 + 4 + 5 + 6 = 17$.

If the two populations have the same distribution, then the sum of the ranks of the first sample and those in the second sample should be the same relative to their sample sizes. Our test statistic is

$$W_n = \sum_{i=1}^{n} i S_i(\mathbf{X}, Y),$$

where $S_i(\mathbf{X}, Y)$ is an indicator function defined as 1 if the ith ranked observation is from the first sample and as 0 if the observation is from the second sample. If there are no ties, then under H_0,

$$\mathbb{E}(W_n) = \frac{n_1(n+1)}{2} \quad \text{and} \quad \mathbb{V}\mathrm{ar}(W_n) = \frac{n_1 n_2 (n+1)}{12}.$$

The statistic W_n achieves its minimum when the first sample is entirely smaller than the second and its maximum when the opposite occurs:

$$\min W_n = \sum_{i=1}^{n_1} i = \frac{n_1(n_1+1)}{2}, \quad \max W_n = \sum_{i=n-n_1+1}^{n} i = \frac{n_1(2n - n_1 + 1)}{2}.$$

The exact distribution of W_n is computed in a tedious but straightforward manner. The probabilities for W_n are symmetric about the value of $\mathbb{E}(W_n) = n_1(n+1)/2$.

Example 7.8 Suppose $n_1 = 2, n_2 = 3$, so that $n = n_1 + n_2 = 5$. There are $\binom{5}{2} = \binom{5}{3} = 10$ distinguishable configurations of the vector $(S_1, S_2, \ldots, S_5)$. The minimum of W_5 is 3 and the maximum is 9. Table 7.2 gives the values for W_5 in this example, along with the configurations of ones in the vector $(S_1, S_2, \ldots, S_5)$ and the probability under H_0. Notice the symmetry in probabilities about $\mathbb{E}(W_5)$.

Let $k_{n_1, n_2}(m)$ be the number of all arrangements of zeroes and ones in $(S_1(\mathbf{X}, \mathbf{Y}), \ldots, S_n(\mathbf{X}, \mathbf{Y}))$ such that $W_n = \sum_{i=1}^{n} i S_i(\mathbf{X}, \mathbf{Y}) = m$. Then the probability distribution

$$P(W_n = m) = \frac{k_{n_1, n_2}(m)}{\binom{n}{n_1}}, \quad \frac{n_1(n_1+1)}{2} \le m \le \frac{n_1(2n - n_1 + 1)}{2},$$

can be used to perform an exact test. Deriving this distribution is no trivial matter, mind you. When n is large, the calculation of exact distribution of W_n is cumbersome.

Table 7.2 Distribution of W_5 when $n_1 = 2$ and $n_2 = 3$.

W_5	Configuration	Probability
3	(1,2)	1/10
4	(1,3)	1/10
5	(1,4), (2,3)	2/10
6	(1,5), (2,4)	2/10
7	(2,5), (3,4)	2/10
8	(3,5)	1/10
9	(4,5)	1/10

The statistic W_n in WSuRT is an example of a *linear rank statistic* (see Section 7.1) for which the normal approximation holds,

$$W_n \sim \mathcal{N}\left(\frac{n_1(n+1)}{2}, \frac{n_1 n_2(n+1)}{12}\right).$$

A better approximation is

$$P(W_n \leq w) \approx \Phi(x) + \phi(x)(x^3 - 3x)\frac{n_1^2 + n_2^2 + n_1 n_2 + n}{20 n_1 n_2(n+1)},$$

where $\phi(x)$ and $\Phi(x)$ are the probability distribution function (PDF) and CDF of a standard normal distribution and $x = (w - \mathbb{E}(W) + 0.5)/\sqrt{\mathbb{V}\mathrm{ar}(W_n)}$. This approximation is satisfactory for $n_1 > 5$ and $n_2 > 5$ if there are no ties.

7.5.1 Ties in the Data

If ties are present, let $t_1, \ldots, t_k$ be the number of different observations among all the observations in the combined sample. The adjustment for ties is needed only in $\mathbb{V}\mathrm{ar}(W_n)$, because $\mathbb{E}(W_n)$ does not change. The variance decreases to

$$\mathbb{V}\mathrm{ar}(W_n) = \frac{n_1 n_2(n+1)}{12} - \frac{n_1 n_2 \sum_{i=1}^{k}(t_i^3 - t_i)}{12n(n+1)}. \tag{7.4}$$

For a proof of (7.4) and more details, see Lehmann (1998).

Example 7.9 Let the combined sample be $\{\ 2\ \boxed{2}\ \boxed{3}\ \boxed{4}\ 4\ 4\ 5\ \}$, where the boxed numbers are observations from the first sample. Then $n = 7, n_1 = 3, n_2 = 4$, and the ranks are $\{1.5\ 1.5\ 3\ 5\ 5\ 5\ 7\}$. The statistic $w = 1.5 + 3 + 5 = 9.5$ has mean $\mathbb{E}(W_n) = n_1(n+1)/2 = 12$. To adjust the variance for the ties, first note that there are $k = 4$ different groups of observations, with $t_1 = 2$, $t_2 = 1$, $t_3 = 3$ and

$t_4 = 1$. With $t_i = 1$, $t_i^3 - t_i = 0$, only the values of $t_i > 1$ (genuine ties) contribute to the adjusting factor in the variance. In this case,

$$\mathbb{V}\mathrm{ar}(W_7) = \frac{3 \cdot 4 \cdot 8}{12} - \frac{3 \cdot 4 \cdot ((8-2) + (27-3))}{12 \cdot 7 \cdot 8} = 8 - 0.5357 = 7.4643.$$

7.6 Mann–Whitney U Test

Like the Wilcoxon test above, the Mann–Whitney test is applied to find differences in two populations and does not assume that the populations are normally distributed. However, if we extend the method to tests involving population means (instead of just $\mathbb{E}(D_{ij}) = P(Y < X)$), we need an additional assumption.

Assumption: The shapes of the two distributions are identical.

This is satisfied if we have $F_X(t) = F_Y(t + \delta)$ for some $\delta \in \mathbb{R}$. Let $X_1, \ldots, X_{n_1}$ and $Y_1, \ldots, Y_{n_2}$ represent two independent samples. Define $D_{ij} = I(Y_j < X_i)$, $i = 1, \ldots, n_1$ and $j = 1, \ldots, n_2$. The Mann–Whitney statistic for testing the equality of distributions for X and Y is the linear rank statistic

$$U = \sum_{i=1}^{n_1} \sum_{j=1}^{n_2} D_{ij}.$$

It turns out that the test using U is equivalent to the test using W_n in Section 7.5.

7.6.1 Equivalence of Mann–Whitney and Wilcoxon Sum Rank Test

Fix i and consider

$$\sum_{j=1}^{n_2} D_{ij} = D_{i1} + D_{i2} + \cdots + D_{i,n_2}. \tag{7.5}$$

The sum in (7.5) is exactly the number of index values j for which $Y_j < X_i$. Apparently, this sum is equal to the rank of the X_i in the combined sample, $r(X_i)$, minus the number of Xs that are $\le X_i$. Denote the number of Xs that are $\le X_i$ by k_i. Then,

$$U = \sum_{i=1}^{n_1} (r(X_i) - k_i) = \sum_{i=1}^{n_1} r(X_i) - (k_1 + k_2 + \cdots + k_{n_1})$$

$$= \sum_{i=1}^{n_1} i S_i(\mathbf{X}, \mathbf{Y}) - \frac{n_1(n_1 + 1)}{2} = W_n - \frac{n_1(n_1 + 1)}{2},$$

because $k_1 + k_2 + \cdots + k_{n_1} = 1 + 2 + \cdots + n_1$. After all this, the Mann–Whitney (U) statistic and the Wilcoxon sum rank statistic (W_n) are equivalent. As a result, the Wilcoxon sum rank test and Mann–Whitney test are often referred simply as the *Wilcoxon-Mann–Whitney* test.

Example 7.10 Let the combined sample be { ⑦ 12 13 ⑮ ⑮ 18 28}, where boxed observations come from sample 1. The statistic U is $0 + 2 + 2 = 4$. On the other hand, $W_7 - 3 \cdot 4/2 = (1 + 4.5 + 4.5) - 6 = 4$.

The R function wmw computes the Wilcoxon–Mann–Whitney test using the same arguments from tests listed above. In the example below, w is the sum of ranks for the first sample, and z is the standardized rank statistic for the case of ties:

```
> source("wmw.r")
> result<-wmw(c(1,2,3,4,5),c(2,4,2,11,1),0)
> result
          R          T          p
28.0000000  0.1063990  0.9575729
```

7.7 Test of Variances

Compared with parametric tests of the mean, statistical tests on population variances based on the assumption of normal distributed populations are less robust. That is, the parametric tests for variances are known to perform quite poorly if the normal assumptions are wrong.

Suppose we have two populations with CDFs F and G and we collect random samples $X_1, ..., X_{n_1} \sim F$ and $Y_1, ..., Y_{n_2} \sim G$ (the same setup used in the Mann–Whitney test). This time, our null hypothesis is

$$H_0 : \sigma_X^2 = \sigma_Y^2$$

versus one of three alternative hypotheses (H_1): $\sigma_X^2 \neq \sigma_Y^2, \sigma_X^2 < \sigma_Y^2, \sigma_X^2 > \sigma_Y^2$. If $\bar{x}$ and $\bar{y}$ are the respective sample means, the test statistic is based on

$\tilde{R}(x_i)$ = rank of $(x_i - \bar{x})^2$ among all $n = n_1 + n_2$ squared differences
$\tilde{R}(y_i)$ = rank of $(y_i - \bar{y})^2$ among all $n = n_1 + n_2$ squared differences
with test statistic

$$T = \sum_{i=1}^{n_1} \tilde{R}(x_i).$$

Assumption: The measurement scale needs to be interval (at least).

7.7.1 Ties in the Data

If there are ties in the data, it is better to use

$$T^* = \frac{T - n_1 V_R}{\sqrt{\frac{n_1 n_2}{n(n-1)} W_R - \frac{n_1 n_2}{n-1} V^2_R}}$$

where

$$V_R = n^{-1} \left[\sum^{n_1} \tilde{R}(x_i)^2 + \sum^{n_2} \tilde{R}(y_i)^2 \right] \quad \text{and}$$

$$W_R = \sum^{n_1} \tilde{R}(x_i)^4 + \sum^{n_2} \tilde{R}(y_i)^4.$$

The critical region for the test corresponds to the direction of the alternative hypothesis. This is called the *Conover test of equal variances*, and tabled quantiles for the null distribution of T are found in Conover and Iman (1978). If we have larger samples ($n_1 \geq 10$, $n_2 \geq 10$), the following normal approximation for T can be used:

$$T \sim \mathcal{N}(\mu_T, \sigma_T^2), \quad \text{with } \mu_T = \frac{n_1(n+1)(2n+1)}{6},$$

$$\sigma_T^2 = \frac{n_1 n_2(n+1)(2n+1)(8n+11)}{180}.$$

For example, with an α-level test, if $H_1 : \sigma_X^2 > \sigma_Y^2$, we reject H_0 if $z_0 = (T - \mu_T)/\sigma_T > z_\alpha$, where z_α is the $1 - \alpha$ quantile of the normal distribution. The test for three or more variances is discussed in Chapter 8, after the Kruskal–Wallis test for testing differences in three or more population medians.

Use the R function `conover.test(x,y,p,alt)` for the test of two variances, where x and y are the samples, p is the sought-after quantile from the null distribution of T, `alt = 1` for the test of $H_1 : \sigma_X^2 > \sigma_Y^2$ (use $p/2$ for the two-sided test), `alt = -1` for the test of $H_1 : \sigma_X^2 < \sigma_Y^2$, and `alt = 0` for the test of $H_1 : \sigma_X^2 \neq \sigma_Y^2$. In the first example below, the test statistic $T = -1.5253$ is inside the region the interval $(-1.6449, 1.6449)$, and we do not reject $H_0 : \sigma_X^2 = \sigma_Y^2$ at level $\alpha = 0.10$:

```
> source("conover.test.r")
>
> x <- c(1, 2, 3, 4, 5);
> y <- c(1, 3, 5, 7, 9);
> conover.test(x,y,0.1,0);
[1] "Tied value(s) exist(s)."
            T        Tstar        pval       Tlower      Tupper        ties
111.2500000  -1.5252798   0.1271893   -1.6448536   1.6448536   1.0000000
>
> x <- c(1, 3, 4, 10, 13, 14, 17, 19, 23, 27);
> y <- c(51, 52, 53, 57, 61, 79, 80, 82, 85, 86);
> conover.test(x,y,0.1,0);
```

```
[1] "Sample sizes are equal to or bigger than 10: n1 >= 10, n2 >= 10."
         T          pval        Tlower        Tupper           ties
6.430000e+02 5.618590e-03 9.645747e+02 1.905425e+03 0.000000e+00
>
> x <- c(1, 3, 4, 10, 13, 14, 17, 19, 23);
> y <- c(51, 52, 53, 57, 61, 79, 80, 82, 85);
> conover.test(x,y,0.1,0);
[1] "Sample sizes are smaller than 10: n1 < 10, n2 < 10."
      T Tlower Tupper   ties
    400    689   1420      0
```

7.8 Walsh Test for Outliers

Suppose that r outliers are suspect, where $r \geq 1$ and fixed. Order observations $X_1, \ldots, X_n$ and obtain the order statistic $X_{(1)} \leq X_{(2)} \leq \cdots \leq X_{(n)}$, and set the significance level α.

We will explain the steps and provide an R implementation, but readers interested in details are directed to Walsh (1962). The Walsh method has the following steps, following Madansky (1988):

Step 1: Calculate $c = \lfloor \sqrt{2n} \rfloor$, where $\lfloor x \rfloor$ is the largest integer $\leq x$, $b^2 = 1/\alpha$, $k = r + c$, and $a = \left(1 + b\sqrt{\frac{c-b^2}{c-1}}\right)/(c - b^2 - 1)$.

Step 2. The smallest r values $X_{(1)} \leq \cdots \leq X_{(r)}$ are outliers if

$$rL = X_{(r)} - (1 + a)X_{(r+1)} + aX_k < 0,$$

and the largest r values $X_{(n-r+1)} \leq \cdots \leq X_{(n)}$ are outliers if

$$rU = X_{(n-r+1)} - (1 + a)X_{(n-r)} + aX_{n-k+1} > 0.$$

The sample size has to satisfy $n \geq \frac{1}{2}\left(1 + \frac{1}{\alpha}\right)^2$. To achieve $\alpha = 0.05$, a sample size of at least 221 is needed. As an outcome, one either rejects no outliers, rejects the smallest r or the largest r, or even both the smallest and largest r, thus potentially rejecting a total of $2r$ observations. In R, the custom function walshnp(data, r, alpha) evaluates data with r and alpha and provides outliers:

```
> source("walshnp.r")
> dat <- rnorm(300,mean=0,sd=2)
> data <- c(-11.26, dat, 13.12)
> walshnp(data)
[1] "Lower outliers are: -11.26"
[1] "Upper outliers are: 13.12"
[1] -11.26  13.12
```

7.9 Exercises

7.1 For the sign test, what is the two-sided *p*-value of the test statistic T if $n = 50$ and $T = 32$?

7.2 With the Spearman correlation statistic, show that when the ranks are opposite, $\hat{\rho} = -1$.

7.3 Diet A was given to a group of 10 overweight boys between the ages of 8 and 10. Diet B was given to another independent group of 8 similar overweight boys. The weight loss is given in the table below. Using WMW test, test the hypothesis that the diets are of comparable effectiveness against the two-sided alternative. Use $\alpha = 5\%$ and normal approximation:

Diet A	7	2	3	−1	4	6	0	1	4	6
Diet B	5	6	4	7	8	9	7	2		

7.4 A psychological study involved the rating of rats along a dominance–submissiveness continuum. To determine the reliability of the ratings, the ranks given by two different observers were tabulated below. Are the ratings agreeable? Explain your answer:

Animal	Rank observer A	Rank observer B	Animal	Rank observer A	Rank observer B
A	12	15	I	6	5
B	2	1	J	9	9
C	3	7	K	7	6
D	1	4	L	10	12
E	4	2	M	15	13
F	5	3	N	8	8
G	14	11	O	13	14
H	11	10	P	16	16

7.5 Two vinophiles, X and Y, were asked to rank $N = 8$ tasted wines from best to worst (rank #1 = highest, rank #8 = lowest). Find the Spearman coefficient of correlation between the experts. If the sample size increased to $N = 80$ and we find $\hat{\rho}$ is 10 times smaller than what you found above, what would the *p*-value be for the two-sided test of hypothesis?

Wine brand	a	b	c	d	e	f	g	h
Expert X	1	2	3	4	5	6	7	8
Expert Y	2	3	1	4	7	8	5	6

7.6 Use the link below to see the results of an experiment on the effect of prior information on the time to fuse random dot stereograms. One group (NV) was given either no information or just verbal information about the shape of the embedded object. A second group (VV) received both verbal information and visual information (e.g. a drawing of the object). Does the median time prove to be greater for the NV group? Compare your results to those from a two-sample t-test.

```
http://lib.stat.cmu.edu/DASL/Datafiles/FusionTime
.html
```

7.7 Derive the exact distribution of the Mann–Whitney U statistic in the case that $n_1 = 4$ and $n_2 = 2$.

7.8 A number of Vietnam combat veterans were discovered to have dangerously high levels of the dioxin 2,3,7,8-TCDD in blood and fat tissue as a result of their exposure to the defoliant Agent Orange. A study published in *Chemosphere (Vol. 20, 1990)* reported on the TCDD levels of 20 Massachusetts Vietnam veterans who were possibly exposed to Agent Orange. The amounts of TCDD (measured in parts per trillion) in blood plasma and fat tissue drawn from each veteran are shown in the table: Is there sufficient evidence of a difference between the distributions of TCDD levels in plasma and fat tissue for Vietnam veterans exposed to Agent Orange?

TCDD levels in plasma	TCDD levels in fat tissue
2.5 3.1 2.1	4.9 5.9 4.4
3.5 3.1 1.8	6.9 7.0 4.2
6.8 3.0 36.0	10.0 5.5 41.0
4.7 6.9 3.3	4.4 7.0 2.9
4.6 1.6 7.2	4.6 1.4 7.7
1.8 20.0 2.0	1.1 11.0 2.5
2.5 4.1	2.3 2.5

Year, month	No. of accidents Friday the 6th	No. of accidents Friday the 13th	Sign	Hospital
1989, October	9	13	–	SWTRHA hospital
1990, July	6	12	–	
1991, September	11	14	–	
1991, December	11	10	+	
1992, March	3	4	–	
1992, November	5	12	–	

7.9 For the two samples in Exercise 7.6, test for equal variances.

7.10 The following two data sets are part of a larger data set from Scanlon et al. (1993). The data analysis in this paper addresses the issues of how superstitions regarding Friday the 13th affect human behavior. Scanlon et al. collected data on shopping patterns and traffic accidents for Fridays the 6th and the 13th between October 1989 and November 1992:

i) The first data set is found online at

```
http://lib.stat.cmu.edu/DASL/Datafiles/Fridaythe
13th.html
```

The data set lists the number of shoppers in nine different supermarkets in southeast England. At the level $\alpha = 10\%$, test the hypothesis that "Friday 13th" affects spending patterns among South Englanders.

ii) The second data set is the number of patients accepted in SWTRHA hospital on dates of Friday the 6th and Friday the 13th. At the level $\alpha = 10\%$, test the hypothesis that the "Friday the 13th" effect is present:

7.11 Professor Inarb claims that 50% of his students in a large class achieve a final score 90 points or and higher. A suspicious student asks 17 randomly selected students from Professor Inarb's class, and they report the following scores:

80	81	87	94	79	78	89	90	92	88	81	79	82	79	77	89	90

Year, month	No. of shoppers Friday the 6th	No. of shoppers Friday the 13th	Sign	Supermarket
1990, July	4942	4882	+	Epsom
1991, September	4895	4736	+	
1991, December	4805	4784	+	
1992, March	4570	4603	−	
1992, November	4506	4629	−	
1990, July	6754	6998	−	Guildford
1991, September	6704	6707	−	
1991, December	5871	5662	+	
1992, March	6026	6162	−	
1992, November	5676	5665	+	
1990, July	3685	3848	−	Dorking
1991, September	3799	3680	+	
1991, December	3563	3554	+	
1992, March	3673	3676	−	
1992, November	3558	3613	−	
1990, July	5751	5993	−	Chichester
1991, September	5367	5320	+	
1991, December	4949	4960	−	
1992, March	5298	5467	−	
1992, November	5199	5092	+	
1990, July	4141	4389	−	Horsham
1991, September	3674	3660	+	
1991, December	3707	3822	−	
1992, March	3633	3730	−	
1992, November	3688	3615	+	
1990, July	4266	4532	−	East Grinstead
1991, September	3954	3964	−	
1991, December	4028	3926	+	
1992, March	3689	3692	−	
1992, November	3920	3853	+	

(Continued)

Year, month	No. of shoppers Friday the 6th	No. of shoppers Friday the 13th	Sign	Supermarket
1990, July	7138	6836	+	Lewisham
1991, September	6568	6363	+	
1991, December	6514	6555	−	
1992, March	6115	6412	−	
1992, November	5325	6099	−	
1990, July	6502	6648	−	Nine Elms
1991, September	6416	6398	+	
1991, December	6422	6503	−	
1992, March	6748	6716	+	
1992, November	7023	7057	−	
1990, July	4083	4277	−	Crystal Palace
1991, September	4107	4334	−	
1991, December	4168	4050	+	
1992, March	4174	4198	−	
1992, November	4079	4105	−	

Test the hypothesis that the Professor Inarb's claim is not consistent with the evidence, i.e. that the 50%-tile (0.5-quantile, median) is not equal to 90. Use $\alpha = 0.05$.

7.12 Why does the moon look bigger on the horizon? Kaufman and Rock (1962) tested 10 subjects in an experimental room with moons on a horizon and straight above. The ratios of the perceived size of the horizon moon and the perceived size of the zenith moon were recorded for each person. Does the horizon moon seem bigger?

Subject	Zenith	Horizon	Subject	Zenith	Horizon
1	1.65	1.73	2	1	1.06
3	2.03	2.03	4	1.25	1.4
5	1.05	0.95	6	1.02	1.13
7	1.67	1.41	8	1.86	1.73
9	1.56	1.63	10	1.73	1.56

7.13 To compare the *t*-test with the WSuRT, set up the following simulation in R: (i) generate $n = 10$ observations from $\mathcal{N}(0,1)$; (ii) for the test of $H_0 : \mu = 1$ versus $H_1 : \mu < 1$, perform a *t*-test at $\alpha = 0.05$; (iii) run an analogous nonparametric test; (iv) repeat this simulation 1000 times, and compare the power of each test by counting the number of times H_0 is rejected; and (v) repeat the entire experiment using a non-normal distribution, and comment on your result.

RELEVANT R FUNCTIONS AND DATA SETS IN THIS CHAPTER

R codes: `conover.test.r`, `sign.test.r`, `wmw.r`, `wilcoxon.signed.r`, `wilcoxon.signed2.r`, `walshnp.r`

R functions: `rank`, `cor`, `binom.test`, `wilcox.test`, `mood.test`, `fligner.test`, `ansari.test`

`exer7.3.csv`, `exer7.5.csv`, `exer7.7.csv`, `exer7.9.csv`, `exer7.11.csv`

References

Arbuthnot, J. (1710), "An Argument for Divine Providence," *Philosophical Transactions*, 27, 186–190.

Arvin, D. V., and Spaeth, R. (1997), "Trends in Indiana's water use 1986–1996 special report," Technical report by State of Indiana Department of Datural Resources, Division of Water.

Conover, W. J. (1999), *Practical Nonparametric Statistics*, New York: Wiley.

Conover, W. J., and Iman, R. L. (1978), "Some Exact Tables for the Squared Ranks Test," *Communications in Statistics*, 5, 491–513.

Hervey, K. (1888), *The Book of Christmas; Descriptive of the Customs, Ceremonies, Traditions, Superstitions, Fun, Feeling, & Festivities of the Christmas Season* ISBN - 978-1162917085.

Hotelling, H., and Pabst, M. (1936), "Rank Correlation and Tests of Significance Involving the Assumption of Normality," *Annals of Mathematical Statistics*, 7, 29–43.

Kaufman, L., and Rock, I. (1962), "The Moon Illusion," *Science*, 136, 953–961.

Kendall, M. G. (1938), "A New Measure of Rank Correlation," *Biometrika*, 30, 81–93.

Lehmann, E. L. (1998), *Nonparametrics: Statistical Methods Based on Ranks*, New Jersey: Prentice Hall.

Madansky, A. (1988), *Prescriptions for Working Statisticians*, New York, NY: Springer.

Scanlon, T. J., Luben, R. N., Scanlon, F. L., and Singleton, N. (1993), "Is Friday the 13th Bad for Your Health?," *BMJ*, 307, 1584–1586.

Stichler, R. D., Richey, G. G., and Mandel, J. (1953), "Measurement of Treadware of Commercial Tires," *Rubber Age*, 2, 73.

Spearman, C. (1904), "The Proof and Measurement for Association Between Two Things," *American Journal of Psychology*, 15, 72–101.

Walsh, J. E. (1962), *Handbook of Nonparametric Statistics I and II*, Princeton, NJ: Van Nostrand.

8

Designed Experiments

Luck is the residue of design.
Branch Rickey,[1] former owner of the Brooklyn Dodgers (1881–1965)

This chapter deals with the nonparametric statistical analysis of designed experiments. The classical parametric methods in analysis of variance (ANOVA), from one-way to multi-way tables, often suffer from a sensitivity to the effects of non-normal data. The nonparametric methods discussed here are much more robust. In most cases, they mimic their parametric counterparts but focus on analyzing ranks instead of response measurements in the experimental outcome. In this way, the chapter represents a continuation of the rank tests presented in Chapter 7.

We cover the *Kruskal–Wallis* (KW) *test* to compare three or more samples in an ANOVA, the *Friedman test* to analyze two-way ANOVA in a "randomized block" design (RBD), and nonparametric tests of variances for three or more samples.

8.1 Kruskal–Wallis Test

The Kruskal–Wallis (KW) test (Kruskal (1952)) is a logical extension of the Wilcoxon–Mann–Whitney test. It is a nonparametric test used to compare three or more samples. It is used to test the null hypothesis that all populations have identical distribution functions against the alternative hypothesis that at least two of the samples differ only with respect to location (median), if at all.

The KW test is the analogue to the F-test used in the one-way ANOVA. While ANOVA tests depends on the assumption that all populations under comparison are independent and normally distributed, the KW test places no such restriction

---

1 Contributed almost nothing to nonparametric statistics, but helped major league baseball develop a minor league system, catalyzed racial integration by signing Jackie Robinson, and even introduced the batting helmet.

Nonparametric Statistics with Applications to Science and Engineering with R, Second Edition.
Paul Kvam, Brani Vidakovic, and Seong-joon Kim.
© 2023 John Wiley & Sons, Inc. Published 2023 by John Wiley & Sons, Inc.

on the comparison. Suppose the data consist of k independent random samples with sample sizes $n_1, \ldots, n_k$. Let $n = n_1 + \cdots + n_k$:

Sample 1	$X_{11},$	$X_{12},$	$\cdots$	X_{1,n_1}
Sample 2	$X_{21},$	$X_{22},$	$\cdots$	X_{2,n_2}
$\vdots$	$\vdots$			
Sample $k-1$	$X_{k-1,1},$	$X_{k-1,2},$	$\cdots$	$X_{k-1,n_{k-1}}$
Sample k	$X_{k1},$	$X_{k2},$	$\cdots$	X_{k,n_k}

Under the null hypothesis, we can claim that all of the k samples are from a common population. The expected sum of ranks for the sample i, $\mathbb{E}(R_i)$, would be n_i times the expected rank for a single observation. That is, $n_i(n+1)/2$, and the variance can be calculated as $\mathbb{V}\text{ar}(R_i) = n_i(n+1)(n-n_i)/12$. One way to test H_0 is to calculate $R_i = \sum_{j=1}^{n_i} r(X_{ij})$ – the total sum of ranks in sample i. The statistic

$$\sum_{i=1}^{k}\left[R_i - \frac{n_i(n+1)}{2}\right]^2 \tag{8.1}$$

will be large if the samples differ, so the idea is to reject H_0 if (8.1) is "too large." However, its distribution is a jumbled mess, even for small samples, so there is little use in pursuing an exact test. Alternatively we can use the normal approximation

$$\frac{R_i - \mathbb{E}(R_i)}{\sqrt{\mathbb{V}\text{ar}(R_i)}} \overset{\text{appr}}{\sim} \mathcal{N}(0,1) \Rightarrow \sum_{i=1}^{k} \frac{\left(R_i - \mathbb{E}(R_i)\right)^2}{\mathbb{V}\text{ar}(R_i)} \overset{\text{appr}}{\sim} \chi^2_{k-1},$$

where the χ^2 statistic has only $k-1$ degrees of freedom due to the fact that only $k-1$ ranks are unique.

Based on this idea, Kruskal and Wallis (1952) proposed the test statistic

$$H' = \frac{1}{S^2}\left[\sum_{i=1}^{k}\frac{R_i^2}{n_i} - \frac{n(n+1)^2}{4}\right], \tag{8.2}$$

where

$$S^2 = \frac{1}{n-1}\left[\sum_{i=1}^{k}\sum_{j=1}^{n_i} r(X_{ij})^2 - \frac{n(n+1)^2}{4}\right].$$

If there are no ties in the data, (8.2) simplifies to

$$H = \frac{12}{n(n+1)}\sum_{i=1}^{k}\frac{1}{n_i}\left[R_i - \frac{n_i(n+1)}{2}\right]^2. \tag{8.3}$$

They showed that this statistic has an approximate χ^2 distribution with $k-1$ degrees of freedom.

The R routine

```
kruskal.test
```

implements the KW test using a vector to represent the responses and another to identify the population from which the response came. Suppose we have the following responses from three treatment groups:

$$(1, 3, 4), \quad (3, 4, 5), \quad (4, 4, 4, 6, 5),$$

which are sample from three populations. The R code for testing the equality of locations of the three populations computes a p-value of 0.1428:

```
> data <- c(1, 3, 4, 3, 4, 5, 4, 4, 4, 6, 5);
> belong <- c(1, 1, 1, 2, 2, 2, 3, 3, 3, 3, 3);
>
> kruskal.test(data,belong)
        Kruskal-Wallis rank sum test
data:  data and belong
Kruskal-Wallis chi-squared = 3.8923, df = 2, p-value = 0.1428
```

Example 8.1 The following data are from a classic agricultural experiment measuring crop yield in four different plots. For simplicity, we identify the treatment (plot) using the integers {1,2,3,4}. The third treatment mean measures far above the rest, and the null hypothesis (the treatment means are equal) is rejected with a p-value less than 0.0002:

```
> yield <- c(83, 91, 94, 89, 89, 96, 91, 92, 90, 84, 91, 90, 81, 83,
+ 84, 83, 88, 91, 89, 101, 100, 91, 93, 96, 95, 94, 81, 78, 82, 81,
+ 77, 79, 81, 80);
> plot <- c(1, 1, 1, 1, 1, 1, 1, 1, 1, 1, 2, 2, 2, 2, 2, 2, 2, 2, 2,
+ 3, 3, 3, 3, 3, 3, 3, 4, 4, 4, 4, 4, 4, 4, 4);
>
> kruskal.test(yield,plot)
        Kruskal-Wallis rank sum test
data:  yield and plot
Kruskal-Wallis chi-squared = 20.3371, df = 3, p-value = 0.0001445
```

8.1.1 Kruskal–Wallis Pairwise Comparisons

Results of the nonparametric ANOVAs determine if there are significant differences in the means among the groups being sampled. The results are not to be used to determine which treatments are best or worst, however. Instead, we employ post hoc tests, called pairwise comparisons, to determine which groups differ from one another.

If the KW test detects overall treatment differences, we can determine if two particular treatment groups (say, i and j) are different at level α if

$$\left| \frac{R_i}{n_i} - \frac{R_j}{n_j} \right| > t_{n-k,\, 1-\alpha/2} \sqrt{\frac{S^2(n-1-H')}{n-k} \cdot \left(\frac{1}{n_i} + \frac{1}{n_j} \right)}. \tag{8.4}$$

Example 8.2   We decided the four crop treatments were statistically different, and it would be natural to find out which ones seem better and which ones seem worse. In the table below, we compute the statistic

$$T = \frac{\left| \frac{R_i}{n_i} - \frac{R_j}{n_j} \right|}{\sqrt{\frac{S^2(n-1-H')}{n-k}\left(\frac{1}{n_i} + \frac{1}{n_j}\right)}},$$

for every combination of $1 \leq i \neq j \leq 4$, and compare it to $t_{30,0.975} = 2.042$:

(i,j)	1	2	3	4
1	0	1.856	1.859	5.169
2	1.856	0	3.570	3.363
3	1.859	3.570	0	6.626
4	5.169	3.363	6.626	0

This shows that the third treatment is the best, but not significantly different from the first treatment, which is second best. Treatment 2, which is third best, is not significantly different from Treatment 1, but is different from Treatment 4 and Treatment 3. These differences are more plainly seen in the box plot in Figure 8.1:

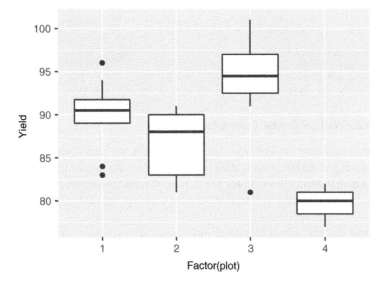

Figure 8.1   Box plot for crop yields.

```
>crop <- data.frame(yield,plot)
>library(ggplot2)
>ggplot(data=crop, aes(x=factor(plot),y=yield)) + geom_boxplot()
```

8.1.2 Jonckheere–Terpstra Ordered Alternative

If we want to see if the treatment effects in a one-way ANOVA have an implicit ordering, we can improve on the Kruskal–Wallace procedure by using the Jonckheere–Terpstra test for independent samples (see Jonckheere (1954), Terpstra (1952)). Naturally, if there exists an a priori ordering of treatments, this test will offer more power than the KW test. With the alternative to the null hypothesis (that all treatment means are the same), the alternative hypothesis will set the first treatment effect smaller (or equal to) than the second, and so on, with at least one strict inequality.

For any two treatment groups (say, i and j), we compute the Mann–Whitney U statistic, so that U_{ij} counts the number of observations in the ith treatment group that are larger than each observation from the jth treatment group. Under the null hypothesis, the Jonckheere–Terpstra test statistic

$$H_{JT} = \sum_{i<j} U_{ij},$$

is approximately normal (if n is sufficiently large) with mean and variance

$$\mathbb{E}(H_{JT}) = \frac{1}{2}\sum_{i<j} n_i n_j \quad \text{and} \quad \mathbb{V}\mathrm{ar}(H_{JT}) = \frac{1}{72}\left(n^2(2n+3) - \sum_i n_i^2(2n_i+3)\right).$$

Large values of the test statistic are considered significant according to the order of treatments described above. The PMCMRplus package in R contains various mean rank sum tests and procedures for multiple comparisons, including jonckheere.test. Arguments for jonckheere.test include a vector of response outcomes and a vector (or factor object) that identifies group identity.

8.2 Friedman Test

Well, first of all, tell me: is there a society that isn't run on greed?

Milton Friedman (1912–2006)

The *Friedman test* is a nonparametric alternative to the RBD in regular ANOVA. It replaces the RBD when the assumptions of normality are in question or when variances are possibly different from population to population. This test uses the ranks of the data rather than their raw values to calculate the test statistic. Because the Friedman test does not make distribution assumptions, it is not as powerful as the standard test if the populations are indeed normal.

Milton Friedman published the first results for this test, which was eventually named after him. He received the Nobel Prize for Economics in 1976, and one of the listed breakthrough publications was his article "The Use of Ranks to Avoid the Assumption of Normality Implicit in the Analysis of Variance," published in 1937 Friedman (1937).

Recall that the RBD design requires repeated measures for each block at each level of treatment. Let X_{ij} represent the experimental outcome of subject (or "block") i with treatment j, where $i = 1, \ldots, b$ and $j = 1, \ldots, k$:

	Treatments			
Blocks	**1**	**2**	**...**	**k**
1	X_{11}	X_{12}	...	X_{1k}
2	X_{21}	X_{22}	...	X_{2k}
⋮	⋮	⋮		⋮
b	X_{b1}	X_{b2}	...	X_{bk}

To form the test statistic, we assign ranks $\{1, 2, \ldots, k\}$ to each *row* in the table of observations. Thus the expected rank of any observation under H_0 is $(k+1)/2$. We next sum all the ranks by columns (by treatments) to obtain $R_j = \sum_{i=1}^{b} r(X_{ij})$, $1 \le j \le k$. If H_0 is true, the expected value for R_j is $\mathbb{E}(R_j) = b(k+1)/2$. The statistic

$$\sum_{j=1}^{k} \left(R_j - \frac{b(k+1)}{2} \right)^2$$

is an intuitive formula to reveal treatment differences. It has expectation $bk(k^2 - 1)/12$ and variance $k^2 b(b-1)(k-1)(k+1)^2/72$. Once normalized to

$$S = \frac{12}{bk(k+1)} \sum_{j=1}^{k} \left(R_j - \frac{b(k+1)}{2} \right)^2, \tag{8.5}$$

it has moments $\mathbb{E}(S) = k - 1$ and $\mathbb{V}\text{ar}(S) = 2(k-1)(b-1)/b \approx 2(k-1)$, which coincide with the first two moments of χ^2_{k-1}. Higher moments of S also approximate well those of χ^2_{k-1} when b is large.

In the case of ties, a modification to S is needed. Let $C = bk(k+1)^2/4$ and $R^* = \sum_{i=1}^{b} \sum_{j=1}^{k} r(X_{ij})^2$. Then,

$$S' = \frac{k-1}{R^* - bk(k+1)^2/4} \left(\sum_{j=1}^{k} R_j^2 - bC \right) \tag{8.6}$$

is also approximately distributed as χ^2_{k-1}.

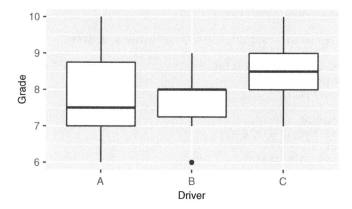

Figure 8.2 Box plot of vehicle performance grades of three cars (A,B,C).

Iman and Davenport (1980) showed that the Friedman test can be improved by replacing the S (or S') statistic with

$$F = \frac{(b-1)S}{b(k-1) - S},$$

which is approximately distributed as $F_{k-1,(b-1)(k-1)}$. Tests based on this approximation are generally superior to those based on chi-square tests that use S (Figure 8.2).

Example 8.3 In an evaluation of vehicle performance, six professional drivers (labeled I, II, III, IV, V, VI) evaluated three cars (A, B, and C) in a randomized order (see Figure 8.2). Their grades concern only the performance of the vehicles and supposedly are not influenced by the vehicle brand name or similar exogenous information. Here are their rankings on the scale 1–10:

Car	I	II	III	IV	V	VI
A	7	6	6	7	7	8
B	8	10	8	9	10	8
C	9	7	8	8	9	9

To use the R function

```
friedman.test,
```

the first input vector represents blocks (drivers), and the second represents treatments (cars):

```
> data <- matrix(c(7,8,9,6,10,7,6,8,8,7,9,8,7,10,9,8,8,9),
+ nrow=6,byrow=TRUE,dimnames=(list(1:6,c("A","B","C")))));
>
> result <- friedman.test(data);
> result
```

```
            Friedman rank sum test
data:   data
Friedman chi-squared = 8.2727, df = 2, p-value = 0.01598
>
> S <- as.numeric(result$statistic);
> b <- dim(data)[1];
> k <- dim(data)[2];
>
> F <- (b-1)*S/(b*(k-1)-S);
> pF <- 1-pf(F,k-1,(b-1)*(k-1));
> print(c(F,pF))
[1] 11.09756098  0.00289101
```

8.2.1 Friedman Pairwise Comparisons

If the *p*-value from the Friedman test is small enough to warrant multiple comparisons of treatments, we can use the following pairwise comparison: consider two treatments *i* and *j* to be different at level α if

$$\left| R_i - R_j \right| > t_{(b-1)(k-1),1-\alpha/2} \sqrt{2 \cdot \frac{bR^* - \sum_{j=1}^{k} R_j^2}{(b-1)(k-1)}}. \tag{8.7}$$

Example 8.4 From Example 8.3, the three cars (A,B,C) are considered significantly different at test level α = 0.01 (if we use the *F*-statistic). We can use the R function

```
friedman.pairwise.comparison(data,i,j,alpha)
```

to make a pairwise comparison between treatment *i* and treatment *j* at level *a*. The output = 1 if the treatments *i* and *j* are different; otherwise it is 0. The Friedman pairwise comparison reveals that car A is rated significantly lower than both car B and car C, but car B and car C are not considered to be different:

```
> source("friedman.pairwise.comparison.r")
>
> apply(data,2,mean) # mean of each treatment
         A        B        C
  6.833333 8.833333 8.333333
> apply(data,2,sd) # standard deviation of each treatment
         A        B        C
  0.7527727 0.9831921 0.8164966
> friedman.pairwise.comparison(data,1,2,0.01)
[1] 1
> friedman.pairwise.comparison(data,1,3,0.01)
[1] 1
> friedman.pairwise.comparison(data,2,3,0.01)
[1] 0
```

An alternative test for k matched populations is the test by Quade (1966), which is an extension of the Wilcoxon signed-rank test. In general, the *Quade test* performs no better than Friedman's test, but slightly better in the case $k = 3$. For that reason, we reference it but will not go over it in any detail.

8.2.2 Page Test for Ordered Alternative

To test the hypothesis that the treatment effects are ordered (e.g. the first treatment effect being the smallest and the kth treatment being the largest), we can apply the ordered alternatives test suggested by Page (1963). The test statistic H_P is based on the weighted combination of rank sums $\sum_{j=1}^{k} jR_j$:

$$H_P = \frac{(12 \sum_{j=1}^{k} jR_j - 3kn(n+1)^2)^2}{kn^2(n^2 - 1)(n + 1)},$$

and we reject the null hypothesis of identical treatment effects in favor of the ordered alternative if H_P is large. Under H_0, H_P has an approximate chi-square distribution with one degree of freedom. The R function `pageTest` is found in the `PMCMRplus` package.

Example 8.5 An example by Sachs (1997) features nine reviewers who are ranking four treatments (labeled B, C, D, and A). To test an implied order among those four treatments (i.e. $\mu_B \leq \mu_C \leq \mu_D \leq \mu_A$ with at least one strict inequality), the Page rank sum test `pageTest` will assume matrix columns indicate treatment groups if data values and groups are not identified in the argument:

```
> pageTest(reviewers, alternative = "greater")
        Page's ordered aligned rank sum test

data:  reviewers
z = 3.06, p-value = 0.001107
alternative hypothesis: greater
```

8.3 Variance Test for Several Populations

In Chapter 7, the test for variances from two populations was achieved with the nonparametric *Conover test*. In this section, the test is extended to three or more populations using a setup similar to that of the KW test. For the hypotheses H_0 (k variances are equal) versus H_1 (some of the variances are different), let $n_i =$ the number of observations sampled from each population and X_{ij} is the jth observation from population i. We denote the following:

- $n = n_1 + \cdots + n_k$
- $\bar{x}_i$ = sample average for ith population
- $R(x_{ij})$ = rank of $(x_{ij} - \bar{x}_i)^2$ among n items
- $T_i = \sum_{j=1}^{n_i} R(x_{ij})^2$
- $\bar{T} = n^{-1} \sum_{j=1}^{k} T_j$
- $V_T = (n-1)^{-1} \left(\sum_i \sum_j R(x_{ij})^4 - n\bar{T}^2 \right)$

Then the test statistic is

$$T = \frac{\sum_{j=1}^{k}(T_j^2/n_j) - n\bar{T}^2}{V_T}. \tag{8.8}$$

Under H_0, T has an approximate χ^2 distribution with $k-1$ degrees of freedom, so we can test for equal variances at level α by rejecting H_0 if $T > \chi_{k-1}^2(1-\alpha)$. Conover (1999) notes that the asymptotic relative efficiency (ARE), relative to the regular test for different variances, is 0.76 (when the data are actually distributed normally). If the data are distributed as *double exponential*, the ARE is over 1.08.

8.3.1 Multiple Comparisons for Variance Test

If H_0 is rejected, we can determine which populations have unequal variances using the following paired comparisons:

$$\left| \frac{T_i}{n_i} - \frac{T_j}{n_j} \right| > \sqrt{ \left(\frac{1}{n_i} + \frac{1}{n_j} \right) V_T \left(\frac{n-1-T}{n-k} \right) } \, t_{n-k}(1-\alpha/2),$$

where $t_{n-k}(\alpha)$ is the α quantile of the t distribution with $n-k$ degrees of freedom. If there are no ties, $\bar{T}$ and V_T are simple constants: $\bar{T} = (n+1)(2n+1)/6$ and $V_T = n(n+1)(2n+1)(8n+11)/180$.

Example 8.6 For the crop data in the Example 8.1, we can apply the variance test and obtain $n = 34$, $T_1 = 3845$, $T_2 = 4631$, $T_3 = 4032$, $T_4 = 1174.5$, and $\bar{T} = 402.51$. The variance term $V_T = \left(\sum_i \sum_j R(x_{ij})^4 - 34(402.51)^2 \right)/33 = 129\,090$ leads to the test statistic

$$T = \frac{\sum_{j=1}^{k}(T_j^2/n_j) - 34(402.51)^2}{V_T} = 4.5086.$$

Using the approximation that $T \sim \chi_3^2$ under the null hypothesis of equal variances, the p-value associated with this test is $P(T > 4.5086) = 0.2115$. There is no strong evidence to conclude the underlying variances for crop yields are significantly different.

8.4   Exercises

> The statistician who supposes that his main contribution to the plan-
> ning of an experiment will involve statistical theory, finds repeatedly
> that he makes his most valuable contribution simply by persuading
> the investigator to explain why he wishes to do the experiment, by per-
> suading him to justify the experimental treatments, and to explain
> why it is that the experiment, when completed, will assist him in his
> research.
>
> Gertrude Mary Cox (1900–1978)

8.1   Show that, when ties are not present, the KW statistic H' in (8.2) coincides
with H in (8.3).

8.2   Generate three samples of size 10 from an exponential distribution with
$\lambda = 0.10$. Perform both the F-test and the KW test to see if there are
treatment differences in the three groups. Repeat this 1000 times, recording
the p-value for both tests. Compare the simulation results by comparing
the two histograms made from these p-values. What do the results mean?

8.3   The data set hypnosis contains data from a study investigating whether hypnosis has the same effect on skin potential (measured in millivolts) for four
emotions (Lehmann, 1975, p. 264). Eight subjects are asked to display fear,
joy, sadness, and calmness under hypnosis. The data are recorded as one
observation per subject for each emotion:

```
1 fear 23.1   1 joy 22.7   1 sadness 22.5   1 calmness 22.6
2 fear 57.6   2 joy 53.2   2 sadness 53.7   2 calmness 53.1
3 fear 10.5   3 joy  9.7   3 sadness 10.8   3 calmness  8.3
4 fear 23.6   4 joy 19.6   4 sadness 21.1   4 calmness 21.6
5 fear 11.9   5 joy 13.8   5 sadness 13.7   5 calmness 13.3
6 fear 54.6   6 joy 47.1   6 sadness 39.2   6 calmness 37.0
7 fear 21.0   7 joy 13.6   7 sadness 13.7   7 calmness 14.8
8 fear 20.3   8 joy 23.6   8 sadness 16.3   8 calmness 14.8
```

8.4   The points-per-game statistics from the 1993 NBA season were analyzed
for basketball players who went to college in four particular ACC schools:
Duke, North Carolina, North Carolina State, and Georgia Tech. We want
to find out if scoring is different for the players from different schools. Can
this be analyzed with a parametric procedure? Why or why not? The classical F-test that assumes normality of the populations yields $F = 0.41$ and H_0
is not rejected. What about the nonparametric procedure?

Duke	UNC	NCSU	GT
7.5	5.5	16.9	7.9
8.7	6.2	4.5	7.8
7.1	13.0	10.5	14.5
18.2	9.7	4.4	6.1
	12.9	4.6	4.0
	5.9	18.7	14.0
	1.9	8.7	
		15.8	

8.5 Some varieties of nematodes (roundworms that live in the soil and are frequently so small they are invisible to the naked eye) feed on the roots of lawn grasses and crops such as strawberries and tomatoes. This pest, which is particularly troublesome in warm climates, can be treated by the application of nematocides. However, because of size of the worms, it is difficult to measure the effectiveness of these pesticides directly. To compare four nematocides, the yields of equal-size plots of one variety of tomatoes were collected. The data (yields in pounds per plot) are shown in the table. Use a nonparametric test to find out which nematocides are different:

Nematocide A	Nematocide B	Nematocide C	Nematocide D
18.6	18.7	19.4	19.0
18.4	19.0	18.9	18.8
18.4	18.9	19.5	18.6
18.5	18.5	19.1	18.7
17.9		18.5	

8.6 An experiment was run to determine whether four specific firing temperatures affect the density of a certain type of brick. The experiment led to the following data. Does the firing temperature affect the density of the bricks?

Temperature	Density					
100	21.8	21.9	21.7	21.7	21.6	21.7
125	21.7	21.4	21.5	21.4		
150	21.9	21.8	21.8	21.8	21.6	21.5
175	21.9	21.7	21.8	21.4		

8.7 A chemist wishes to test the effect of four chemical agents on the strength of a particular type of cloth. Because there might be variability from one bolt to another, the chemist decides to use a RBD, with the bolts of cloth considered as blocks. She selects five bolts and applies all four chemicals in random order to each bolt. The resulting tensile strengths follow. How do the effects of the chemical agents differ?

Chemical	Bolt no. 1	Bolt no. 2	Bolt no. 3	Bolt no. 4	Bolt no. 5
1	73	68	74	71	67
2	73	67	75	72	70
3	75	68	78	73	68
4	73	71	75	75	69

8.8 The venerable auction house of Snootly & Snobs will soon be putting three fine seventeenth- and eighteenth-century violins, A, B, and C, up for bidding. A certain musical arts foundation, wishing to determine which of these instruments to add to its collection, arranges to have them played by each of 10 concert violinists. The players are blindfolded, so that they cannot tell which violin is which; each plays the violins in a randomly determined sequence (BCA, ACB, etc.)

The violinists are not informed that the instruments are classic masterworks; all they know is that they are playing three different violins. After each violin is played, the player rates the instrument on a 10-point scale of overall excellence (1 = lowest, 10 = highest). The players are told that they can also give fractional ratings, such as 6.2 or 4.5, if they wish. The results are shown in the table below. For the sake of consistency, the $n = 10$ players are listed as "subjects":

Violin	Subject									
	1	2	3	4	5	6	7	8	9	10
A	9	9.5	5	7.5	9.5	7.5	8	7	8.5	6
B	7	6.5	7	7.5	5	8	6	6.5	7	7
C	6	8	4	6	7	6.5	6	4	6.5	3

8.9 From Exercise 8.5, test to see if the underlying variances for the four plot yields are the same. Use a test level of $\alpha = 0.05$.

RELEVANT R FUNCTIONS AND DATA SETS IN THIS CHAPTER

http://www.npstat.org/

R code: friedman.pairwise.comparison.r

R functions: kruskal.test, friedman.test

exer8.3.csv, exer8.4.csv, exer8.5.csv, exer8.6.csv, exer8.7.csv, exer8.8.csv

---

References

Conover, W. J. (1999), *Practical Nonparametric Statistics*, New York: Wiley.

Friedman, M. (1937), "The Use of Ranks to Avoid the Assumption of Normality Implicit in the Analysis of Variance," *Journal of the American Statistical Association*, 32, 675–701.

Iman, R. L., and Davenport, J. M. (1980), "Approximations of the Critical Region of the Friedman Statistic," *Communications in Statistics A: Theory and Methods*, 9, 571–595.

Jonckheere, A. R. (1954), "A Distribution-Free K-Sample Test Against Ordered Alternatives," *Biometrika*, 41, 133–145.

Kruskal, W. H. (1952), "A Nonparametric Test for the Several Sample Problem," *Annals of Mathematical Statistics*, 23, 525–540.

Kruskal, W. H., and Wallis, W. A. (1952), "Use of Ranks in One-Criterion Variance Analysis," *Journal of the American Statistical Association*, 47, 583–621.

Lehmann, E. L. (1975), *Testing Statistical Hypotheses*, New York: Wiley.

Page, E. B. (1963), "Ordered Hypotheses for Multiple Treatments: A Significance Test for Linear Ranks," *Journal of the American Statistical Association*, 58, 216–230.

Quade, D. (1966), "On the Analysis of Variance for the K-Sample Population," *Annals of Mathematical Statistics*, 37, 1747–1785.

Sachs, L. (1997), *Angewandte Statistik*, Berlin: Springer-Verlag.

Terpstra, T. J. (1952). "The Asymptotic Normality and Consistency of Kendall's Test Against Trend, When Ties Are Present in One Ranking," *Indagationes Mathematicae*, 14, 327–333.

9

Categorical Data

*Ignorance more frequently begets confidence than does knowledge: it is those
who know little, and not those who know much, who so positively assert that
this or that problem will never be solved by science.*

Charles Darwin (1843–1927)

A *categorical* variable is a variable that is nominal or ordinal in scale. Ordinal
variables have more information than nominal ones because their levels can be
ordered. For example, an automobile could be categorized in an ordinal scale
(compact, mid-size, large) or a nominal scale (Honda, Tesla, Audi). Opposed to
interval data, which are quantitative, nominal data are *qualitative*, so comparisons
between the variables cannot be described mathematically. Ordinal variables are
more useful than nominal ones because they can possibly be ranked, yet they are
not quite quantitative. Categorical data analysis is seemingly ubiquitous in statistical practice, and we encourage readers who are interested in a more comprehensive coverage to consult monographs by Agresti (2012) and Simonoff (2003).

At the turn of the nineteenth century, while probabilists in Russia, France,
and other parts of the world were hastening the development of statistical theory
through probability, British academic researchers achieved great methodological
developments in statistics through applications in the biological sciences. This
was due in part from the gush of research following Charles Darwin's publication
of *The Origin of Species* in 1859. Darwin's theories helped to catalyze research
in the variations of traits within species, and this strongly affected the growth of
applied statistics and biometrics. It was several years later that the genetics dis-
coveries of Gregor Mendel (1822–1844), published in 1865 in a relatively obscure
Austrian journal, were "rediscovered" in light of these new theories of evolution.

When it comes to the development of statistical methods, two individuals are
dominant from this era: Karl Pearson and R. A. Fisher. Both were cantankerous
researchers influenced by William S. Gosset, the man who derived the (Student's)

Nonparametric Statistics with Applications to Science and Engineering with R, Second Edition.
Paul Kvam, Brani Vidakovic, and Seong-joon Kim.
© 2023 John Wiley & Sons, Inc. Published 2023 by John Wiley & Sons, Inc.

t distribution. Karl Pearson, in particular, contributed seminal results to the study of categorical data, including the chi-square test of statistical significance (Pearson, 1900). Fisher used Mendel's theories as a framework for the research of biological inheritance.[1] Both researchers were motivated by problems in heredity, and both played an interesting role in its promotion.

Although their research was undoubtedly brilliant, racial bigotry strongly prevailed in Western society during this colonial period, and scientists were hardly exceptional in this regard. Fisher, an upper-class British conservative and intellectual, theorized the decline of Western civilization due to the diminished fertility of the upper classes. Pearson, his rival, was a staunch socialist yet ironically advocated a "war on inferior races," which he often associated with the working class. Pearson said, "no degenerate and feeble stock will ever be converted into healthy and sound stock by the accumulated effects of education, good laws and sanitary surroundings."

9.1 Chi-Square and Goodness-of-Fit

Pearson's chi-square statistic found immediate applications in biometry, genetics, and other life sciences. It is introduced in the most rudimentary science courses. For instance, if you are at a party and you meet a typical college graduate of the social sciences, it is likely one of the few things they remember about the required statistics class they suffered through in college is the term "chi-square."

To motivate the chi-square statistic, let $X_1, X_2, \ldots, X_n$ be a sample from any distribution. As in Chapter 6, we would like to test the goodness-of-fit hypothesis

$$H_0 : F_X(x) = F_0(x).$$

Let the domain of the distribution $D = (a, b)$ be split into r non-overlapping intervals, $I_1 = (a, x_1]$, $I_2 = (x_1, x_2]$, ..., $I_r = (x_{r-1}, b)$. Such intervals have (theoretical) probabilities $p_1 = F_0(x_1) - F_0(a)$, $p_2 = F_0(x_2) - F_0(x_1)$, ..., $p_r = F_0(b) - F_0(x_{r-1})$, under H_0.

Let $n_1, n_2, \ldots, n_r$ be observed frequencies of intervals $I_1, I_2, \ldots, I_r$. In this notation, n_1 is the number of elements of the sample $X_1, \ldots, X_n$ that fall into the interval I_1. Of course, $n_1 + \cdots + n_r = n$ because the intervals are a partition of the domain of the sample. The discrepancy between observed frequencies n_i and theoretical frequencies np_i is the rationale for forming the statistic

$$X^2 = \sum_{i=1}^{r} \frac{(n_i - np_i)^2}{np_i} \tag{9.1}$$

---

1 Actually, Fisher (1918) showed statistically that Mendel's data were probably fudged a little to support the theory for his new genetic model. See Section 9.2.

that has a chi-square (χ^2) distribution with $r - 1$ degrees of freedom. Large values of X^2 are critical for H_0. Alternative representations include

$$X^2 = \sum_{i=1}^{r} \frac{n_i^2}{np_i} - n \quad \text{and} \quad X^2 = n \left[\sum_{i=1}^{r} \left(\frac{\hat{p}_i}{p_i} \right) \hat{p}_i - 1 \right],$$

where $\hat{p}_i = n_i / n$.

In some experiments, the distribution under H_0 cannot be *fully* specified; for example, one might conjecture the data are generated from a Poisson distribution without knowing the exact value of the Poisson rate parameter λ. In cases like this, the unknown parameters are estimated using the sample.

Suppose that k parameters are estimated to fully specify F_0. Then, the resulting statistic in (9.1) has a χ^2 distribution with $r - k - 1$ degrees of freedom. A degree of freedom is lost with the estimation of a parameter. In fairness, if we estimated a parameter and then inserted it into the hypothesis without further acknowledgment, the hypothesis will undoubtedly fit the data at least and any alternative hypothesis we could construct with a known parameter. So the lost degree of freedom represents a form of handicapping.

There is no orthodoxy in selecting the categories or even the number of categories to use. If possible, make the categories approximately equal in probability. Practitioners may want to arrange interval selection so that all $np_i > 1$ and that at least 80% of the np_i's exceed 5. The rule of thumb is $n \geq 10$, $r \geq 3$, $n^2/r \geq 10$, and $np_i \geq 0.25$.

As mentioned in Chapter 6, the chi-square test is not altogether efficient for testing known continuous distributions, especially compared with individualized tests such as Shapiro–Wilk or Anderson–Darling. Its advantage is manifest with discrete data and special distributions that cannot be fit in a Kolmogorov-type statistical test.

Example 9.1 *Mendel's Data.*

In 1865, Mendel (1866) discovered a basic genetic code by breeding green and yellow peas in an experiment. Because the yellow pea gene is dominant, the first generation hybrids all appeared yellow, but the second generation hybrids were about 75% yellow and 25% green. The green color reappears in the second generation because there is a 25% chance that two peas, both having a yellow and green gene, will contribute the green gene to the next hybrid seed. In another pea experiment[2] (model illustrated in Figure 9.1) that considered both color and texture traits, the outcomes from repeated experiments came out as in Table 9.1.

The statistical analysis shows a strong agreement with the hypothesized outcome with a *p*-value of 0.9166. While this, by itself, is not sufficient proof to consider

---

2 See Section 16.2 for more detail on probability models in basic genetics.

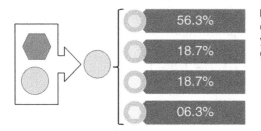

Figure 9.1   Genetic model for a dihybrid cross between round, yellow peas and wrinkled, green peas.

Table 9.1   Mendel's data.

Type of pea	Observed number	Expected number
Smooth yellow	315	313
Wrinkled yellow	101	104
Smooth green	108	104
Wrinkled green	32	35

foul play, Fisher noted this kind of result in a sequence of several experiments. His "meta-analysis" (see Chapter 6) revealed a *p*-value around 0.00013 (Figure 9.2):

```
> o <- c(315, 101, 108, 32);
> th <- c(313, 104, 104, 35);
>
> sum((o-th)^2/th)
[1] 0.510307
>
> 1-pchisq(0.510307,4-1)
[1] 0.9166212
```

Example 9.2   *Horse-Kick Fatalities.*
During the latter part of the nineteenth century, Prussian officials collected information on the hazards that horses posed to cavalry soldiers. A total of 10 cavalry corps were monitored over a period of 20 years. Recorded for each year and each corps was X, the number of fatalities due to kicks. Table 9.2 shows the distribution of X for these 200 "corps-years."

Altogether there were 122 fatalities ($109(0) + 65(1) + 22(2) + 3(3) + 1(4)$), meaning that the observed fatality *rate* was $122/200 = 0.61$ fatalities per corps-year. A Poisson model for X with a mean of $\mu = 0.61$ was proposed by von Bortkiewicz (1898). Table 9.2 shows the expected frequency corresponding to $x = 0,1,\ldots,$

24   Zweites Kapitel. & 12.

	75	76	77	78	79	80	81	82	83	84	85	86	87	88	89	90	91	92	93	94
G	–	2	2	1	–	–	1	1	–	3	–	2	1	–	–	1	–	1	–	1
I	–	–	–	2	–	3	–	2	–	–	–	1	1	1	–	2	–	3	1	–
II	–	–	–	2	–	2	–	–	1	1	–	–	2	1	1	–	–	2	–	–
III	–	–	–	1	1	1	2	–	2	–	–	–	1	–	1	2	1	–	–	–
IV	–	1	–	1	1	1	1	–	–	–	–	1	–	–	–	–	1	1	–	–
V	–	–	–	–	2	1	–	–	1	–	–	1	–	1	1	1	1	1	1	–
VI	–	–	1	–	2	–	–	1	2	–	1	1	3	1	1	1	–	3	–	–
VII	1	–	1	–	–	–	1	–	1	1	–	–	2	–	–	2	1	–	2	–
VIII	1	–	–	–	1	–	–	1	–	–	–	–	1	–	–	–	1	1	–	1
IX	–	–	–	–	–	2	1	1	1	–	2	1	1	–	1	2	–	1	–	–
X	–	–	1	1	–	1	–	2	–	2	–	–	–	–	2	1	3	–	1	1
XI	–	–	–	–	2	4	–	1	3	–	1	1	1	1	2	1	3	1	3	1
XIV	1	1	2	1	1	3	–	4	–	1	–	3	2	1	–	2	1	1	–	–
XV	–	1	–	–	–	–	–	1	–	1	1	–	–	–	2	2	–	–	–	–

Figure 9.2 Original data of horse-kick fatalities from von Bortkiewicz (1898).

Table 9.2 Horse-kick fatalities data.

x	Observed number of corps-years	Expected number of corps-years
0	109	108.7
1	65	66.3
2	22	20.2
3	3	4.1
4	1	0.7
	200	200

assuming the Poisson model for X was correct. The agreement between the observed and the expected frequencies is remarkable. The R procedure below shows that the resulting X^2 statistic $= 0.6104$:

```
> o <- c(109, 65, 22, 3, 1);
> th <- c(108.7, 66.3, 20.2, 4.1, 0.7);
>
> sum( (o-th)^2/th)
[1] 0.6104076
>
> 1-pchisq(0.6104076,4-1)
[1] 0.8940457
```

It should occur to you that the hypothesized mean of $\mu = 0.61$ was estimated from the data, so the original model should include a reduced degree of freedom in the chi-square test statistic. That is, if the Poisson distribution is correctly specified, the statistic should be distributed χ^2 with three degrees of freedom, so the p-value is computed $P(W > 0.6104) = 0.8940$.

Example 9.3 *Benford's Law.*
Benford's law (Benford, 1938; Hill, 1998) concerns relative frequencies of leading digits of various data sets, numerical tables, accounting data, etc. This, also called *the first digit law*, states that in numbers from many sources, the leading digit 1 occurs much more often than the others (namely, about 30% of the time). Furthermore, the higher the digit, the less likely it is to occur as the leading digit of a number. This applies to figures related to the natural world or of social significance, be it numbers taken from electricity bills, newspaper articles, street addresses, stock prices, population numbers, death rates, areas or lengths of rivers, or physical and mathematical constants.

To be precise, Benford's law states that the leading digit n, $(n = 1, \ldots, 9)$ occurs with probability $P(n) = \log_{10}(n + 1) - \log_{10}(n)$, approximated to three digits in the table below:

Digit n	1	2	3	4	5	6	7	8	9
$P(n)$	0.301	0.176	0.125	0.097	0.079	0.067	0.058	0.051	0.046

The table below lists the distribution of the leading digit for all 307 numbers appearing in a particular issue of *Reader's Digest*. With p-value of 0.8719, the support for H_0 (the first digits in *Reader's Digest* are distributed according to Benford's law) is strong:

Digit	1	2	3	4	5	6	7	8	9
count	103	57	38	23	20	21	17	15	13

The agreement between the observed digit frequencies and Benford's distribution is good. The R calculation shows that the resulting X^2 statistic is 3.8322. Under H_0, X^2 is distributed as χ_8^2, and more extreme values of X^2 are quite likely. The p-value is almost 90%:

```
> x <- c(103, 57, 38, 23, 20, 21, 17, 15, 13);
> e <- 307*c(0.301, 0.176, 0.125, 0.097, 0.079, 0.067,
+ 0.058, 0.051, 0.046);
>
> sum( (x-e)^2/e)
[1] 3.832247
>
> 1-pchisq(3.832247, 8)
[1] 0.8719326
```

9.2 Contingency Tables: Testing for Homogeneity and Independence

Suppose there are m populations (more specifically, m levels of factor A: $(R_1, \ldots, R_m)$) under consideration. Furthermore, each observation can be classified in a different ways, according to another factor B, which has k levels $(C_1, \ldots, C_k)$. Let n_{ij} be the number of all observations at the ith level of A and jth level of B. We seek to find out if the populations (from A) and treatments (from B) are independent. If we treat the levels of A as population groups and the levels of B as treatment groups, there are

$$n_{i.} = \sum_{j=1}^{k} n_{ij}$$

observations in population i, where $i = 1, \ldots, m$. Each of the treatment groups is represented

$$n_{.j} = \sum_{i=1}^{n} n_{ij}$$

times, and the total number of observations is

$$n_{1.} + \cdots + n_{m.} = n_{..}.$$

The following table summarizes the above description:

	C_1	C_2	$\cdots$	C_k	**Total**
R_1	n_{11}	n_{12}		n_{1k}	$n_{1.}$
R_2	n_{21}	n_{22}		n_{2k}	$n_{2.}$
R_m	n_{m1}	n_{m2}		n_{mk}	$n_{m.}$
Total	$n_{.1}$	$n_{.2}$		$n_{.k}$	$n_{..}$

We are interested in testing independence of factors A and B, represented by their respective levels $R_1, \ldots, R_m$ and $C_1, \ldots, C_k$, on the basis of observed frequencies n_{ij}. Recall the definition of independence of component random variables X and Y in the random vector (X, Y):

$$P(X = x_i, Y = y_j) = P(X = x_i) \cdot P(Y = y_j).$$

Assume that the random variable ξ is to be classified. Under the hypothesis of independence, the cell probabilities $P(\xi \in R_i \cap C_j)$ should be equal to the product of probabilities $P(\xi \in R_i) \cdot P(\xi \in C_j)$. Thus, to test the independence of factors A and B, we should evaluate how different the sample counterparts of cell probabilities

$$\frac{n_{ij}}{n_{..}}$$

are from the product of marginal probability estimators:

$$\frac{n_{i\cdot}}{n_{\cdot\cdot}} \cdot \frac{n_{\cdot j}}{n_{\cdot\cdot}}$$

or, equivalently, how different the observed frequencies, n_{ij}, are from the expected (under the hypothesis of independence) frequencies

$$\hat{n}_{ij} = n_{\cdot\cdot} \frac{n_{i\cdot}}{n_{\cdot\cdot}} \frac{n_{\cdot j}}{n_{\cdot\cdot}} = \frac{n_{i\cdot} \cdot n_{\cdot j}}{n_{\cdot\cdot}}.$$

The measure of discrepancy is defined as

$$X^2 = \sum_{i=1}^{m} \sum_{j=1}^{k} \frac{(n_{ij} - \hat{n}_{ij})^2}{\hat{n}_{ij}}, \tag{9.2}$$

and under the assumption of independence, the statistic in (9.2) has a χ^2 distribution with $(m-1)(k-1)$ degrees of freedom. Here is the rationale: the observed frequencies n_{ij} are distributed as multinomial $\mathcal{M}n(n_{\cdot\cdot}; \theta_{11}, \dots, \theta_{mk})$, where $\theta_{ij} = P(\xi \in R_i \cap C_j)$:

	C_1	C_2	$\cdots$	C_k	Total
R_1	θ_{11}	θ_{12}		θ_{1k}	$\theta_{1\cdot}$
R_2	θ_{21}	θ_{22}		θ_{2k}	$\theta_{2\cdot}$
R_m	θ_{m1}	θ_{m2}		θ_{mk}	$\theta_{m\cdot}$
Total	$\theta_{\cdot 1}$	$\theta_{\cdot 2}$		$\theta_{\cdot k}$	1

The corresponding likelihood is

$$L = \prod_{i=1}^{m} \prod_{j=1}^{k} (\theta_{ij})^{n_{ij}}, \qquad \sum_{i,j} \theta_{ij} = 1.$$

The null hypothesis of independence states that for any pair i, j, the cell probability is the product of marginal probabilities, $\theta_{ij} = \theta_{i\cdot} \cdot \theta_{\cdot j}$. Under H_0, the likelihood becomes

$$L = \prod_{i=1}^{m} \prod_{j=1}^{n} (\theta_{i\cdot} \cdot \theta_{\cdot j})^{n_{ij}}, \quad \text{where } \sum_{i} \theta_{i\cdot} = \sum_{j} \theta_{\cdot j} = 1.$$

If the estimators of $\theta_{i\cdot}$ and $\theta_{\cdot j}$ are $\hat{\theta}_{i\cdot} = n_{i\cdot}/n_{\cdot\cdot}$ and $\hat{\theta}_{\cdot j} = n_{\cdot j}/n_{\cdot\cdot}$, respectively, then, under H_0, the observed frequency n_{ij} should be compared with its theoretical counterpart:

$$\hat{n}_{ij} = \hat{\theta}_{ij} n_{\cdot\cdot} = \frac{n_{i\cdot} n_{\cdot j}}{n_{\cdot\cdot}}.$$

As the n_{ij} are binomially distributed, they can be approximated by the normal distribution, and the χ^2 forms when those values are squared. The statistic is based on $(m-1)+(k-1)$ estimated parameters, $\theta_{i\cdot}$, $i = 1, \ldots, m-1$, and $\theta_{\cdot j}$, $j = 1, \ldots, k-1$. The remaining parameters are determined: $\theta_{m\cdot} = 1 - \sum_{i=1}^{m-1} \theta_{i\cdot}$, $\theta_{\cdot n} = 1 - \sum_{n=1}^{k-1} \theta_{\cdot j}$. Thus, the chi-square statistic

$$\chi^2 = \sum_{i=1}^{m}\sum_{j=1}^{k} \frac{(n_{ij} - \hat{n}_{ij})^2}{\hat{n}_{ij}}$$

has $mk - 1 - (m-1) - (k-1) = (m-1)(k-1)$ degrees of freedom.

Pearson first developed this test but mistakenly used $mk - 1$ degrees of freedom. It was Fisher (1922), who later deduced the correct degrees of freedom, $(m-1)$ $(k-1)$. This probably did not help to mitigate the antagonism in their professional relationship!

Example 9.4 *Icelandic Dolphins.*
From Rasmussen, Wahlberg and Miller (2004), groups of dolphins were observed off the coast in Iceland, and their frequency of observation was recorded along with the time of day and the perceived activity of the dolphins at that time. Table 9.3 provides the data. To see if the activity is independent of the time of day, the R function

```
tablerxc
```

takes the input table X and computes the χ^2 statistic, its associated p-value, and a table of expected values under the assumption of independence. The R function chisq.test also provides the same results. In this example, the activity and time of day appear to be dependent (Figure 9.3):

```
> source("tablerxc.r")
> tab<- matrix(c(6,6,14,13,28,4,0,56,38,5,9,10),nrow=4,
+ dimnames=list(c("Morning","Noon","Afternoon","Evening"),
+ c("Traveling","Feeding","Socializing")))
> tablerxc(tab)
$chisq
[1] 68.46457
```

Table 9.3 Observed groups of dolphins, including *time of day* and *activity*.

Time of day	Traveling	Feeding	Socializing
Morning	6	28	38
Noon	6	4	5
Afternoon	14	0	9
Evening	13	56	10

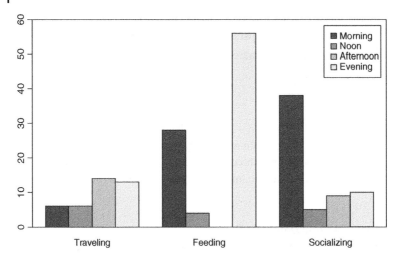

Figure 9.3 Barplot of dolphin's data.

```
$Pvalue
[1] 8.438805e-13

$exp
           [,1]        [,2]        [,3]
[1,]  14.857143  33.523810  23.619048
[2,]   3.095238   6.984127   4.920635
[3,]   4.746032  10.708995   7.544974
[4,]  16.301587  36.783069  25.915344
>
> chisq.test(tab)
        Pearson's Chi-squared test

data:  tab
X-squared = 68.4646, df = 6, p-value = 8.439e-13
```

9.2.1 Relative Risk

Statistically speaking, US soldiers have less of a chance of dying from all causes in Iraq than citizens have of being murdered in California, which is roughly the same geographical size. California has more than 2300 homicides each year, which means about 6.6 murders each day. Meanwhile, US troops have been in Iraq for 160 days, which means they're incurring about 1.7 deaths, including illness and accidents each day.[3]

Brit Hume, Fox News, August 2003.

---

3 By not taking the total population of each group into account, Hume failed to note the relative risk of death (Section.2) to a soldier in Iraq was 65 times higher than the murder rate in California.

In simple 2×2 tables, the comparison of two proportions might be more important if those proportions veer toward zero or one. For example, a procedure that decreases production errors from 5% to 2% could be much more valuable than one that decreases errors in another process from 45% to 42%. This is the *relative* in relative risk.

For example, if the rate of murder in California is compared with the death rate of US military personnel in Iraq in 2003, the data from a 2×2 table

	Killed	Not killed	Total
California	6.6	37 999,993.4	38 000 000
Iraq	1.7	149 998.3	150 000
Total	8.3	38,149,981.7	

can be summarized by relative risk, which is defined here as the risk of death in Iraq (for US soldiers) divided by the risk of murder for citizens of California. For example, McWilliams and Piotrowski (2005) determined the rate of 6.6 Californian homicide victims (out of 38 000 000 at risk) per day. On the other hand, there were 1.7 average daily military related deaths in Iraq (with 150 000 soldiers at risk):

$$\frac{\theta_{11}}{\theta_{11} + \theta_{12}} \left(\frac{\theta_{21}}{\theta_{21} + \theta_{22}} \right)^{-1} = \frac{1.7}{150\,000} \left(\frac{6.6}{38\,000\,000} \right)^{-1} = 65.25.$$

Fixed Marginal Totals. The categorical analysis above was developed based on assuming that each observation is to be classified according to the stochastic nature of the two factors. It is actually common, however, to have either row or column totals fixed. If row totals are fixed, for example, we are observing n_j observations distributed into k bins and essentially comparing multinomial observations. In this case we are testing differences in the multinomial parameter sets. However, if we look at the experiment this way (where n_j is fixed), the test statistic and rejection region remain the same. This is also true if *both* row and column totals are fixed. This is less common; for example, if $m = k = 2$, this is essentially Fisher's exact test.

9.3 Fisher Exact Test

Along with Pearson, R. A. Fisher contributed important new methods for analyzing categorical data. Pearson and Fisher both recognized that the statistical methods of their time were not adequate for small categorized samples, but their disagreements are more well known. In 1922, Pearson used his position as editor of

Biometrika to attack Fisher's use of the chi-squared test. Fisher attacked Pearson with equal fierceness. While at University College, Fisher continued to criticize Pearson's ideas even after his passing. With Pearson's son Egon also holding a chair there, the departmental politics were awkward, to say the least.

Along with his original concept of maximum likelihood estimation, Fisher pioneered research in small sample analysis, including a simple categorical data test that bears his name (*Fisher exact test*). Fisher (1966) described a test based on the claims of a British woman who said she could taste a cup of tea, with milk added, and identify whether the milk or tea was added to the cup first. She was tested with eight cups, of which she knew four had the tea added first and four had the milk added first. The results are listed below:

First poured	Lady's guess		Total
	Tea	Milk	
Tea	3	1	4
Milk	1	3	4
Total	4	4	

Both *marginal totals* are fixed at four, so if X is the number of times the woman guessed tea was poured first when, in truth, tea *was* poured first, then X determines the whole table, and under the null hypothesis (that she is just guessing), X has a hypergeometric distribution with probability mass function (PMF)

$$p_X(x) = \frac{\binom{4}{x}\binom{4}{4-x}}{\binom{8}{4}}.$$

To see this more easily, count the number of ways to choose x cups from the correct 4 and the remaining $4 - x$ cups from the incorrect 4, and divide by the total number of ways to choose 4 cups from the 8 total. The lady guessed correctly $x = 3$ times. In this case, because the only better guess is all four, the p-value is $P(X = 3) + P(X = 4) = 0.229 + 0.014 = 0.243$. Because the sample is so small, not much can be said of the experimental results.

In general, the Fisher exact test is based on the null hypothesis that two factors, each with two factor levels, are independent, conditional on fixing marginal frequencies for *both* factors (e.g. the number of times tea was poured first and the number of times the lady guesses that tea was poured first).

9.4 McNemar Test

Quinn McNemar's expertise in statistics and psychometrics led to an influential textbook titled *Psychological Statistics*. The McNemar test (McNemar, 1947) is a simple way to test *marginal homogeneity* in 2×2 tables. This is not a regular contingency table, so the usual analysis of contingency tables would not be applicable.

Consider such a table that, for instance, summarizes agreement between two evaluators choosing only two grades 0 and 1, so in the table below, a represents the number of times that both evaluators graded an outcome with 0. The marginal totals, unlike the Fisher exact test, are not fixed:

	0	1	Total
0	a	b	$a + b$
1	c	d	$c + d$
Total	$a + c$	$b + d$	$a + b + c + d$

Marginal homogeneity (i.e. the graders give the same proportion of zeros and ones, on average) implies that row totals should be close to the corresponding column totals, or

$$a + b \approx a + c, \tag{9.3}$$
$$c + d \approx b + d.$$

More formally, suppose that a matched pair of Bernoulli random variables (X, Y) is to be classified into a table

	0	1	Marginal
0	θ_{00}	θ_{01}	$\theta_{0 \cdot}$
1	θ_{10}	θ_{11}	$\theta_{1 \cdot}$
Marginal	$\theta_{\cdot 0}$	$\theta_{\cdot 1}$	1

in which $\theta_{ij} = P(X = i, Y = j)$, $\theta_{i \cdot} = P(X = i)$, and $\theta_{\cdot j} = P(Y = j)$, for $i, j \in \{0,1\}$. The null hypothesis H_0 can be expressed as a hypothesis of symmetry

$$H_0 : \quad \theta_{01} = P(X = 0, Y = 1) = P(X = 1, Y = 0) = \theta_{10}, \tag{9.4}$$

but after adding $\theta_{00} = P(X = 0, Y = 0)$ or $\theta_{11} = P(X = 1, Y = 1)$ to the both sides in (9.4), we get H_0 in the form of marginal homogeneity:

$$H_0 : \quad \theta_{0.} = P(X = 0) = P(Y = 0) = \theta_{.0}, \quad \text{or equivalently}$$
$$H_0 : \quad \theta_{1.} = P(X = 1) = P(Y = 1) = \theta_{.1} \ .$$

As a and d on both sides of (9.3) cancel out, it implies $b \approx c$. A sensible test statistic for testing H_0 might depend on how much b and c differ. The values of a and d are called ties and do not contribute to the testing of H_0.

When $b + c > 20$, the McNemar statistic is calculated as

$$X^2 = \frac{(b - c)^2}{b + c},$$

which has a χ^2 distribution with one degree of freedom. Some authors recommend a version of the McNemar test with a correction for discontinuity, calculated as $X^2 = (|b - c| - 1)^2 / (b + c)$, but there is no consensus among experts that this statistic is better.

If $b + c < 20$, a simple statistics

$$T = b$$

can be used. If H_0 is true, $T \sim Bin(b + c, 1/2)$, and testing is as in the sign test. In some sense, where the standard two-sample paired t-test is for normally distributed responses, the McNemar test is for paired binary responses.

Example 9.5 A study by Johnson and Johnson (1972) involved 85 patients with Hodgkin's disease. Hodgkin's disease is a cancer of the lymphatic system; it is known also as a lymphoma. Each patient in the study had a sibling who did not have the disease. In 26 of these pairs, both individuals had a tonsillectomy (T). In 37 pairs, neither of the siblings had a tonsillectomy (N). In 15 pairs, only the individual with Hodgkin's disease had a tonsillectomy, and in 7 pairs, only the non-Hodgkin's disease sibling had a tonsillectomy:

	Sibling/T	Sibling/N	Total
Patient/T	26	15	41
Patient/N	7	37	44
Total	33	52	85

The pairs (X_i, Y_i), $i = 1, \ldots, 85$ represent siblings – one of which is a patient with Hodgkin's disease (X) and the second without the disease (Y). Each of the siblings is also classified (as $T = 1$ or $N = 0$) with respect to having a tonsillectomy:

	$Y = 1$	$Y = 0$
$X = 1$	26	15
$X = 0$	7	37

The test we are interested in is based on $H_0 : P(X = 1) = P(Y = 1)$, i.e. that the probabilities of siblings having a tonsillectomy are the same with and without the disease. Because $b + c > 20$, the statistic of choice is

$$\chi^2 = \frac{(b - c)^2}{b + c} = 8^2/(7 + 15) = 2.9091.$$

The p-value is $p = P(W \geq 2.9091) = 0.0881$, where $W \sim \chi_1^2$. Under H_0, $T = 15$ is a realization of a binomial $Bin(22, 0.5)$ random variable, and the p value is $2 \cdot P(T \geq 15) = 2 \cdot P(T > 14) = 0.1338$, that is,

```
> 2*(1-pbinom(14,22,0.5))
[1] 0.1338005
```

With such a high p-value, there is scant evidence to reject the null hypothesis of homogeneity of the two groups of patients with respect to having a tonsillectomy.

9.5 Cochran's Test

Cochran's (1950) test is essentially a randomized block design (RBD), as described in Chapter 8, but the responses are dichotomous. That is, each treatment-block combination receives a 0 or 1 response.

If there are only two treatments, the experimental outcome is equivalent to McNemar's test with marginal totals equaling the number of blocks. To see this, consider the last example as a collection of dichotomous outcomes; each of the 85 patients are initially classified into two blocks depending on whether the patient had or had not received a tonsillectomy. The response is 0 if the patient's sibling did not have a tonsillectomy and 1 if they did.

Example 9.6 Consider the software debugging data in Table 9.4. Here the software reviewers (A,B,C,D,E) represent five blocks, and the 27 bugs are considered to be treatments. Let the column totals be denoted $\{C_1, \ldots, C_5\}$ and row totals as $\{R_1, \ldots, R_{27}\}$. We are essentially testing H_0 where treatments (software bugs) have an equal chance of being discovered versus H_a where some software bugs are more prevalent (or easily found) than others. The test statistic is

$$T_C = \frac{\sum_{j=1}^{m} \left(C_j - \frac{n}{m} \right)^2}{\left(\frac{\sum_{i=1}^{k} R_i(m - R_i)}{m(m-1)} \right)},$$

Table 9.4 Five reviewers found 27 issues in software example as in Gilb and Graham (1993).

A	B	C	D	E	A	B	C	D	E
1	1	1	1	1	0	0	1	0	0
1	0	1	0	1	0	0	1	0	0
1	1	1	0	1	0	0	0	1	0
1	0	1	1	1	1	1	1	0	1
1	0	1	1	1	0	0	1	0	1
1	0	1	1	1	1	0	0	0	0
1	1	1	1	1	0	1	0	0	0
1	1	1	1	1	1	0	1	1	1
0	0	1	0	0	0	0	0	0	0
1	0	1	0	0	0	0	0	0	0
0	1	0	0	0	0	1	0	0	1
1	0	0	1	1	1	0	0	0	0
1	0	1	0	1	1	0	0	0	0
0	0	1	0	1					

where $n = \sum C_j = \sum R_i$, $m = 5$ (blocks) and $k = 27$ treatments (software bugs). Under H_0, T_C has an approximate chi-square distribution with $m - 1$ degrees of freedom. In this example, $T_C = 13.725$, corresponding to a test p-value of 0.00822:

```
> d <- data.frame(
+ A=c(1,1,1,1,1,1,1,1,0,1,0,1,1,0,0,0,0,1,0,1,0,1,0,0,0,1,1),
+ B=c(1,0,1,0,0,0,1,1,0,0,1,0,0,0,0,0,0,1,0,0,1,0,0,0,1,0,0),
+ C=c(1,1,1,1,1,1,1,1,1,1,0,0,1,1,1,1,0,1,1,0,0,1,0,0,0,0,0),
+ D=c(1,0,0,1,1,1,1,1,0,0,0,1,0,0,0,0,1,0,0,0,0,1,0,0,0,0,0),
+ E=c(1,1,1,1,1,1,1,1,0,0,0,1,1,1,0,0,0,1,1,0,0,1,0,0,1,0,0))
>
> d2 <- data.frame(bug=as.vector(t(d)),reviewer=rep(LETTERS[1:5],
+ 27),treatment=as.vector(sapply(1:27,rep,times=5)))
>
> test <- cochran.qtest(bug~reviewer|treatment,data=d2)
> print(c(stat=test$statistic,p.value=test$p.value))
      stat.Q       p.value
13.725490196  0.008224734
>
> C <- apply(d,2,sum);R<-apply(d,1,sum);m<-5
> TC <- sum(((C-sum(C)/m)^2)/(sum(R*(m-R))/(m*(m-1))))
> print(c(TC=TC,p.value=1-pchisq(TC,5-1)))
          TC       p.value
13.725490196  0.008224734
```

9.6   Mantel–Haenszel Test

Suppose that k independent classifications into a 2×2 table are observed. We could denote the ith such table by

x_i	$r_i - x_i$	r_i
$c_i - x_i$	$n_i - r_i - c_i + x_i$	$n_i - r_i$
c_i	$n_i - c_i$	n_i

It is assumed that the marginal totals (r_i, n_i or just n_i) are fixed in advance and that the sampling was carried out until such fixed marginal totals are satisfied. If each of the k tables represent an independent study of the same classifications, the Mantel–Haenszel Test essentially pools the studies together in a "meta-analysis" that combines all experimental outcomes into a single statistic (Mantel and Haenszel 1959). For more about nonparametric approaches to this kind of problem, see the section on meta-analysis in Chapter 6.

For the ith table, p_{1i} is the proportion of subjects from the first row falling in the first column, and likewise, p_{2i} is the proportion of subjects from the second row falling in the first column. The hypothesis of interest here is if the population proportions p_{1i} and p_{2i} coincide over all k experiments.

Suppose that in experiment i there are n_i observations. All items can be categorized as type 1 (r_i of them) or type 2 ($n_i - r_i$ of them). If c_i items are selected from the total of n_i items, the probability that exactly x_i of the selected items are of the type 1 is

$$\frac{\binom{r_i}{x_i} \cdot \binom{n_i - r_i}{c_i - x_i}}{\binom{n_i}{c_i}}. \tag{9.5}$$

Likewise, all items can be categorized as type A (c_i of them) or type B ($n_i - c_i$ of them). If r_i items are selected from the total of n_i items, the probability that exactly x_i of the selected are of the type A is

$$\frac{\binom{c_i}{x_i} \cdot \binom{n_i - c_i}{r_i - x_i}}{\binom{n_i}{r_i}}. \tag{9.6}$$

Of course these two probabilities are equal, i.e.

$$\frac{\binom{r_i}{x_i} \cdot \binom{n_i - r_i}{c_i - x_i}}{\binom{n_i}{c_i}} = \frac{\binom{c_i}{x_i} \cdot \binom{n_i - c_i}{r_i - x_i}}{\binom{n_i}{r_i}}.$$

These are hypergeometric probabilities with mean and variance

$$\frac{r_i \cdot c_i}{n_i}, \quad \text{and} \quad \frac{r_i \cdot c_i \cdot (n_i - r_i) \cdot (n_i - c_i)}{n_i^2(n_i - 1)},$$

respectively. The k experiments are independent, and the statistic

$$T = \frac{\sum_{i=1}^{k} x_i - \sum_{i=1}^{k} \frac{r_i c_i}{n_i}}{\sqrt{\sum_{i=1}^{k} \frac{r_i \cdot c_i \cdot (n_i - r_i) \cdot (n_i - c_i)}{n_i^2(n_i - 1)}}} \tag{9.7}$$

is approximately normal (if n_i is large, the distributions of the x_i's are close to binomial and thus the normal approximation holds. In addition, summing over k independent experiments makes the normal approximation more accurate.) Large values of $|T|$ indicate that the proportions change across the k experiments.

Example 9.7   The three 2×2 tables provide classification of people from three Chinese cities, Zhengzhou, Taiyuan, and Nanchang, with respect to smoking habits and incidence of lung cancer (Liu, 1992):

	Zhengzhou			Taiyuan			Nanchang		
Cancer diagnosis:	Yes	No	Total	Yes	No	Total	Yes	No	Total
Smoker	182	156	338	60	99	159	104	89	193
Nonsmoker	72	98	170	11	43	54	21	36	57
Total	254	254	508	71	142	213	125	125	250

We can apply the Mantel–Haenszel Test to decide if the proportions of cancer incidence for smokers and nonsmokers coincide for the three cities, i.e. $H_0 : p_{1i} = p_{2i}$ where p_{1i} is the proportion of incidence of cancer among smokers in the city i and p_{2i} is the proportion of incidence of cancer among nonsmokers in the city i, $i = 1, 2, 3$. We use the two-sided alternative, $H_1 : p_{1i} \neq p_{2i}$ for some $i \in \{1, 2, 3\}$ and fix the type-I error rate at $\alpha = 0.10$.

From the tables, $\sum_i x_i = 182 + 60 + 104 = 346$. Also, $\sum_i r_i c_i / n_i = 338 \cdot 254/508 + 159 \cdot 71/213 + 193 \cdot 125/250 = 169 + 53 + 96.5 = 318.5$. To compute T in (9.7),

$$\sum_i \frac{r_i \, c_i \, (n_i - r_i) \, (n_i - c_i)}{n_i^2 \, (n_i - 1)} = \frac{338 \cdot 254 \cdot 170 \cdot 254}{508^2 \cdot 507} + \frac{159 \cdot 71 \cdot 54 \cdot 142}{213^2 \cdot 212}$$

$$+ \frac{193 \cdot 125 \cdot 57 \cdot 125}{250^2 \cdot 249}$$

$$= 28.33333 + 9 + 11.04518 = 48.37851.$$

Therefore,

$$T = \frac{\sum_i x_i - \sum_i \frac{r_i \, c_i}{n_i}}{\sqrt{\sum_i \frac{r_i \, c_i \, (n_i - r_i) \, (n_i - c_i)}{n_i^2 \, (n_i - 1)}}} = \frac{346 - 318.5}{\sqrt{48.37851}} \approx 3.95.$$

Because T is approximately $\mathcal{N}(0,1)$, the p-value (via R) is

```
> source("mantel.haenszel.r")
> dat <- cbind(c(182,72,60,11,104,21),c(156,98,99,43,89,36))
> mantel.haenszel(dat)
        stat        Pvalue
3.953725e+00 7.694392e-05
```

In this case, there is clear evidence that the differences in cancer rates are not constant across the three cities.

9.7 Central Limit Theorem for Multinomial Probabilities

Let $E_1, E_2, \ldots, E_r$ be events that have probabilities $p_1, p_2, \ldots, p_r$; $\sum_i p_i = 1$. Suppose that in n independent trials, the event E_i appears n_i times ($n_1 + \cdots + n_r = n$). Consider

$$\zeta^{(n)} = \left(\sqrt{\frac{n}{p_1}} \left(\frac{n_1}{n} - p_1 \right), \ldots, \sqrt{\frac{n}{p_r}} \left(\frac{n_r}{n} - p_r \right) \right).$$

The vector $\zeta^{(n)}$ can be represented as

$$\zeta^{(n)} = \frac{1}{\sqrt{n}} \sum_{j=1}^{n} \xi^{(j)},$$

where components $\xi^{(j)}$ are given by $p_i^{-1/2}[\mathbf{1}(E_i) - p_i]$, $i = 1, \ldots, r$. Vectors $\xi^{(j)}$ are independently and identically distributed (i.i.d.), with $E(\xi_i^{(j)}) = p_i^{-1}(E\mathbf{1}(E_i) - p_i) = 0$, $E(\xi_{\xi_i}^{(j)})^2 = (p_i^{-1})p_i(1 - p_i) = 1 - p_i$, and $E(\xi_i^{(j)} \xi_\ell^{(j)}) = (p_i p_\ell)^{-1/2}(E\mathbf{1}(E_i)\mathbf{1}(E_\ell) - p_i p_\ell) = -\sqrt{p_i p_\ell}$, $i \neq \ell$.

Result. When $n \to \infty$, the random vector $\zeta^{(n)}$ is asymptotically normal with mean 0 and the covariance matrix:

$$\Sigma = \begin{bmatrix} 1 - p_1 & -\sqrt{p_1 p_2} & \cdots & -\sqrt{p_1 p_r} \\ -\sqrt{p_2 p_1} & 1 - p_2 & \cdots & -\sqrt{p_2 p_r} \\ & & \cdots & \\ -\sqrt{p_r p_1} & -\sqrt{p_r p_2} & \cdots & 1 - p_r \end{bmatrix} = I - zz',$$

where I is the $r \times r$ identity matrix and $z = (\sqrt{p_1} \ \ \sqrt{p_2} \ , \ldots , \ \sqrt{p_r})'$. The matrix Σ is singular. Indeed, $\Sigma z = z - z(z'z) = 0$, due to $z'z = 1$.

As a consequence, $\lambda = 0$ is characteristic value of Σ corresponding to a characteristic vector z. Because $|\zeta^{(n)}|^2$ is a continuous function of $\zeta^{(n)}$, its limiting distribution is the same as $|\zeta|^2$, where $|\zeta|^2$ is distributed as χ^2 with $r - 1$ degrees of freedom.

This is more clear if we consider the following argument. Let Ξ be an orthogonal matrix with the first row equal to $(\sqrt{p_1}, \ldots, \sqrt{p_r})$ and the rest being arbitrary, but subject to orthogonality of Ξ. Let $\eta = \Xi\zeta$. Then $E\eta = 0$ and $\Sigma_\eta = E\eta\eta' = E(\Xi\zeta)(\Xi\zeta)' = \Xi E\zeta\zeta'\Xi' = \Xi\Sigma\Xi' = I - (\Xi z)(\Xi z)'$, because $\Xi' = \Xi^{-1}$. It follows that $\Xi z = (1, 0, 0, \ldots, 0)$ and $(\Xi z)(\Xi z)'$ are a matrix with element at the position $(1,1)$ as the only nonzero element. Thus,

$$\Sigma_\eta = I - (\Xi z)(\Xi z)' = \begin{bmatrix} 0 & 0 & 0 & \ldots & 0 \\ 0 & 1 & 0 & \ldots & 0 \\ 0 & 0 & 1 & \ldots & 0 \\ & & \ldots & & \\ 0 & 0 & 0 & \ldots & 1 \end{bmatrix},$$

and $\eta_1 = 0$, $w.p.1$; $\eta_2, \ldots, \eta_r$ are i.i.d. $\mathcal{N}(0,1)$. The orthogonal transformation preserves the L_2 norm,

$$|\zeta|^2 = |\eta|^2 = \sum_{i=2}^{r} \eta_i^2 \overset{d}{=} \chi_{r-1}^2.$$

But, $|\zeta^{(n)}|^2 = \sum_{i=1}^{r} \frac{(n_i - np_i)^2}{np_i} \overset{d}{\to} |\zeta|^2$.

9.8 Simpson's Paradox

Simpson's paradox is an example of changing the favorability of marginal proportions in a set of contingency tables due to aggregation of classes. In this case the manner of classification can be thought as a "lurking variable" causing seemingly paradoxical reversal of the inequalities in the marginal proportions when they are aggregated. Mathematically, there is no paradox – the set of vectors cannot be ordered in the traditional fashion. We can show Simpson's paradox through a two-factor example.

In 1964, the US Congress passed the "Civil Rights Act," which prohibited discrimination based on race, color, religion, sex, and national origin. It also stopped racial segregation in schools and public accommodations and made employment discrimination illegal. The paradox is apparent when we consider which party (Democratic or Republican) showed more support for the bill. In Figure 9.4, we see that Republicans voted in favor of the bill at a higher rate than the Democrats ($138/172 = 0.8023$ is greater than $152/248 = 0.6129$).

What the chart in Figure 9.5 does not show you is that there is another factor that is more important than party in determining how a congress person voted on

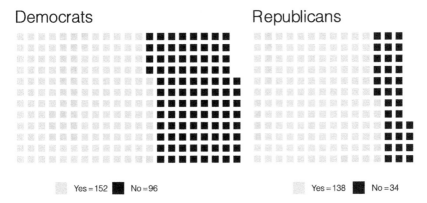

Figure 9.4 Waffle chart for showing party split in voting for 1964 Civil Rights Act.

this bill, but it is strongly intertwined with party. If we divide up congress based on geography, specifically on whether the congress person represents a state tied to the "South" (the rebellious confederate states from the US civil war) or the "North," we will see that the factor of geography is much more important in determining votes.

Figure 9.5 show how Republicans and Democrats voted for the Civil Rights Act, but it is first broken into two categories (North versus South) that determine where the elected official is from. If we consider just members of congress from northern states, then Democrats voted in favor of the Civil Rights Act (145/154 = 0.9416) in a higher proportion than Republicans (138/162 = 0.8519). In the South, most Democratic representatives voted against the bill, but all 10 Republicans voted against it. The paradox occurs because in 1964, elected White Southerners were predominantly registered in the Democratic party (at this point in time, the Republican party was still perceived as the party of Abraham Lincoln, which was not an appealing party attribute to a white southern voter):

```
> nrows <- 10
> DN <- waffle( c('Yes=145' = 145, 'No=9' = 9, 162-145-9),
+ rows = nrows, colors = c("black", "gray","white"),
title = 'Democrats: North', legend_pos="bottom")
> RN <- waffle(c('Yes=138' = 138, 'No=24' = 25, 0),
+ rows = nrows, colors = c("black", "gray","white"),
title = 'Republicans: North', legend_pos="bottom")
> DS <- waffle( c('Yes=7' = 7, 'No=87' = 87, 162-87-7),
+ rows = nrows, colors = c("black", "gray","white"),
title = 'Democrats: South', legend_pos="bottom")
> RS <- waffle( c('Yes=0' = 0, 'No=10' = 10, 162-10),
+ rows = nrows, colors = c("black", "gray","white"),
title = 'Republicans: South', legend_pos="bottom
> grid.arrange(DN,RN,DS,RS, ncol=2)
```

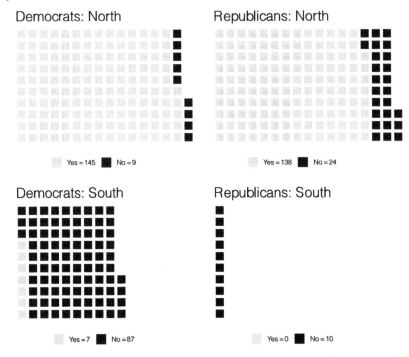

Figure 9.5 Waffle chart for showing demographic and party splits in voting for 1964 Civil Rights Act.

9.9 Exercises

9.1 Duke University has always been known for its great school spirit, especially when it comes to men's basketball. One way that school enthusiasm is shown is by donning Duke paraphernalia including shirts, hats, shorts, and sweat shirts. A class of Duke students explored possible links between school spirit (measured by the number of students wearing paraphernalia) and some other attributes. It was hypothesized that males would wear Duke clothes more frequently than females. The data were collected on the Bryan Center walkway starting at 12:00 pm on 10 different days. Each day 50 men and 50 women were tallied. Do the data bear out this claim?

	Duke paraphernalia	No Duke paraphernalia	Total
Male	131	369	500
Female	52	448	500
Total	183	817	1000

9.2 Gene Siskel and Roger Ebert hosted the most famous movie review shows in history. Below are their respective judgments on 43 films that were released in 1995. Each critic gives his judgment with a "thumbs up" or "thumbs down." Do they have the same likelihood of giving a movie a positive rating?

		Ebert's review	
		Thumbs up	**Thumbs down**
Siskel's	Thumbs up	18	6
Review	Thumbs down	9	10

9.3 Bickel, Hammel, and O'Connell (1975) investigated whether there was any evidence of gender bias in graduate admissions at the University of California at Berkeley. The table below comes from their cross-classification of 4526 applications to graduate programs in 1973 by gender (male or female), admission (whether or not the applicant was admitted to the program), and program (A, B, C, D, E, or F). What does the data reveal?

A: Admit	**Male**	**Female**	**B: Admit**	**Male**	**Female**
Admitted	512	89	Admitted	353	17
Rejected	313	19	Rejected	207	8

C: Admit	**Male**	**Female**
Admitted	120	202
Rejected	205	391

D: Admit	**Male**	**Female**	**E: Admit**	**Male**	**Female**
Admitted	138	131	Admitted	53	94
Rejected	279	244	Rejected	138	299

F: Admit	**Male**	**Female**
Admitted	22	24
Rejected	351	317

9.4 When an epidemic of severe intestinal disease occurred among workers in a plant in South Bend, Indiana, doctors said that the illness resulted from infection with the amoeba *Entamoeba histolytica* (Cohen, 1973). There are actually two races of these amoebas, large and small, and the large ones were believed to be causing the disease. Doctors suspected that the presence of the small ones might help people resist infection by the large ones. To check on this, public health officials chose a random sample of 138 apparently healthy workers and determined if they were infected with either the large or small amoebas. The table below gives the resulting data. Is the presence of the large race independent of the presence of the small one?

	Large race		
Small race	Present	Absent	Total
Present	12	23	35
Absent	35	68	103
Total	47	91	138

9.5 A study was designed to test whether or not aggression is a function of anonymity. The study was conducted as a field experiment on Halloween; 300 children were observed unobtrusively as they made their rounds. Of these 300 children, 173 wore masks that completely covered their faces while 127 wore no masks. It was found that 101 children in the masked group displayed aggressive or antisocial behavior versus 36 children in unmasked group. What conclusion can be drawn? State your conclusion in terminology of the problem using $\alpha = 0.01$.

9.6 Deathbed scenes in which a dying mother or father holds to life until after the long-absent son returns home and dies immediately after are all too familiar in movies. Do such things happen in everyday life? Are some people able to postpone their death until after an anticipated event takes place? It is believed that famous people do so with respect to their birthdays to which they attach some importance. A study by David P. Phillips (in Tanur, 1972, pp. 52–65) seems to be consistent with the notion. Phillips obtained data[4] on months of birth and death of 1251 famous Americans; the deaths were classified by the time period between the birth

---

4 Three hundred forty-eight were people listed in *Four Hundred Notable Americans*, and 903 are listed as foremost families in three volumes of *Who Was Who* for the years 1951–1960, 1943–1950, and 1897–1942.

dates and death dates as shown in the table below. What do the data suggest?

b	e	f	o	r	e	Birth	a	f	t	e	r
6	5	4	3	2	1	Month	1	2	3	4	5
90	100	87	96	101	86	119		118	121	114	113	106

9.7 Using a calculator, mimic the R results for X^2 from Benford's law example (from p. 172). Here are some theoretical frequencies rounded to two decimal places:

92.41	54.06	•	29.75	24.31	•	•	15.72	14.06

Use χ^2 tables and compare X^2 with the critical χ^2 quantile at $\alpha = 0.05$.

9.8 Assume that a contingency table has two rows and two columns with frequencies of a and b in the first row and frequencies of c and d in the second row:

(a) Verify that the χ^2 test statistic can be expressed as

$$\chi^2 = \frac{(a+b+c+d)(ad-bc)^2}{(a+b)(c+d)(b+d)(a+c)}.$$

(b) Let $\hat{p}_1 = a/(a+c)$ and $\hat{p}_2 = b/(b+d)$. Show that the test statistic

$$z = \frac{(\hat{p}_1 - \hat{p}_2) - 0}{\sqrt{\frac{\bar{p}\bar{q}}{n_{.1}} + \frac{\bar{p}\bar{q}}{n_{.2}}}}, \quad \text{where } \bar{p} = \frac{a+b}{a+b+c+d}$$

and $\bar{q} = 1 - \bar{p}$, coincides with χ^2 from (a).

9.9 Generate a sample of size $n = 216$ from $\mathcal{N}(0,1)$. Select intervals by partitioning $\mathbb{R}$ at points $-2.7, -2.2, -2, -1.7, -1.5, -1.2, -1, -0.8, -0.5, -0.3, 0,$ $0.2, 0.4, 0.9, 1, 1.4, 1.6, 1.9, 2, 2.5,$ and 2.8. Using a χ^2-test, confirm the normality of the sample. Repeat this procedure using sample contaminated by the Cauchy distribution in the following way: `0.95*rnorm(1)+ 0.05*rcauchy(1)`.

9.10 It is well known that when the arrival times of customers constitute a Poisson process with the rate λt, the inter-arrival times follow an exponential distribution with density $f(t) = \lambda e^{-\lambda t}$, $t \geq 0, \lambda > 0$. It is often of interest to establish that the process is Poisson because many theoretical results are available for such processes, ubiquitous in the domain of industrial engineering.

In the following example, $n = 109$ inter-arrival times of an arrival process were recorded, averaged ($\bar{x} = 2.5$), and categorized into time intervals as follows:

Interval	[0,1)	[1,2)	[2,3)	[3,4)	[4,5)	[5,6)	[6,∞)
Frequency	34	20	16	15	9	7	8

Test the hypothesis that the process described with the above inter-arrival times is Poisson, at level $\alpha = 0.05$. You must first estimate λ from the data.

9.11 In a long study of heart disease, the day of the week on which 63 seemingly healthy men died was recorded. These men had no history of disease and died suddenly:

Day of week	Monday	Tuesday	Wednesday	Thursday	Friday	Satday	Sunday
No. of deaths	22	7	6	13	5	4	6

(i) Test the hypothesis that these men were just as likely to die on one day as on any other. Use $\alpha = 0.05$. (ii) Explain in words what constitutes Type II error in the above testing.

9.12 Write a R function mcnemar.r. If $b + c \geq 20$, use the χ^2 approximation. If $b + c < 20$, use exact binomial p-values. You will need pchisq and pbinom. Use your program to solve Exercise 9.4.

9.13 Doucet et al. (1999) compared applications to different primary care programs at Tulane University. The "medicine/pediatrics" program students are trained in both primary care specialties. The results for 148 survey responses, in the table below, are broken down by race. Does ethnicity seem to be a factor in program choice?

Ethnicity	Medical school applicants		
	Medicine	Pediatrics	Medicine/pediatrics
White	30	35	19
Black	11	6	9
Hispanic	3	9	6
Asian	9	3	8

9.14 The Donner party is the name given to a group of emigrants, including the families of George Donner and his brother Jacob, who became trapped in the Sierra Nevada mountains during the winter of 1846–1847. Nearly half of the party died. The experience has become legendary as one of the most spectacular episodes in the record of Western migration in the United States. In total, of the 89 men, women, and children in the Donner party, 48 survived, and 41 died. The following table gives the numbers of males/females according their survival status:

	Male	Female
Died	32	9
Survived	23	25

Test the hypothesis that in the population of consisting of members of Donner's Party, the gender and survival status were independent. Use $\alpha = 0.05$. The following table gives the numbers of males/females who survived according to their age (children/adults). Test the hypothesis that in the population of consisting of surviving members of Donner's Party, the gender and age were independent. Use $\alpha = 0.05$:

	Adult	Children
Male	7	16
Female	10	15

Interesting facts are as follows (not needed for the solution):

- Two-thirds of the women survived; two-thirds of the men died.
- Four girls aged three and under died; two survived. No girls between the ages of 4 and 16 died.
- Four boys aged three and under died; none survived. Six boys between the ages of 4 and 16 died.
- All the adult males who survived the entrapment (Breen, Eddy, Foster, Keseberg) were fathers.
- All the bachelors (single males over age 21) who were trapped in the Sierra died. Jean-Baptiste Trudeau and Noah James survived the entrapment but were only about 16 years old and are not considered bachelors.

9.15 West of Tokyo lies a large alluvial plain, dotted by a network of farming villages. Matui (1968) analyzed the position of the 911 houses making up one of those villages. The area studied was a rectangle, 3 km × 4 km. A grid was superimposed over a map of the village, dividing its 12 km2 into 1200 plots,

Figure 9.6 (a) Matrix of 1200 plots (30 × 40). Lighter color corresponds to higher number of houses. (b) Histogram of number of houses per plot.

each 100 m on a side. The number of houses on each of those plots was recorded in a 30 × 40 matrix of data. Test the hypothesis that the distribution of number of houses per plot is Poisson. Use $\alpha = 0.05$:

Number	0	1	2	3	4	≥ 5
Frequency	584	398	168	35	9	6

Hint: assume that parameter $\lambda = 0.76$ (approximately the ratio 911/1200). Find theoretical frequencies first. For example, the theoretical frequency for number = 2 is $np_2 = 1200 \times 0.76^2/2! \times \exp\{-0.76\} = 162.0745$, while the observed frequency is 168. Subtract an additional degree of freedom because λ is estimated from the data (Figure 9.6).

9.16 A poll was conducted to determine if perceptions of the hazards of smoking were dependent on whether or not the person smoked. One hundred people were randomly selected and surveyed. The results are given below:

	Very dangerous [code 0]	dangerous [code 1]	Somewhat dangerous [code 2]	Not Dangerous [code 3]
Smokers	11 (18.13)	15 (15.19)	14 (9.80)	9 ( )
Nonsmokers	26 (18.87)	16 ( )	6 ( )	3 (6.12)

(a) Test the hypothesis that smoking status does not affect perception of the dangers of smoking at $\alpha = 0.05$ (five theoretical/expected frequencies are given in the parentheses).

(b) Observed frequencies of perceptions of danger [codes] for smokers are

[code 0]	[code 1]	[code 2]	[code 3]
11	15	14	9

Are the codes coming from a discrete uniform distribution (i.e. each code is equally likely)? Use $\alpha = 0.01$.

9.17 Radelet (1981) investigated the relationship between race and whether criminals (convicted of homicide) receive the death penalty (versus a lesser sentence) for regional Florida court cases during 1976–1977. The table lists the death sentence frequencies categorized by the defendant's race and the (murder) victim's race. We see the importance of the victim's race in death penalty decisions. African-Americans were sentenced to death more often if the victim was Caucasian (17.5% versus 12.6%) or African-American (5.8% to 0.0%).[5] Show how Simpson's paradox works if we consider aggregating over the factor of victim's race:

Race of defendant	Race of victim	Death penalty	Lesser sentence
Caucasian	Caucasian	19	132
	African-American	0	9
African-American	Caucasian	11	52
	African-American	6	97

RELEVANT R FUNCTIONS AND DATA SETS IN THIS CHAPTER

R codes: `mantel.haenszel.r`, `tablerxc.r`

R functions: `chisq.test`, `cochran.qtest`, `grid.arrange`, `waffle`

R package: `gridExtra`, `waffle`, `survival`, `RVAideMemoire`

---

5 Note that other covariate information about the defendant and victim, such as income or wealth, might have led to similar results.

References

Agresti, A. (2012), *Categorical Data Analysis*, Third Edition, New York: Wiley.

Benford, F. (1938), "The Law of Anomalous Numbers," *Proceedings of the American Philosophical Society*, 78, 551.

Bickel, P. J., Hammel, E. A., and O'Connell, J. W. (1975), "Sex Biasin Graduate Admissions: Data from Berkeley," *Science*, 187, 398–404.

Cochran, W. G. (1950), "The Comparison of Percentages in Matched Samples," *Biometrika*, 37, 256–266.

Cohen, J. E. (1973), "Independence of Amoebas," in *Statistics by Example: Weighing Chances*, Eds. F. Mosteller, R. S. Pieters, W. H. Kruskal, G. R. Rising, R. F. Link, Link, with the assistance of R. Carlson, and M. Zelinka Reading, MA: Addison-Wesley, 72.

Darwin, C. (1859), *The Origin of Species by Means of Natural Selection*, First Edition, London, UK: Murray.

Doucet, H., Shah, M. K., Cummings, T. L., and Kahm, M. J. (1999), "Comparison of Internal Medicine, Pediatric and Medicine/Pediatrics Applicants and Factors Influencing Career Choices," *Southern Medical Journal*, 92, 296–299.

Fisher, R. A. (1918), "The Correlation Between Relatives on the Supposition of Mendelian Inheritance," *Philosophical Transactions of the Royal Society of Edinburgh*, 52, 399–433.

Fisher, R. A. (1922), "On the Interpretation of Chi-Square from Contingency Tables, and the Calculation of P," *Journal of the Royal Statistical Society*, 85, 87–94.

Fisher, R. A. (1966), *The Design of Experiments*, Eighth Edition, Edinburgh, UK: Oliver and Boyd.

Gilb, T., and Graham, D. (1993), *Software Inspection*, Reading, MA: Addison-Wesley.

Hill, T. (1998), "The First Digit Phenomenon," *American Scientist*, 86, 358.

Johnson, S., and Johnson, R. (1972), "Tonsillectomy History in Hodgkin's Disease," *New England Journal of Medicine*, 287, 1122–1125.

Liu, Z. (1992), "Smoking and Lung Cancer in China: Combined Analysis of Eight Case-Control Studies," *International Journal of Epidemiology*, 21, 197–201.

Mantel, N., and Haenszel, W. (1959), "Statistical Aspects of the Analysis of Data from Retrospective Studies of Disease," *Journal of the National Cancer Institute*, 22, 719–729.

Matui, I. (1968), "Statistical Study of the Distribution of Scattered Villages in Two Regions of the Tonami Plain, Toyama Prefecture," in *Spatial Patterns*, Eds. B. J. L. Berry and D. F. Marble, Englewood Clifs, NJ: Prentice-Hall, 149–158.

Mendel, J. G. (1866), "Versuche über Pflanzenhybriden," *Verhandlungen des naturforschenden Vereines zu Brünn*, 4 (1865), 3–47.

McNemar, Q. (1947), "A Note on the Sampling Error of the Difference Between Correlated Proportions or Percentages," *Psychometrika*, 12, 153–157.

McWilliams, W. C., and Piotrowski, H. (2005), *The World Since 1945: A History of International Relations*, Boulder, CO: Lynne Rienner Publishers.

Pearson, K. (1900), "On the Criterion that a Given System of Deviations from the Probable in the Case of a Correlated System of Variables is such that it can be Reasonably Supposed to have Arisen from Random Sampling," *Philosophical Magazine*, 50, 157–175.

Radelet, M. (1981), "Racial Characteristics and the Imposition of the Death Penalty," *American Sociological Review*, 46, 918–927.

Rasmussen, M. H., Wahlberg, M., Miller, L. A. (2004), "Estimated transmission beam pattern of clicks recorded from free-ranging white-beaked dolphins (Lagenorhynchus albirostris)," *The Journal of the Acoustical Society of America*, 116, 1826–1831.

Simonoff, J. S. (2003), *Analyzing Categorical Data*, New York: Springer-Verlag.

Tanur, J. M., Ed. (1972), *Statistics: A Guide to the Unknown*, San Francisco, CA: Holden-Day.

von Bortkiewicz, L. (1898), *Das Gesetz der Kleinen Zahlen*, Leipzig, Germany: Teubner.

10

Estimating Distribution Functions

The harder you fight to hold on to specific assumptions, the more likely there's gold in letting go of them.

John Seely Brown (1997), former Chief Scientist at Xerox Corporation

10.1 Introduction

Let $X_1, X_2, \ldots, X_n$ be a sample from a population with continuous cumulative distribution function (CDF) F. In Chapter 3, we defined the *empirical (cumulative) distribution function* (EDF) based on a random sample as

$$F_n(x) = \frac{1}{n} \sum_{i=1}^{n} \mathbf{1}(X_i \leq x).$$

Because $F_n(x)$, for a fixed x, has a sampling distribution directly related to the binomial distribution, its properties are readily apparent, and it is easy to work with as an estimating function.

The EDF provides a sound estimator for the CDF, but not through any methodology that can be extended to general estimation problems in nonparametric statistics. For example, what if the sample is right truncated? Or censored? What if the sample observations are not independent or identically distributed? In standard statistical analysis, the method of *maximum likelihood* provides a general methodology for achieving inference procedures on unknown parameters, but in the nonparametric case, the unknown parameter is the function $F(x)$ (or, equivalently, the survival function $S(x) = 1 - F(x)$). Essentially, there are an infinite number of parameters. In the Section 10.2 we develop a general formula for estimating the distribution function for non-i.i.d. (independently

Nonparametric Statistics with Applications to Science and Engineering with R, Second Edition.
Paul Kvam, Brani Vidakovic, and Seong-joon Kim.
© 2023 John Wiley & Sons, Inc. Published 2023 by John Wiley & Sons, Inc.

and identically distributed) samples. Specifically, the Kaplan–Meier estimator is constructed to estimate $F(x)$ when censoring is observed in the data.

This theme continues in Chapter 11 where we introduce *Density Estimation* as a practical alternative to estimating the CDF. Unlike the cumulative distribution, the density function provides a better visual summary of how the random variable is distributed. Corresponding to the EDF, the *empirical density function* is a discrete uniform probability distribution on the observed data, and its graph does not explain much about the distribution of the data. The properties of the more refined density estimators in Chapter 11 are not so easily discerned, but it will give the researcher a smoother and visually more interesting estimator to work with.

In medical research, survival analysis is the study of lifetime distributions along with associated factors that affect survival rates. The time event might be an organism's death, or perhaps the occurrence or recurrence of a disease or symptom.

10.2 Nonparametric Maximum Likelihood

As a counterpart to the parametric likelihood, we define the nonparametric likelihood of the sample $X_1, \ldots, X_n$ as

$$L(F) = \prod_{i=1}^{n} \left(F(x_i) - F(x_i^-) \right), \tag{10.1}$$

where $F(x_i^-)$ is defined as $P(X < x_i)$. This framework was first introduced by Kiefer and Wolfowitz (1956).

One serious problem with this definition is that $L(F) = 0$ if F is continuous, which we might assume about the data. In order for L to be positive, the argument (F) must put positive weight (or probability mass) on every one of the observations in the sample. Even if we know F is continuous, the nonparametric maximum likelihood estimator (NPMLE) must be noncontinuous at the points of the data.

For a reasonable class of estimators, we consider nondecreasing functions F that can have discrete and continuous components. Let $p_i = F(X_{i:n}) - F(X_{i-1:n})$, where $F(X_{0:n})$ is defined to be 0. We know that $p_j > 0$ is required, or else $L(F) = 0$. We also know that $p_1 + \cdots + p_n = 1$, because if the sum is less than one, there would be probability mass assigned outside the set $x_1, \ldots, x_n$. That would be impractical because if we reassigned that residual probability mass (say, $q = 1 - p_1 - \cdots - p_n > 0$) to any one of the values x_i, the likelihood $L(F)$ would increase in the term $F(x_i) - F(x_i^-) = p_i + q$. So the NPMLE assigns probability mass not only to every observation but *only* to that set; hence the likelihood can be equivalently expressed as

$$L(p_1, \ldots, p_n) = \prod_{i=1}^{n} p_i,$$

which, under the constraint that $\sum p_i = 1$, is the *multinomial* likelihood. The NPMLE is easily computed as $\hat{p}_i = 1/n$, $i = 1, \dots, n$. Note that this solution is quite intuitive – it places equal "importance" on all n of the observations, and it satisfies the constraint given above that $\sum p_i = 1$. This essentially proves the following theorem.

Theorem 10.1 Let $X_1, \dots, X_n$ be a random sample generated from F. For any distribution function F_0, the nonparametric likelihood $L(F_0) \leq L(F_n)$, so that the empirical distribution function is the NPMLE.

10.3 Kaplan–Meier Estimator

The nonparametric likelihood can be generalized to all sorts of observed data sets beyond a simple i.i.d. sample. The most commonly observed phenomenon outside the i.i.d. case involves *censoring*. To describe censoring, we will consider $X \geq 0$, because most problems involving censoring consist of lifetime measurements (e.g. time until failure).

Definition 10.1 Suppose X is a lifetime measurement. X is **right censored** at time t if we know the failure time occurred *after* time t, but the actual time is unknown. X is **left censored** at time t if we know the failure time occurred *before* time t, but the actual time is unknown.

Definition 10.2 **Type-I censoring** occurs when n items on test are stopped at a fixed time t_0, at which time all surviving test items are taken off test and are right censored.

Definition 10.3 **Type-II censoring** occurs when n items $(X_1, \dots, X_n)$ on test are stopped after a prefixed number of them (say, $k \leq n$) have failed, leaving the remaining items to be right censored at the random time $t = X_{k:n}$.

Type-I censoring is a common problem in drug treatment experiments based on human trials; if a patient receiving an experimental drug is known to survive up to a time t but leaves the study (and humans are known to leave such clinical trials much more frequently than lab mice), the lifetime is right censored.

Suppose we have a sample of possibly right-censored values. We will assume the random variables represent lifetimes (or "occurrence times"). The sample is summarized as $\{(X_i, \delta_i), i = 1, \dots, n\}$, where X_i is a time measurement and δ_i equals 1 if the X_i represents the lifetime and equals 0 if X_i is a (right) censoring time. If $\delta_i = 1$, X_i contributes $dF(x_i) \equiv F(x_i) - F(x_i^-)$ to the likelihood (as it does in the i.i.d. case). If $\delta_i = 0$, we know only that the lifetime surpassed time X_i, so this event contributes $1 - F(x_i)$ to the likelihood. Then

$$L(F) = \prod_{i=1}^{n} \left(1 - F(x_i)\right)^{1-\delta_i} \left(dF(x_i)\right)^{\delta_i}. \tag{10.2}$$

The argument about the NPMLE has changed from (10.1). In this case, no probability mass need be assigned to a value X_i for which $\delta_i = 0$, because in that case, $dF(X_i)$ does not appear in the likelihood. Furthermore, the accumulated probability mass of the NPMLE on the observed data does not necessarily sum to one, because if the largest value of X_i is a censored observation, the term $S(X_i) = 1 - F(X_i)$ will only be positive if probability mass is assigned to a point or interval to the right of X_i.

Let p_i be the probability mass assigned to $X_{i:n}$. This new notation allows for positive probability mass (call it p_{n+1}) that can be assigned to some arbitrary point or interval after the last observation $X_{n:n}$. Let $\tilde{\delta}_i$ be the censoring indicator associated with $X_{i:n}$. Note that even though $X_{1:n} < \cdots < X_{n:n}$ are ordered, the set $(\tilde{\delta}_1,\ldots,\tilde{\delta}_n)$ is not necessarily so ($\tilde{\delta}_i$ is called a *concomitant*).

If $\tilde{\delta}_i = 1$, the likelihood is clearly maximized by setting probability mass (say, p_i) on $X_{i:n}$. If $\tilde{\delta}_i = 0$, some mass will be assigned to the right of $X_{i:n}$, which has interval probability $p_{i+1} + \cdots + p_{n+1}$. The likelihood based on censored data is expressed as

$$L(p_1,\ldots,p_{n+1}) = \prod_{i=1}^{n} p_i^{\tilde{\delta}_i} \left(\sum_{j=i+1}^{n+1} p_j\right)^{1-\tilde{\delta}_i}.$$

Instead of maximizing the likelihood in terms of $(p_1,\ldots,p_{n+1})$, it will prove to be much easier using the transformation

$$\lambda_i = \frac{p_i}{\sum_{j=i}^{n+1} p_j}.$$

This is a convenient one-to-one mapping where

$$\sum_{j=i}^{n+1} p_j = \prod_{j=1}^{i-1}(1-\lambda_j), \quad p_i = \lambda_i \prod_{j=1}^{i-1}(1-\lambda_j).$$

The likelihood simplifies to

$$L(\lambda_1,\ldots,\lambda_{n+1}) = \prod_{i=1}^{n} \left(\left(\lambda_i \prod_{j=1}^{i-1}(1-\lambda_j)\right)^{\tilde{\delta}_i} \left(\prod_{j=1}^{i}(1-\lambda_j)\right)^{1-\tilde{\delta}_i} \right)$$

$$= \left(\prod_{i=1}^{n} \lambda_i^{\tilde{\delta}_i}(1-\lambda_i)^{1-\tilde{\delta}_i}\right) \left(\prod_{i=1}^{n-1}(1-\lambda_i)^{n-i}\right)$$

$$= \prod_{i=1}^{n} \left(\frac{\lambda_i}{1-\lambda_i}\right)^{\tilde{\delta}_i} (1-\lambda_i)^{n-i+1}.$$

As a function of $(\lambda_1, \ldots, \lambda_{n+1})$, L is maximized at $\hat{\lambda}_i = \tilde{\delta}_i/(n - i + 1)$, $i = 1, \ldots,$ $n + 1$. Equivalently,

$$\hat{p}_i = \frac{\tilde{\delta}_i}{n - i + 1} \prod_{j=1}^{i-1} \left(1 - \frac{\tilde{\delta}_j}{n - j + 1} \right).$$

The NPMLE of the distribution function (denoted $F_{KM}(x)$) can be expressed as a sum in p_i. For example, at the observed order statistics, we see that

$$S_{KM}(x_{i:n}) \equiv 1 - F_{KM}(x_{i:n}) = \prod_{j=1}^{i} \left(1 - \frac{1}{n - j + 1} \right)^{\tilde{\delta}_j} \tag{10.3}$$

$$= \prod_{j=1}^{i} \left(1 - \frac{\tilde{\delta}_j}{n - j + 1} \right).$$

This is the *Kaplan–Meier* nonparametric estimator, developed by Kaplan and Meier (1958) for censored lifetime data analysis. It has been one of the most influential developments in the past century; their paper is the most cited paper in statistics (Stigler, 1994). E. L. Kaplan (1920–2002) and Paul Meier (1924–2011) never actually met during this time, but they both submitted their idea of the "product limit estimator" to the *Journal of the American Statistical Association* at approximately the same time, so their joint results were amalgamated through letter correspondence.

For non-censored observations, the Kaplan–Meier estimator is identical to the regular maximum likelihood estimator (MLE). The difference occurs when there is a censored observation – then the Kaplan–Meier estimator takes the "weight" normally assigned to that observation and distributes it evenly among all observed values to the right of the observation. This is intuitive because we know that the true value of the censored observation must be somewhere to the right of the censored value, but we do not have any more information about what the exact value should be.

The estimator is easily extended to sets of data that have potential tied values. If we define d_j = number of failures at x_j and m_j = number of observations that had survived up to x_j^-, then

$$F_{KM}(t) = 1 - \prod_{x_j \leq t} \left(1 - \frac{d_j}{m_j} \right). \tag{10.4}$$

Example 10.1 Muenchow (1986) tested whether male or female flowers (of *Western White Clematis*) were equally attractive to insects. The data in Table 10.1

Table 10.1   Waiting times for insects to visit flowers.

Male flowers			Female flowers		
1	9	27	1	19	57
1	9	27	2	23	59
2	9	30	4	23	67
2	11	31	4	26	71
4	11	35	5	28	75
4	14	36	6	29	75[a]
5	14	40	7	29	78[a]
5	14	43	7	29	81
6	16	54	8	30	90[a]
6	16	61	8	32	94[a]
6	17	68	8	35	96
7	17	69	9	35	96[a]
7	18	70	14	37	100[a]
8	19	83	15	39	102[a]
8	19	95	18	43	105[a]
8	19	102[a]	18	56	
		104[a]			

a) Waiting times for insects to visit flowers. Right censoring indicated by *.

represent waiting times (in minutes), which includes censored data. We can use the R functions Surv and survfit (from the R package survival) to construct nonparametric estimators from the data. The survfit function creates survival curves for the Kaplan–Meier estimator, while the Surv function creates an object for storing the response variable (Figure 10.1):

```
> library(survival)
>
> male <- c(1,1,2,2,4,4,5,5,6,6,6,7,7,8,8,8,
+ 9,9,9,11,11,14,14,14,16,16,17,17,18,19,19,19,
+ 27,27,30,31,35,36,40,43,54,61,68,69,70,83,95,102,104)
> male.event <- c(rep(1,47),0,0)
> # denoting '1' if failure, denoting '0' if censored.
>
> female <- c(1,2,4,4,5,6,7,7,8,8,8,9,14,15,18,18,
+ 19,23,23,26,28,29,29,29,30,32,35,35,37,39,43,56,
+ 57,59,67,71,75,75,78,81,90,94,96,96,100,102,105);
> female.event <- c(rep(1,32),1,1,1,1,1,0,0,1,0,0,1,0,0,0,0);
```

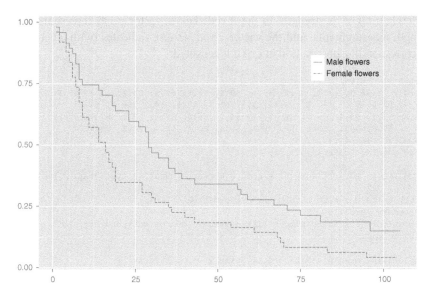

Figure 10.1 Kaplan–Meier estimator for waiting times (solid line for male flowers and dashed line for female flowers).

```
>
> male.fit <- survfit(Surv(time=male,event=male.event)~1,
+          type="kaplan-meier");
> female.fit <- survfit(Surv(time=female,event=female.event)~1,
+          type="kaplan-meier");
> # try "summary(male.fit)" and "attributes(male.fit)"
> # to obtain more information
>
> dat <- data.frame(x=c(male.fit$time,female.fit$time),
+    y=c(male.fit$surv,female.fit$surv),group=c(rep("Male Flowers",
+    length(male.fit$time)),rep("Female Flowers",length(female.fit$time))))
> p <- ggplot(aes(x=x,y=y,group=group,lty=group,col=group),data=dat)
+    + geom_step()
> p <- p + theme(legend.position=c(0.8,0.8),legend.background=
+    element_rect(fill=NA),
+    legend.title=element_blank()) + xlab("") + ylab("")
> print(p)
```

You can obtain information on the sample means, medians, and extrema using the R function `summary`. The `attributes` function provides variable names for all the stored information inside the object.

Example 10.2 Data from Crowder et al. (1991) lists strength measurements (in coded units) for 48 pieces of weathered cord. Seven of the pieces of cord were damaged and yielded strength measurements that are considered right censored. That is, because the damaged cord was taken off test, we know only the lower limit of its strength. In the R code below, vector `cord` represents the strength measurements, and the vector `cord.event` indicates (with a zero) if the corresponding observation in `cord` is censored:

```
> library(survival)
> cord <- c(36.3, 41.7, 43.9, 49.9, 50.1, 50.8, 51.9, 52.1, 52.3, 52.3,
+ 52.4, 52.6, 52.7, 53.1, 53.6, 53.6, 53.9, 53.9, 54.1, 54.6, 54.8,
+ 54.8, 55.1, 55.4, 55.9, 56, 56.1, 56.5, 56.9, 57.1, 57.1, 57.3,
+ 57.7, 57.8, 58.1, 58.9, 59, 59.1, 59.6, 60.4, 60.7, 26.8, 29.6,
+ 33.4, 35, 40, 41.9, 42.5)
> cord.event <- c(rep(1,41),rep(0,7))
>
> cord.fit <- survfit(Surv(time=cord,event=cord.event)~1,type="kaplan-meier")
>
> # try below codes to obtain more information
> # summary(cord.fit)
> # with(cord.fit,cbind(time,n.risk,n.event,n.censor,surv,std.err,
    lower,upper))
>
> ggplot()+geom_step(aes(x=cord.fit$time,y=cord.fit$surv))+
+ geom_step(aes(x=cord.fit$time,y=cord.fit$lower),lty=2)+
+ geom_step(aes(x=cord.fit$time,y=cord.fit$upper),lty=2)
```

The table below shows how the Kaplan–Meier estimator is calculated using the formula in (10.4) for the first 16 measurements, which includes 7 censored observations. Figure 10.2 shows the estimated survival function for the cord strength data:

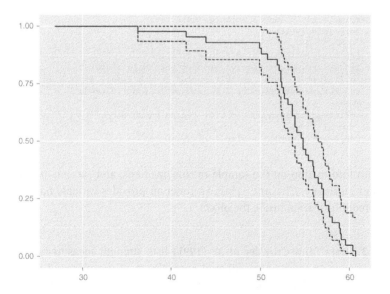

Figure 10.2 Kaplan–Meier estimator cord strength (in coded units).

Uncensored	x_j	m_j	d_j	$\frac{m_j-d_j}{m_j}$	$1 - F_{KM}(x_j)$
	26.8	48	0	1.000	1.000
	29.6	47	0	1.000	1.000
	33.4	46	0	1.000	1.000
	35.0	45	0	1.000	1.000
1	36.3	44	1	0.977	0.977
	40.0	43	0	1.000	0.977
2	41.7	42	1	0.976	0.954
	41.9	41	0	1.000	0.954
	42.5	40	0	1.000	0.954
3	43.9	39	1	0.974	0.930
4	49.9	38	1	0.974	0.905
5	50.1	37	1	0.973	0.881
6	50.8	36	1	0.972	0.856
7	51.9	35	1	0.971	0.832
8	52.1	34	1	0.971	0.807
9	52.3	33	2	0.939	0.758
⋮	⋮	⋮	⋮	⋮	⋮

Example 10.3 Consider observing the lifetime of a series system. Recall that a series system is a system of $k \geq 1$ components that fails at the time the first component fails. Suppose we observe n different systems that are each made of k_i identical components ($i = 1, \ldots, n$) with lifetime distribution F. The lifetime data is denoted $(x_1, \ldots, x_n)$. Further suppose there is (random) right censoring, and $\delta_i = I$ (x_i represents a lifetime measurement). How do we estimate F?

If $F(x)$ is continuous with derivative $f(x)$, then the ith system's survival function is $S(x)^{k_i}$, and its corresponding likelihood is

$$\ell_i(F) = k_i(1 - F(x))^{k_i-1}f(x).$$

It is easier to express the full likelihood in terms of $S(x) = 1 - F(x)$:

$$L(S) = \prod_{i=1}^{n}\left(k_i\big(S(x_i)\big)^{k_i-1}f(x_i)\right)^{\delta_i}\left(S(x_i)^{k_i}\right)^{1-\delta_i},$$

where $1 - \delta$ indicates censoring.

To make the likelihood more easy to solve, let us examine the ordered sample $y_i = x_{i:n}$ so we observe $y_1 < y_2 < \cdots < y_n$. Let $\tilde{k}_i$ and $\tilde{\delta}_i$ represent the size of the series system and the censoring indicator for y_i. Note that $\tilde{k}_i$ and $\tilde{\delta}_i$ are concomitants of y_i.

The likelihood, now as a function of $(y_1, \ldots, y_n)$, is expressed as

$$\tilde{L}(S) = \prod_{i=1}^{n} \left(\tilde{k}_i \left(S(y_i) \right)^{\tilde{k}_i - 1} f(y_i) \right)^{\tilde{\delta}_i} \left(S(y_i)^{\tilde{k}_i} \right)^{1 - \tilde{\delta}_i}$$

$$\propto \prod_{i=1}^{n} f(y_i)^{\tilde{\delta}_i} S(y_i)^{\tilde{k}_i - \tilde{\delta}_i}.$$

For estimating F nonparametrically, it is again clear that $\hat{F}$ (or $\hat{S}$) will be a step function with jumps occurring only at points of observed system failure. With this in mind, let $S_i = S(y_i)$ and $\alpha_i = S_i / S_{i-1}$. Then $f_i = S_{i-1} - S_i = \prod_{r=1}^{i-1} \alpha_r (1 - \alpha_i)$. If we let $\tau_j = \tilde{k}_j + \cdots + \tilde{k}_n$, the likelihood can be expressed simply (see Exercise 10.5) as

$$\tilde{L}(S) = \prod_{i=1}^{n} \alpha_i^{\tau_i - \tilde{\delta}_i} \left(1 - \alpha_i \right)^{\tilde{\delta}_i},$$

and the NPMLE for $S(x)$, in terms of the ordered system lifetimes, is

$$\hat{S}(y_i) = \prod_{r=1}^{i} \left(\frac{\tau_r - \tilde{\delta}_r}{\tau_r} \right).$$

Note the special case in which $k_i = 1$ for all i; we end up with the Kaplan–Meier estimator.

10.4 Confidence Interval for *F*

Like all estimators, $\hat{F}(x)$ is only as good as its measurement of uncertainty. Confidence intervals can be constructed for $F(x)$ just as they are for regular parameters, but a typical inference procedure refers to a *pointwise* confidence interval about $F(x)$ where x is fixed.

A simple, approximate $1 - \alpha$ confidence interval can be constructed using a normal approximation

$$\hat{F}(x) \pm z_{1-\alpha/2} \hat{\sigma}_{\hat{F}},$$

where $\hat{\sigma}_{\hat{F}}$ is our estimate of the standard deviation of $\hat{F}(x)$. If we have an i.i.d. sample, $\hat{F} = F_n$, and $\sigma_{F_n}^2 = F(x)[1 - F(x)]/n$, so that

$$\hat{\sigma}_{\hat{F}}^2 = F_n(x)[1 - F_n(x)]/n.$$

Recall that $nF_n(x)$ is distributed as binomial $Bin(n, F(x))$ and an exact interval for $F(x)$ can be constructed using the bounding procedure for the binomial parameter p in Chapter 3.

In the case of right censoring, a confidence interval can be based on the Kaplan–Meier estimator, but the variance of $F_{KM}(x)$ does not have a simple form. Greenwood's formula (Greenwood, 1926), originally concocted for grouped data, can be applied to construct a $1 - \alpha$ confidence interval for the survival function ($S = 1 - F$) under right censoring:

$$S_{KM}(t_i) \pm z_{\alpha/2} \hat{\sigma}_{KM}(t_i),$$

where

$$\hat{\sigma}^2_{KM}(t_i) = \hat{\sigma}^2(S_{KM}(t_i)) = S_{KM}(t_i)^2 \sum_{t_j \leq t_i} \frac{d_j}{m_j(m_j - d_j)}.$$

It is important to remember these are *pointwise* confidence intervals, based on fixed values of t in $F(t)$. Simultaneous confidence bands are a more recent phenomenon and apply as a confidence statement for F across all values of t for which $0 < F(t) < 1$. Nair (1984) showed that the confidence bands by Hall and Wellner (1980) work well in various settings, even though they are based on large-sample approximations. An approximate $1 - \alpha$ confidence band for $S(t)$, for values of t less than the largest observed failure time, is

$$S_{KM}(t) \pm \sqrt{-\frac{1}{2n} \ln\left(\frac{\alpha}{2}\right)} S_{KM}(t)\left(1 + \hat{\sigma}^2_{KM}(t)\right).$$

This interval is based on rough approximation for an infinite series, and a slightly better approximation can be obtained using numerical procedures suggested in Nair (1984). Along with the Kaplan–Meier estimator of the distribution of cord strength, Figure 10.2 also shows a 95% simultaneous confidence band. The pointwise confidence interval at $t = 50$ units is $(0.8121, 0.9934)$. The confidence band, on the other hand, is $(0.7078, 1.0000)$. Note that for small strength values, the band reflects a significant amount of uncertainty in $F_{KM}(x)$.

10.5 Plug-in Principle

With an i.i.d. sample, the EDF serves not only as an estimator for the underlying distribution of the data, but through the EDF, any particular parameter θ of the distribution can also be estimated. Suppose the parameter has a particular functional relationship with the distribution function F:

$$\theta = \theta(F).$$

Examples are easy to construct. The population mean, for example, can be expressed as

$$\mu = \mu(F) = \int_{-\infty}^{\infty} x dF(x)$$

and variance is

$$\sigma^2 = \sigma^2(F) = \int_{-\infty}^{\infty} (x - \mu)^2 \, dF(x).$$

As F_n is the sample analog to F, so $\theta(F_n)$ can serve as a sample-based estimator for θ. This is the idea of the *plug-in principle*. The estimator for the population mean is

$$\hat{\mu} = \mu(F_n) = \int_{-\infty}^{\infty} x \, dF_n(x) = \sum_{x} x_i dF_n(x_i) = \bar{x}.$$

Obviously, the plug-in principle is not necessary for simply estimating the mean, but it is reassuring to see it produce a result that is consistent with standard estimating techniques.

Example 10.4   The quantile x_p can be expressed as a function of F: $x_p = \inf\{x : \int_x^{\infty} dF(x) \leq 1 - p\}$. The sample equivalent is the value $\hat{x}_p = \inf\{x : \int_x^{\infty} dF_n(x) \leq 1 - p\}$. If F is continuous, then we have $x_p = F^{-1}(p)$, and $F_n(\hat{x}_p) = p$ is solved uniquely. If F is discrete, $\hat{x}_p$ is the smallest value of x for which

$$n^{-1} \sum_{i=1}^{n} 1(x \leq x_i) \leq 1 - p,$$

or, equivalently, the smallest order statistic $x_{i:n}$ for which $i/n \leq p$, i.e. $(i + 1)/n > p$. For example, with the flower data in Table 10.1, the median waiting times are easily estimated as the smallest values (x) for which $F_{KM}(x) \leq 1/2$, which are 16 (for the male flowers) and 29 (for the female flowers).

If the data are not i.i.d., the NPMLE $\hat{F}$ can be plugged in for F in $\theta(F)$. This is a key selling point to the plug-in principle; it can be used to formulate estimators where we might have no set rule to estimate them. Depending on the sample, $\hat{F}$ might be the EDF or the Kaplan–Meier estimator. The plug-in technique is simple, and it will form a basis for estimating uncertainty using resampling techniques in Chapter 15.

Example 10.5   To find the average cord strength from the censored data, for example, it would be imprudent to merely average the data, as the censored observations represent a lower bound on the data. Hence, the true mean will be underestimated. By using the plug-in principle, we will get a more accurate estimate; the code below estimates the mean cord strength as 54.1946. The sample mean, ignoring the censoring indicator, is 51.4438:

```
> svtime <- cord.fit$time;
> if(min(svtime)>0){
+ skm <- cord.fit$surv;
+ skm1 <- c(1, skm);
+ dx <- c(svtime[1],diff(svtime),0);
+ svtime2 <- c(0, svtime);
+ svtime3 <- c(svtime,svtime[length(svtime)])
+ mu.hat <- sum(skm1 * dx);
+ print(mu.hat);
+ }else{
+ cdf <- 1-cord.fit$surv;
+ df <- c(0,diff(cdf),1);
+ svtime2 <- c(svtime,0);
+ mu.hat <- sum(svtime2*df);
+ print(mu.hat);
+ }
[1] 54.19459
```

10.6   Semi-Parametric Inference

The *proportional hazards* model for lifetime data relates two populations according to a common underlying hazard rate. Suppose $r_0(t)$ is a baseline hazard rate, where $r(t) = f(t)/(1 - F(t))$. In reliability theory, $r(t)$ is called the *failure rate*. For some covariate x that is observed along with the lifetime, the positive function of $\Psi(x)$ describes how the level of x can change the failure rate (and thus the lifetime distribution):

$$r(t;x) = r_0(t)\Psi(x).$$

This is termed a *semi-parametric model* because $r_0(t)$ is usually left unspecified (and thus a candidate for nonparametric estimation), whereas $\Psi(x)$ is a known positive function, at least up to some possibly unknown parameters. Recall that the CDF is related to the failure rate as

$$\int_{-\infty}^{x} r(u) \, du \equiv R(u) = -\ln S(x),$$

where $S(x) = 1 - F(x)$ is called the survivor function. $R(t)$ is called the *cumulative failure rate* in reliability and life testing. In this case, $S_0(t)$ is the baseline survivor function and relates to the lifetime affected by $\Psi(x)$ as we suggest Statistical Models and Methods for Lifetime Data (1982) by Lawless

$$S(t;x) = S_0(t)^{\Psi(x)}.$$

The most commonly used proportional hazards model used in survival analysis is called the *Cox model* (named after Cox 1972), which has the form

$$r(t;x) = r_0(t)\, e^{x'\beta}.$$

With this model, the (vector) parameter β is left unspecified and must be estimated. Suppose the baseline hazard function of two different populations are related by proportional hazards as $r_1(t) = r_0(t)\lambda$ and $r_2(t) = r_0(t)\theta$. Then if T_1 and T_2 represent lifetimes from these two populations,

$$P(T_1 < T_2) = \frac{\lambda}{\lambda + \theta}.$$

The probability does not depend at all on the underlying baseline hazard (or survivor) function. With this convenient setup, nonparametric estimation of $S(t)$ is possible through maximizing the nonparametric likelihood. Suppose n possibly right-censored observations $(x_1, \ldots, x_n)$ from $F = 1 - S$ are observed. Let ξ_i represent the number of observations at risk just before time x_i. Then, if $\delta_i = 1$ indicates the lifetime was observed at x_i,

$$L(\beta) = \prod_{i=1}^{n} \left(\frac{e^{x_i'\beta}}{\sum_{j \in \xi_i} e^{x_j'\beta}} \right)^{\delta_i}.$$

In general, the likelihood must be solved numerically. For a thorough study of inference with a semi-parametric model, we suggest *Statistical Models and Methods for Lifetime Data* by Lawless. This area of research is paramount in survival analysis.

Related to the proportional hazard model is the *accelerated lifetime model* used in engineering. In this case, the baseline survivor function $S_0(t)$ can represent the lifetime of a test product under usage conditions. In an accelerated life test, and additional stress is put on the test unit, such as high or low temperature, high voltage, high humidity, etc. This stress is characterized through the function $\Psi(x)$, and the survivor function of the stressed test item is

$$S(t;x) = S_0(t\Psi(x)).$$

Accelerated life testing is an important tool in product development, especially for electronics manufacturers who produce gadgets that are expected to last several years on test. By increasing the voltage in a particular way, as one example, the lifetimes can be shortened to hours. The key is how much faith the manufacturer has on the known acceleration function $\Psi(x)$.

In R, the `survival` package offers the function `coxph`, which computes Cox proportional hazards estimator for input data, much in the same way the `survfit` computes the Kaplan–Meier estimator.

10.7 Empirical Processes

> It is a mistake to think you can solve any major problems just with potatoes.
>
> Douglas Adams (1952–2001)

If we express the sample as $X_1(\omega), \dots, X_n(\omega)$, we note that $F_n(x)$ is both a function of x and $\omega \in \Omega$. From this, the EDF can be treated as a random process. The Glivenko–Cantelli theorem from Chapter 3 states that the EDF $F_n(x)$ converges to $F(x)$ (i) almost surely (as random variable, x fixed) and (ii) uniformly in x (as a function of x with ω fixed). This can be expressed as

$$P\left(\omega \;\middle|\; \lim_{n \to \infty} \sup_x |F_n(x) - F(x)| = 0 \right) = 1.$$

Let $W(x)$ be a standard Brownian motion process. It is defined as a stochastic process for which $W(0) = 0$, $W(t) \sim \mathcal{N}(0, t)$, $W(t)$ has independent increments, and the paths of $W(t)$ are continuous. A Brownian bridge is defined as $B(t) = W(t) - tW(1)$, $0 \le t \le 1$. Both ends of a Brownian Bridge, $B(0)$ and $B(1)$, are tied to 0, and this property motivates the name. A Brownian motion $W(x)$ has covariance function $\gamma(t, s) = t \wedge s = \min(t, s)$. This is because $\mathbb{E}(W(t)) = 0$, $\mathbb{V}\mathrm{ar}(W(t)) = s$, for $s < t$, $\mathrm{Cov}(W(t), W(s)) = \mathrm{Cov}(W(s), (W(t) - W(s)) + W(s))$ and W has independent increments.

Define the random process $B_n(x) = \sqrt{n}(F_n(x) - F(x))$. This process converges to a Brownian bridge process, $B(x)$, in the sense that all finite dimensional distributions of $B_n(x)$ (defined by a selection of $x_1, \dots, x_m$) converge to the corresponding finite dimensional distribution of a Brownian bridge $B(x)$.

Using this, one can show that a Brownian bridge has mean zero and covariance function $\gamma(t, s) = t \wedge s - ts$. If $s < t$, $\gamma(s, t) = s(1 - t)$. For $s < t$, $\gamma(s, t) = \mathbb{E}(W(s) - sW(1))(W(t) - tW(1)) = \cdots = s - st$. Because the Brownian bridge is a Gaussian process, it is uniquely determined by its second-order properties. The covariance function $\gamma(t, s)$ for the process $\sqrt{n}(F_n(t) - F(t))$ is

$$\gamma(t, s) = \mathbb{E}\left[\sqrt{n}(F_n(t) - F(t)) \cdot \sqrt{n}(F_n(s) - F(s)) \right]$$

$$= n\mathbb{E}(F_n(t) - F(t))(F_n(s) - F(s)) = \frac{1}{n}(F(t) \wedge F(s) - F(t)F(s)).$$

Proof:

$$\mathbb{E}\gamma(t,s) = \mathbb{E}\left[\left(\frac{1}{n}\sum_i (1(X_i < t) - F(t))\right) \cdot \left(\frac{1}{n}\sum_j (1(X_j < s) - F(s))\right)\right]$$

$$= \frac{1}{n^2}\mathbb{E}\left[\sum_{i,j}(1(X_i < t) - F(t))(1(X_j < s) - F(s))\right]$$

$$= \frac{1}{n}\mathbb{E}(1(X_1 < t) - F(t))(1(X_1 < s) - F(s))$$

$$= \frac{1}{n}\mathbb{E}\left[1(X_1 < t \wedge s) - F(t)1(X_1 < s) - F(s)1(X_1 < t) + F(t)F(s)\right]$$

$$= \frac{1}{n}(F(t \wedge s) - F(t)F(s)).$$

This result is independent of F, as long as F is continuous, as the sample $X_1, \ldots, X_n$ could be transformed to uniform: $Y_1 = F(X_1), \ldots, Y_n = F(X_n)$. Let $G_n(t)$ be the empirical distribution based on $Y_1, \ldots, Y_n$. For the uniform distribution, the covariance is $\gamma(t,s) = t \wedge s - ts$, which is exactly the correlation function of the Brownian bridge. This leads to the following result:

Theorem 10.2 *The random process $\sqrt{n}(F_n(x) - F(x))$ converges in distribution to the Brownian bridge process.*

10.8 Empirical Likelihood

In Chapter 3 we defined the likelihood ratio based on the likelihood function $L(\theta) = \prod f(x_i; \theta)$, where $X_1, \ldots, X_n$ were i.i.d. with density function $f(x; \theta)$. The likelihood ratio function

$$R(\theta_0) = \frac{L(\theta_0)}{\sup_\theta L(\theta)} \tag{10.5}$$

allows us to construct efficient tests and confidence intervals for the parameter θ. In this chapter we extend the likelihood ratio to nonparametric inference, although it is assumed that the research interest lies in some parameter $\theta = \theta(F)$, where $F(x)$ is the unknown CDF.

The likelihood ratio extends naturally to nonparametric estimation. If we focus on the nonparametric likelihood from the beginning of this chapter, from an i.i.d. sample of $X_1, \ldots, X_n$ generated from $F(x)$,

$$L(F) = \prod_{i=1}^n dF(x_i) = \prod_{i=1}^n \left(F(x_i) - F(x_i^-)\right).$$

The likelihood ratio corresponding to this would be $R(F) = L(F)/L(F_n)$, where F_n is the empirical distribution function. $R(F)$ is called the *empirical likelihood ratio* (ELR). In terms of F, this ratio does not directly help us creating confidence intervals. All we know is that for any CDF F, $R(F) \le 1$, reaching its maximum only for $F = F_n$. This means we are considering only functions F that assign mass on the values $X_i = x_i$, $i = 1, \dots, n$, and R is reduced to function of $n - 1$ parameters $R(p_1, \dots, p_{n-1})$ where $p_i = dF(x_i)$ and $\sum p_i = 1$.

It is more helpful to think of the problem in terms of an unknown parameter of interest $\theta = \theta(F)$. Recall the *plug-in principle* can be applied to estimate θ with $\hat{\theta} = \theta(F_n)$. For example, $\mu = \int x \, dF(x)$ was merely the sample mean, i.e. $\int x \, dF_n(x) = \bar{x}$. We will focus on the mean as our first example to better understand the empirical likelihood.

10.8.1 Confidence Interval for the Mean

Suppose we have an i.i.d. sample $X_1, \dots, X_n$ generated from an unknown distribution $F(x)$. In the case $\mu(F) = \int x \, dF(x)$, define the set $C_{\mathbf{p}}(\mu)$ on $\mathbf{p} = (p_1, \dots, p_n)$ as

$$C_{\mathbf{p}}(\mu) = \left\{ \mathbf{p} : \sum_{i=1}^n p_i x_i = \mu, p_i \ge 0, \ i = 1, \dots, n, \ \sum_{i=1}^n p_i = 1 \right\}.$$

The empirical likelihood associated with μ maximizes $L(\mu)$ over $C_{\mathbf{p}}(\mu)$. The restriction $\sum p_i x_i = \mu$ is called the *structural constraint*. The ELR is this empirical likelihood divided by the unconstrained NPMLE, which is just $L(1/n, \dots, 1/n) = n^{-n}$. If we can find a set of solutions to the empirical likelihood, Owen (1988) showed that

$$X^2 = -2 \log \ R(\mu) = -2 \log \left(\sup_{\mathbf{p} \in C_{\mathbf{p}}} \prod_{i=1}^n n p_i \right)$$

is approximately distributed χ_1^2 if μ is correctly specified, so a nonparametric confidence interval for μ can be formed using the values of $-2 \log \ R(\mu)$.

R software is available to help: el.cen.EM function in emplik package computes the empirical likelihood for a specific mean, allowing the user to iterate to make a curve for $R(\mu)$. Computing $R(\mu)$ is no simple matter; we can proceed with Lagrange multipliers to maximize $\sum p_i x_i$ subject to $\sum p_i = 1$ and $\sum \ln(n p_i) = \ln(r_0)$.

Example 10.6 Recall Exercise 6.2. Fuller et al. (1994) examined polished window strength data to estimate the lifetime for a glass airplane window. The units are ksi (or 1000 psi). The R code below constructs the empirical likelihood for the

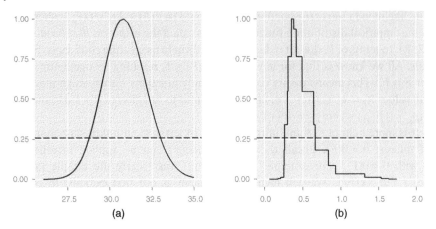

Figure 10.3 Empirical likelihood ratio as a function of (a) the mean and (b) the median (for different samples).

mean glass strength, which is plotted in Figure 10.3a. In this case, a 90% confidence interval for μ is constructed by using the value of r_0 so that $-2\ln r_0 < \chi_1^2(0.90) = 2.7055$, or $r_0 > 0.2585$. The confidence interval is computed as (28.78 ksi, 33.02 ksi):

```
> library(emplik);
> x <- c(18.83, 20.8, 21.657, 23.03, 23.23, 24.05, 24.321, 25.5,
+        25.52, 25.8, 26.69, 26.77, 26.78, 27.05, 27.67, 29.9,
+        31.11, 33.2, 33.73, 33.76, 33.89, 34.76, 35.75, 35.91,
+        36.98, 37.08, 37.09, 39.58, 44.045, 45.29, 45.381);
> n <- length(x);
> mu <- seq(min(x)+0.05,max(x)-0.05,by=0.05);
> ELR.mu <- rep(0,length(mu));
>
> for(i in 1:length(mu)){
+ tmp <- el.cen.EM(x,rep(1,n),mu=mu[i]);
+ ELR.mu[i] <- exp(-tmp$"-2LLR"/2)
+ }
>
> p <- ggplot()+geom_line(aes(x=mu,y=ELR.mu)) + xlim(c(26,35))
> p <- p + geom_abline(intercept=exp(-2.7055/2),slope=0,lty=2)
> print(p)
```

Owen's extension of Wilk's theorem for parametric likelihood ratios is valid for other functions of F, including the variance, quantiles, and more. To construct R for the median, we need only change the structural constraint from $\sum p_i x_i = \mu$ to $\sum p_i \operatorname{sign}(x_i - x_{0.50}) = 0$.

10.8.2 Confidence Interval for the Median

In general, computing $R(x)$ is difficult. For the case of estimating a population quantile, however, the optimizing becomes rather easy. For example, suppose that n_1 observations out of n are less than the population median $x_{0.50}$ and $n_2 = n - n_1$ observations are greater than $x_{0.50}$. Under the constraint $\hat{x}_{0.50} = x_{0.50}$, the nonparametric likelihood estimator assigns mass $(2n_1)^{-1}$ to each observation less than $x_{0.50}$ and assigns mass $(2n_2)^{-1}$ to each observation to the right of $x_{0.50}$, leaving us with

$$R(x_{0.50}; n_1, n_2) = \left(\frac{n}{2n_1}\right)^{n_1} \left(\frac{n}{2n_2}\right)^{n_2}.$$

Example 10.7 Figure 10.3b, based on the R code below, shows the empirical likelihood for the median based on 30 randomly generated numbers from the exponential distribution (with $\mu = 1$ and $x_{0.50} = -\ln(0.5) = 0.6931$). A 90% confidence interval for $x_{0.50}$, again based on $r_0 > 0.2585$, is $(0.2628, 0.6449)$:

```
> n2 <- 30;
> x2 <- rexp(n2,1);
> y <- sort(x2);
> m1 <- seq(1,n2); m2<-n2-m1;
> R <- ((0.5*n2/m1)^m1)*((0.5*n2/m2)^m2);
>
> ggplot() + geom_step(aes(x=y,y=R))+xlim(c(0,2)) +
+ geom_abline(intercept=exp(-2.7055/2),slope=0,lty=2)
```

For general problems, computing the empirical likelihood is no easy matter, and to really utilize the method fully, more advanced study is needed. This section provides a modest introduction to let you know what is possible using the empirical likelihood. Students interested in further pursuing this method are recommended to read Owen's book *Empirical Likelihood* (Owen 2001).

10.9 Exercises

10.1 With an i.i.d. sample of n measurements, use the plug-in principle to derive an estimator for population variance.

10.2 Show that, without censoring, the Kaplan–Meier estimator listed in (10.4) reduces to

$$\hat{F}(t) = \frac{1}{n}\sum_{t_j \le t} d_j.$$

10.3 Twelve people were interviewed and asked how many years they stayed at their first job. Three people are still employed at their first job and have been there for 1.5, 3.0, and 6.2 years. The others reported the following data for years at first job: 0.4, 0.9, 1.1, 1.9, 2.0, 3.3, 5.3, 5.8, and 14.0. Using hand calculations, compute a nonparametric estimator for the distribution of T = time spent (in years) at first job. Verify your hand calculations using R. According to your estimator, what is the estimated probability that a person stays at their job for less than four years? Construct a 95% confidence interval for this estimate.

10.4 Using the estimator in Exercise 10.2, use the plug-in principle to compute the underlying mean number of years a person stays at their first job. Compare it to the faulty estimators based on using (a) only the non-censored items and (b) the censored times but ignoring the censoring mechanism.

10.5 Consider Example 10.3, where we observe series-system lifetimes of a series system. We observe n different systems that are each made of k_i identical components ($i = 1, \ldots, n$) with lifetime distribution F. The lifetime data is denoted $(x_1, \ldots, x_n)$ and are possibly right censored. Show that if we let $\tau_j = \tilde{k}_j + \ldots + \tilde{k}_n$, the likelihood can be expressed as (10.3), and solve for the NPMLE.

10.6 Suppose the i.i.d. sequence $X_1, \ldots, X_n \sim \mathcal{E}xp(\lambda)$ represents failure times in a component reliability test. Along with the failure times, suppose we also observe m i.i.d. right-censored times $Y_1, \ldots, Y_m$. Find the maximum likelihood estimator for λ, and show that it is based on $T = \sum^n X_i + \sum^m Y_i$, which is called the *total time on test* statistic.

10.7 Suppose we observe m different k-out-of-n systems and each system contains i.i.d. components (with distribution F) and the ith system contains n_i components. Set up the nonparametric likelihood function for F based on the n system lifetimes (but do not solve the likelihood).

10.8 Go to the link below to download survival times for 87 people with lupus nephritis. They were followed for 15+ or more years after an initial renal biopsy. The *duration* variable indicates how long the patient had the disease before the biopsy; construct the Kaplan–Meier estimator for survival, ignoring the duration variable.

`http://lib.stat.cmu.edu/datasets/lupus`

10.9 Recall Exercise 6.3 based on 100 measurements of the speed of light in air. Use empirical likelihood to construct a 90% confidence interval for the mean and median.

`http://www.itl.nist.gov/div898/strd/univ/data/`
`Michelso.dat`

10.10 Suppose the ELR for the mean was equal to $R(\mu) = \mu \mathbf{1}(0 \leq \mu \leq 1) + (2 - \mu)\mathbf{1}(1 \leq \mu \leq 2)$. Find a 95% confidence interval for μ.

10.11 The data set `kidney` is from the `survival` package and contains recurrence times (in days) to infection for kidney patients using portable dialysis equipment. Run the following code in R, and use the `summary` function to compare the probability or recurrence at 30 and 60 days:

```
kidney.fit <- survfit(Surv(time=kidney$time,
    + event=kidney$status) ~ 1, type="kaplan-meier")
```

10.12 The *receiver operating characteristic* (ROC) curve is a statistical tool to compare diagnostic tests. Suppose we have a sample of measurements (scores) $X_1, \ldots, X_n$ from a diseased population $F(x)$ and a sample of $Y_1, \ldots, Y_m$ from a healthy population $G(y)$. The healthy population has lower scores, so an observation is categorized as being diseased if it exceeds a given threshold value, e.g. if $X > c$. Then the rate of false-positive results would be $P(Y > c)$. The ROC curve is defined as the plot of $R(p) = F(G^{-1}(p))$. The ROC estimator can be computed using the plug-in principle:

$$\hat{R}(p) = F_n(G_m^{-1}(p)).$$

A common test to see if the diagnostic test is effective is to see if $R(p)$ remains well above 0.5 for $0 \leq p \leq 1$. The *area under the curve* (AUC) is defined as

$$\text{AUC} = \int_0^1 R(p) \, dp.$$

Show that $\text{AUC} = P(X \leq Y)$ and that, by using the plug-in principle, the sample estimator of the AUC is equivalent to the Mann–Whitney two-sample test statistic.

References

Brown, J. S. (1997), *What It Means to Lead*, Fast Company, 7, New York: Mansueto Ventures, LLC.

Cox, D. R. (1972), "Regression Models and Life Tables," *Journal of the Royal Statistical Society (B)*, 34, 187–220.

Crowder, M. J., Kimber, A. C., Smith, R. L., and Sweeting, T. J. (1991), *Statistical Analysis of Reliability Data*, London: Chapman & Hall.

Fuller Jr., E. R., Frieman, S. W., Quinn, J. B., Quinn, G. D., and Carter, W. C. (1994), "Fracture Mechanics Approach to the Design of Glass Aircraft Windows: A Case Study," *SPIE Proceedings*, Vol. 2286, Bellingham, WA: Society of Photo-Optical Instrumentation Engineers (SPIE).

Greenwood, M. (1926), "The Natural Duration of Cancer," in *Reports on Public Health and Medical Subjects*, 33, London: H. M. Stationery Office.

Hall, W. J., and Wellner, J. A. (1980), "Confidence Bands for a Survival Curve," *Biometrika*, 67, 133–143.

Kaplan, E. L., and Meier, P. (1958), "Nonparametric Estimation from Incomplete Observations," *Journal of the American Statistical Association*, 53, 457–481.

Kiefer, J., and Wolfowitz, J. (1956), "Consistency of the Maximum Likelihood Estimator in the Presence of Infinitely Many Incidental Parameters," *Annals of Mathematical Statistics*, 27, 887–906.

Lawless, J. F. (1982), *Statistical Models and Methods for Lifetime Data*, New York: Wiley.

Muenchow, G. (1986), "Ecological Use of Failure Time Analysis," *Ecology*, 67, 246–250.

Nair, V. N. (1984), "Confidence Bands for Survival Functions with Censored Data: A Comparative Study," *Technometrics*, 26, 265–275.

Owen, A. B. (1988), "Empirical Likelihood Ratio Confidence Intervals for a Single Functional," *Biometrika*, 75, 237–249.

Owen, A. B. (2001), *Empirical Likelihood*, Boca Raton, FL: Chapman & Hall/CRC.

Stigler, S. M. (1994), "Citations Patterns in the Journals of Statistics and Probability," *Statistical Science*, 9, 94–108.

11

Density Estimation

George McFly: Lorraine, my density has brought me to you.
Lorraine Baines: What?
George McFly: Oh, what I meant to say was…
Lorraine Baines: Wait a minute, don't I know you from somewhere?
George McFly: Yes. Yes. I'm George, George McFly.
I'm your density. I mean… your destiny.

From the movie *Back to the Future*, 1985

Probability density estimation goes hand in hand with nonparametric estimation of the cumulative distribution function discussed in Chapter 10. There, we noted that the density function provides a better visual summary of how the random variable is distributed across its support. Symmetry, skewness, disperseness, and unimodality are just a few of the properties that are ascertained when we visually scrutinize a probability density plot.

Recall, for continuous i.i.d. data, the *empirical density function* places probability mass $1/n$ on each of the observations. While the plot of the empirical *distribution* function (EDF) emulates the underlying distribution function, for continuous distributions, the empirical density function takes no shape beside the changing frequency of discrete jumps of $1/n$ across the domain of the underlying distribution – see Figure 11.2a.

11.1   Histogram

The histogram provides a quick picture of the underlying density by weighting fixed intervals according the their relative frequency in the data. Pearson (1895) coined the term for this empirical plot of the data, but its history goes as far back as the eighteenth century. William Playfair (1786) is credited with the first

Nonparametric Statistics with Applications to Science and Engineering with R, Second Edition.
Paul Kvam, Brani Vidakovic, and Seong-joon Kim.
© 2023 John Wiley & Sons, Inc. Published 2023 by John Wiley & Sons, Inc.

Figure 11.1 Playfair's 1786 bar chart of wheat prices in England.

appearance of a bar chart (see Figure 11.1) that plotted the price of wheat in England through the seventeenth and eighteenth centuries.

In R, the function hist() in the R's base graphic system will create a histogram using the input vector x. Also advanced graphic system such as ggplot2 package produces the same result using geom_histogram() function. Figure 11.2 shows (a) the empirical density function where vertical bars represent Dirac's point masses at the observations and (b) a histogram for a set of 30 generated $\mathcal{N}(0,1)$ random variables. Obviously, by aggregating observations within the disjoint intervals, we get a better, *smoother* visual construction of the frequency distribution of the sample:

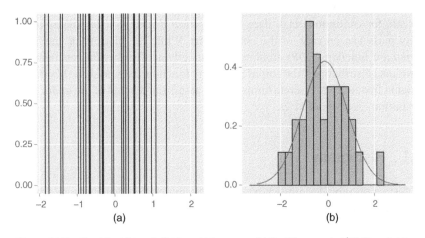

Figure 11.2 Empirical "density" (a) and histogram (b) for 30 normal $\mathcal{N}(0,1)$ variables.

```
> x1 <- rnorm(30)
> x2 <- seq(-3,3,by=0.01)
>
> ggplot() + geom_histogram(aes(x=x1),fill="gray",col="black",
+ binwidth=0.3)
>
> ggplot() + geom_histogram(aes(x=x1,y=..density..),fill="gray",col="black",
+ binwidth=0.3) + geom_line(aes(x=x2,y=dnorm(xx,mean(x),sd(x))),lwd=2)
>
> # R Base graphics
> hist(x)
>
> hist(x,freq=FALSE,col="gray")
> lines(xx,dnorm(xx,mean(x),sd(x)),type="l",col=2,lwd=2)
```

The histogram represents a rudimentary smoothing operation that provides the user a way of visualizing the true empirical density of the sample. Still, this simple plot is primitive and depends on the subjective choices the user makes for bin widths and number of bins. With larger data sets, we can increase the number of bins while still keeping average bin frequency at a reasonable number, say, 5 or more. If the underlying data are continuous, the histogram appears less discrete as the sample size (and number of bins) grows, but with smaller samples, the graph of binned frequency counts will not pick up the nuances of the underlying distribution.

The R codes below plot a histogram with n bins (or k bin width) along with the best fitting normal density curve. Figure 11.3 shows how the appearance of continuity changes as the histogram becomes more refined (with more bins of smaller bin width). Of course, we do not have such luxury with smaller- or medium-sized

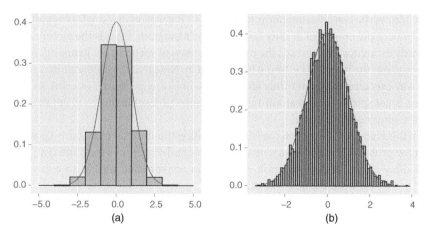

Figure 11.3 Histograms with normal fit of 5000 generated variables using (a) 10 bins and (b) 50 bins.

data sets and are more likely left to ponder the question of underlying normality with a sample of size 30, as in Figure 11.2b:

```
> x <- rnorm(5000)
> xx<-seq(-3,3,by=0.01)
> ggplot() + geom_histogram(aes(x=x,y=..density..),fill="gray",col="black",
+ binwidth=1) + geom_line(aes(x=xx,y=dnorm(xx,mean(x),sd(x))))
>
> ggplot() + geom_histogram(aes(x=x,y=..density..),fill="gray",col="black",
+ binwidth=0.1) + geom_line(aes(x=xx,y=dnorm(xx,mean(x),sd(x))))
>
> # R base graphics
> hist(x,freq=FALSE,col="gray",nclass=10)
> xx<-seq(-3,3,by=0.01)
> lines(xx,dnorm(xx,mean(x),sd(x)),type="l",col=2,lwd=2)
>
> hist(x,freq=FALSE,col="gray",nclass=50)
> xx<-seq(-3,3,by=0.01)
> lines(xx,dnorm(xx,mean(x),sd(x)),type="l",col=2,lwd=2)
```

If you have a dearth of scruples, the histogram provides for you many opportunities to mislead your audience, as you can make the distribution of the data appear differently by choosing your own bin widths centered at a set of points arbitrarily left to your own choosing. If you are completely untrustworthy, you might even consider making bins of unequal length. That is sure to support a conjectured but otherwise unsupportable thesis with your data and might jump-start a promising career for you in politics.

11.2 Kernel and Bandwidth

The idea of the *density estimator* is to spread out the weight of a single observation in a plot of the empirical density function. The histogram, then, is the picture of a density estimator that spreads the probability mass of each sample item *uniformly* throughout the interval (i.e. bin) it is observed in. Note that the observations are in no way expected to be uniformly spread out within any particular interval, so the mass is not spread equally around the observation unless it happens to fall exactly in the center of the interval.

In this chapter, we focus on the kernel density estimator that more fairly spreads out the probability mass of each observation, not arbitrarily in a fixed interval, but smoothly around the observation, typically in a symmetric way. With a sample $X_1, \ldots, X_n$, we write the density estimator

$$\hat{f}(x) = \frac{1}{nh_n} \sum_{i=1}^{n} K\left(\frac{x - x_i}{h_n}\right), \tag{11.1}$$

for $X_i = x_i, i = 1, \ldots, n$. The *kernel function K* represents how the probability mass is assigned, so for the histogram, it is just a constant in any particular interval. The smoothing function h_n is a positive sequence of bandwidths analogous to the bin width in a histogram.

The kernel function K has five important properties:

1. $K(x) \geq 0 \ \forall x$
2. $K(x) = K(-x)$ for $x > 0$
3. $\int K(u)du = 1$
4. $\int uK(u)du = 0$
5. $\int u^2K(u)du = \sigma_K^2 < \infty$

Figure 11.4 shows four basic kernel functions:

1. Normal (or Gaussian) kernel $K(x) = \phi(x)$.
2. Triangular kernel $K(x) = c^{-2}(c - |x|) \ \mathbf{1}(-c < x < c), \ c > 0$.
3. Epanechnikov kernel (described below).
4. Box kernel, $K(x) = \mathbf{1}(-c < x < c)/(2c), \ c > 0$.

While K controls the shape, h_n controls the spread of the kernel. The accuracy of a density estimator can be evaluated using the mean integrated squared error, defined as

$$\text{MISE} = \mathbb{E} \left(\int (f(x) - \hat{f}(x))^2 \, dx \right)$$

$$= \int \text{Bias}^2(\hat{f}(x))dx + \int \mathbb{V}\text{ar}(\hat{f}(x))dx. \tag{11.2}$$

To find a density estimator that minimizes the MISE under the five mentioned constraints, we also will assume that $f(x)$ is continuous (and twice differentiable), $h_n \to 0$ and $nh_n \to \infty$ as $n \to \infty$. Under these conditions, it can be shown that

$$\text{Bias}(\hat{f}(x)) = \frac{h_n^2 \sigma_K^2}{2}f''(x) + O(h_n^4) \text{ and}$$

$$\mathbb{V}\text{ar}(\hat{f}(x)) = \frac{f(x)R(K)}{nh_n} + O(n^{-1}), \tag{11.3}$$

where $R(g) = \int g(u)^2 \, du$.

We determine (and minimize) the MISE by our choice of h_n. From the equations in (11.3), we see that there is a tradeoff. Choosing h_n to reduce bias will increase the variance, and vice versa. The choice of bandwidth is important in the construction of $\hat{f}(x)$. If h is chosen to be small, the subtle nuances in the main part of the density will be highlighted, but the tail of the distribution will be unseemly bumpy. If h is

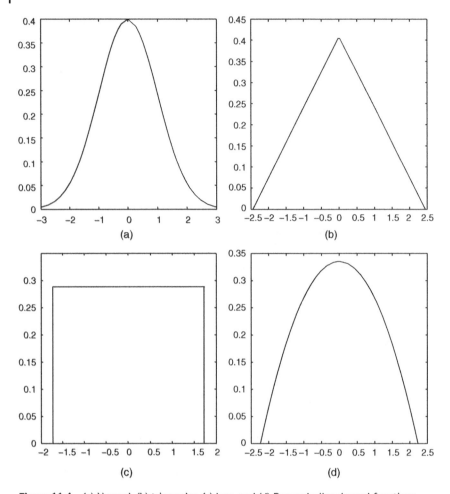

Figure 11.4 (a) Normal, (b) triangular, (c) box, and (d) Epanechnikov kernel functions.

chosen large, the tails of the distribution are better handled, but we fail to see important characteristics in the middle quartiles of the data.

By substituting in the bias and variance in the formula for (11.2), we minimize MISE with

$$h_n^* = \left(\frac{R(K)}{\sigma_K^4 R(f'')} \right)^{1/5} n^{-1/5}.$$

At this point, we can still choose $K(x)$ and insert a "representative" density for $f(x)$ to solve for the bandwidth. Epanechnikov (1969) showed that, upon substituting in $f(x) = \phi(x)$, the kernel that minimizes MISE is

$$K_E(x) = \begin{cases} \frac{3}{4}(1-x^2), & |x| \le 1, \\ 0, & |x| > 1. \end{cases}$$

The resulting bandwidth becomes $h_n^* \approx 1.06\hat{\sigma}n^{-1/5}$, where $\hat{\sigma}$ is the sample standard deviation. This choice relies on the approximation of σ for $f(x)$. Alternative approaches, including cross-validation, lead to slightly different answers.

Adaptive kernels were derived to alleviate this problem. If we use a more general smoothing function tied to the density at x_j, we could generalize the density estimator as

$$\hat{f}(x) = \frac{1}{n}\sum_{i=1}^{n}\frac{1}{h_{n,i}}K\left(\frac{x-x_i}{h_{n,i}}\right). \tag{11.4}$$

This is an advanced topic in density estimation, and we will not further pursue learning more about optimal estimators based on adaptive kernels here. We will also leave out details about estimator limit properties and instead point out that if h_n is a decreasing function of n, under some mild regularity conditions, $|\hat{f}(x) - f(x)| \xrightarrow{P} 0$. For details and more advanced topics in density estimation, see Silverman (1986) and Efromovich (1999).

The (univariate) density estimator from R called

```
density(data)
```

is illustrated in Figure 11.5 using a sample of seven observations. The default estimate is based on a Gaussian kernel; to use another kernel, just enter "rectangular," "triangular," or "epanechnikov" (see code below). Figure 11.5 shows how the normal kernel compares with the (a) rectangular, (2) triangle, and (c) Epanechnikov kernels. Figure 11.6 shows the density estimator using the same data based on the normal kernel, but using five different bandwidth selectors. Note the optimal

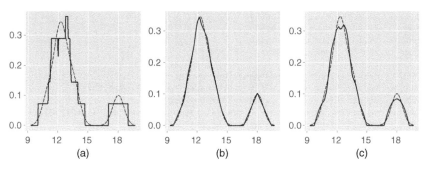

Figure 11.5 Density estimation for sample of size $n = 7$ using various kernels: (all) Gaussian, (a) rectangular, (b) triangular, and (c) Epanechnikov.

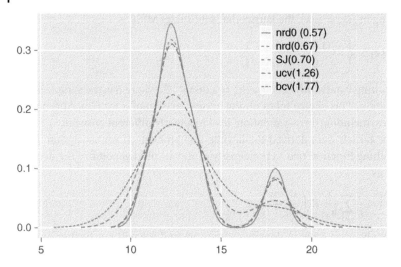

Figure 11.6 Density estimation for sample of size $n = 7$ using various bandwidth selectors.

bandwidth (0.5689) for the default selector (bw.nrd0) can be found by looking the result in the command line:

```
> data1 <- c(11,12,12.2,12.3,13,13.7,18);
> data2 <- c(50,21,25.5,40.0,41,47.6,39);
> ker1 <- density(data1,kernel="gaussian");
> ker2 <- density(data1,kernel="rectangular");
> ker1
Call:

          density.default(x = data1, kernel = "gaussian")

Data: data1 (7 obs.);    Bandwidth 'bw' = 0.5689
        x                   y
 Min.    : 9.293    Min.    :0.000162
 1st Qu.:11.897    1st Qu.:0.008094
 Median :14.500    Median :0.056540
 Mean   :14.500    Mean    :0.095902
 3rd Qu.:17.103    3rd Qu.:0.154429
 Max.:19.707       Max.:0.345574


> fit <- density(data1,kernel="gaussian",bw="nrd0")
> dat <- data.frame(x=fit$x,y=fit$y,group=rep("nrd0 (0.57)",512))
> fit <- density(data1,kernel="gaussian",bw="nrd")
> dat <- rbind(dat,data.frame(x=fit$x,y=fit$y,group=rep("nrd (0.67)",512)))
> fit <- density(data1,kernel="gaussian",bw="SJ")
> dat <- rbind(dat,data.frame(x=fit$x,y=fit$y,group=rep("SJ (0.70)",512)))
> fit <- density(data1,kernel="gaussian",bw="ucv")
> dat <- rbind(dat,data.frame(x=fit$x,y=fit$y,group=rep("ucv (1.26)",512)))
> fit <- density(data1,kernel="gaussian",bw="bcv")
> dat <- rbind(dat,data.frame(x=fit$x,y=fit$y,group=rep("bcv (1.77)",512)))
>
> ggplot(aes(x=x,y=y,group=group,col=group,shape=group,lty=group),data=dat) +
+ geom_line(lwd=0.6) + theme(legend.position=c(0.71,0.8),legend.background=
+ element_rect(fill=NA),legend.title=element_blank())+xlab("")+ylab("")
```

Example 11.1 Radiation Measurements. In some situations, the experimenter might prefer to subjectively decide on a proper bandwidth instead of the objective choice of bandwidth that minimizes MISE. If outliers and subtle changes in the probability distribution are crucial in the model, a more jagged density estimator (with a smaller bandwidth) might be preferred to the optimal one. In Davies and Gather (1993), 2001 radiation measurements were taken from a balloon at a height of 100 ft. Outliers occur when the balloon rotates, causing the balloon's ropes to block direct radiation from the Sun to the measuring device. Figure 11.7 shows two density estimates of the raw data, one based on a narrow bandwidth and the other more smooth density based on a bandwidth 10 times larger (0.01–0.1). Both densities are based upon a normal (Gaussian) kernel. While the more jagged estimator does show the mode and skew of the density as clearly as the smoother estimator, outliers are more easily discerned:

```
> balloon <- read.table("balloon.txt");
> ker1 <- density(balloon[,1],bw=0.01);
> ker2 <- density(balloon[,1],bw=0.1);
>
> p <- ggplot() + geom_line(aes(x=ker1$x,y=ker1$y))
> p <- p + geom_line(aes(x=ker2$x,y=ker2$y),lty=2)
> p <- p + xlim(c(1.5,2.5)) + ylim(c(0,5))
> print(p)
```

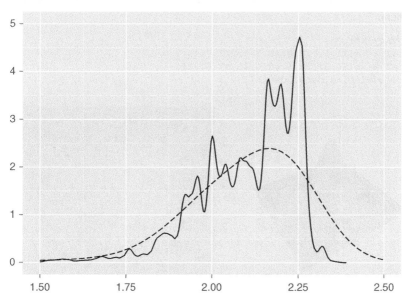

Figure 11.7 Density estimation for 2001 radiation measurements using bandwidths `band = 0.1` (dashed line) and `band=0.01` (solid line).

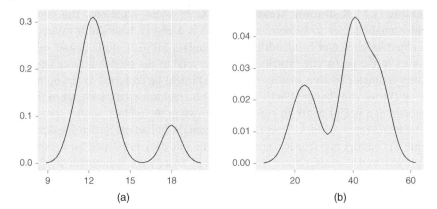

Figure 11.8 (a) Univariate density estimator for first variable. (b) Univariate density estimator for second variable.

11.2.1 Bivariate Density Estimators

To plot density estimators for bivariate data, two-dimensional density estimates can be constructed using R function kde in ks package, noting that both x and y, the vectors designating plotting points for the density, must be of the same size.

In Figure 11.8, (univariate) density estimates are plotted for the seven observations (data1, data2). In Figure 11.9, R functions persp and image are used to produce three-dimensional plots for the seven bivariate observations (coupled together):

```
> library(ks)
> data1 <- c(11,12,12.2,12.3,13,13.7,18);
> data2 <- c(50,21,25.5,40.0,41,47.6,39);
> ker<-kde(cbind(data1,data2))
```

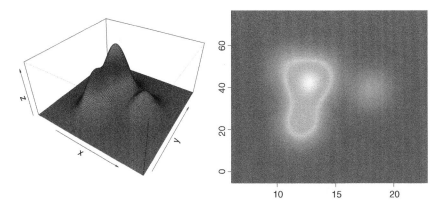

Figure 11.9 Bivariate Density estimation for sample of size *n* = 7.

```
>
> par(mfrow=c(1,2),mar=c(3,3,1,1))
> persp(x,y,z,theta=33,phi=35,shade=0.1,expand=0.5,lwd=0.2,cex=0.6)
> image(x,y,z,col=gray((32:0)/32));box()
```

11.3 Exercises

11.1 Which of the following serve as kernel functions for a density estimator? Prove your assertion one way or the other:
a) $K(x) = \mathbf{1}(-1 < x < 1)/2$.
b) $K(x) = \mathbf{1}(0 < x < 1)$.
c) $K(x) = 1/x$.
d) $K(x) = \frac{3}{2}(2x + 1)(1 - 2x)\,\mathbf{1}\left(-\frac{1}{2} < x < \frac{1}{2}\right)$.
e) $K(x) = 0.75(1 - x^2)\,\mathbf{1}(-1 < x < 1)$.

11.2 With a data set of 12, 15, 16, 20, estimate $p^* = P$ (observation is less than 15) using a density estimator based on a normal (Gaussian) kernel with $h_n = \sqrt{3/n}$. Use hand calculations instead of the R function.

11.3 Generate 12 observations from a mixture distribution, where half of the observations are from $\mathcal{N}(0,1)$ and the other half are from $\mathcal{N}(1, 0.64)$. Use the R function `density` to create a density estimator. Change bandwidth to see its effect on the estimator. Repeat this procedure using 24 observations instead of 12.

11.4 Suppose you have chosen kernel function $K(x)$ and smoothing function h_n to construct your density estimator, where $-\infty < K(x) < \infty$. What should you do if you encounter a right censored observation? For example, suppose the right-censored observation is ranked m lowest out of n, $m \le n - 1$.

11.5 Recall Exercise 6.3 based on 100 measurements of the speed of light in air. In that chapter, we tested the data for normality. Use the same data to construct a density estimator that you feel gives the best visual display of the information provided by the data. What parameters did you choose? The data can be downloaded from

http://www.itl.nist.gov/div898/strd/univ/data/
Michelso.dat

11.6 Go back to Exercise 10.7, where a link is provided to download right-censored survival times for 87 people with lupus nephritis. Construct a density estimator for the survival, ignoring the duration variable.

http://lib.stat.cmu.edu/datasets/lupus

References

Davies, L., and Gather, U. (1993), "The Identification of Multiple Outliers" (discussion paper), *Journal of the American Statistical Association*, 88, 782–792.

Efromovich, S. (1999), *Nonparametric Curve Estimation: Methods, Theory and Applications*, New York: Springer-Verlag.

Epanechnikov, V. A. (1969), "Nonparametric Estimation of a Multivariate Probability Density," *Theory of Probability and its Applications*, 14, 153–158.

Pearson, K. (1895), "Contributions to the Mathematical Theory of Evolution II," *Philosophical Transactions of the Royal Society of London (A)*, 186, 343–414.

Playfair, W. (1786), *Commercial and Political Atlas: Representing, by Copper-Plate Charts, the Progress of the Commerce, Revenues, Expenditure, and Debts of England, during the Whole of the Eighteenth Century*. London: Corry.

Silverman, B. (1986), *Density Estimation for Statistics and Data Analysis*, New York: Chapman & Hall.

12

Beyond Linear Regression

Essentially, all models are wrong, but some models are useful.

George Box, from *Empirical Model-Building and Response Surfaces*

Statistical methods using linear regression are based on the assumptions that errors, and hence the regression responses, are normally distributed. Variable transformations increase the scope and applicability of linear regression toward real applications, but many modeling problems cannot fit in the confines of these model assumptions.

In some cases, the methods for linear regression are robust to minor violations of these assumptions. This has been shown in diagnostic methods and simulation. In examples where the assumptions are more seriously violated; however, estimation and prediction based on the regression model are biased. Some *residuals* (measured difference between the response and the model's estimate of the response) can be overly large in this case and wield a large influence on the estimated model. The observations associated with large residuals are called outliers, which cause error variances to inflate and reduce the power of the inferences made.

In other applications, parametric regression techniques are inadequate in capturing the true relationship between the response and the set of predictors. General "curve fitting" techniques for such data problems are introduced in Chapter 13, where the model of the regression is unspecified and not necessarily linear.

In this chapter, we look at simple alternatives to basic least-squares regression. These estimators are constructed to be less sensitive to the outliers that can affect regular regression estimators. *Robust* regression estimators are made specifically for this purpose. Nonparametric or *rank* regression relies more on the order relations in the paired data rather than the actual data measurements, and *isotonic* regression represents a nonparametric regression model with simple constraints built in, such as the response being monotone with respect to one or more inputs.

Nonparametric Statistics with Applications to Science and Engineering with R, Second Edition.
Paul Kvam, Brani Vidakovic, and Seong-joon Kim.
© 2023 John Wiley & Sons, Inc. Published 2023 by John Wiley & Sons, Inc.

Finally, we overview generalized linear models that, although parametric, encompass some nonparametric methods, such as contingency tables, for example.

12.1 Least-Squares Regression

Before we introduce the less familiar tools of nonparametric regression, we will first review basic linear regression that is taught in introductory statistics courses. Ordinary least-squares regression is synonymous with parametric regression only because of the way the errors in the model are treated. In the simple linear regression case, we observe n independent pairs (X_i, Y_i), where the linear regression of Y on X is the conditional expectation $\mathbb{E}(Y|X)$. A characterizing property of normally distributed X and Y is that the conditional expectation is linear, that is, $\mathbb{E}(Y|X) = \beta_0 + \beta_1 X$.

Standard least-squares regression estimates are based on minimizing squared errors $\sum_i (Y_i - \hat{Y}_i)^2 = \sum_i (Y_i - [\beta_0 + \beta_1 X_i])^2$ with respect to the parameters β_1 and β_0. The least-squares solutions are

$$\hat{\beta}_1 = \frac{\sum_{i=1}^{n}(X_i - \bar{X})(Y_i - \bar{Y})}{\sum_{i=1}^{n}(X_i - \bar{X})^2} \tag{12.1}$$

$$= \frac{\sum_{i=1}^{n}(X_i Y_i - n\bar{X}\bar{Y})}{\sum_{i=1}^{n} X_i^2 - n\bar{X}^2}.$$

$$\hat{\beta}_0 = \bar{Y} - \hat{\beta}_1 \bar{X}. \tag{12.2}$$

This solution is familiar from elementary parametric regression. In fact, $(\hat{\beta}_0, \hat{\beta}_1)$ are the MLEs of (β_0, β_1) in the case the errors are normally distributed. However, with the minimized least-squares approach (treating the sum of squares as a "loss function"), no such assumptions were needed, so the model is essentially nonparametric. However, in ordinary regression, the distributional properties of $\hat{\beta}_0$ and $\hat{\beta}_1$ that are used in constructing tests of hypothesis and confidence intervals are pinned to assuming these errors are homogenous and normal.

12.2 Rank Regression

The truest nonparametric method for modeling bivariate data is Spearman's correlation coefficient that has no specified model (between X and Y) and no assumed distributions on the errors. Regression methods, by their nature, require additional model assumptions to relate a random variable X to Y via a function for the regression of $\mathbb{E}(Y|X)$. The technique discussed here is nonparametric except for the chosen regression model; error distributions are left to be arbitrary. Here we assume the linear model

$$Y_i = \beta_0 + \beta_1 x_i, \quad i = 1, \ldots, n$$

is appropriate, and, using the squared errors as a loss function, we compute $\hat{\beta}_0$ and $\hat{\beta}_1$ as in (12.2) and (12.1) as the least-squares solution.

Suppose we are interested in testing H_0 that the population slope is equal to β_{10} against the three possible alternatives, $H_1 : \beta_1 > \beta_{10}$, $H_1 : \beta_1 < \beta_{10}$, and $H_1 : \beta_1 \neq \beta_{10}$. Recall that in standard least-squares regression, the Pearson coefficient of linear correlation ($\hat{\rho}$) between the Xs and Ys is connected to β_1 via

$$\hat{\rho} = \hat{\beta}_1 \cdot \frac{\sqrt{\sum_i X_i^2 - n(\bar{X})^2}}{\sqrt{\sum_i Y_i^2 - n(\bar{Y})^2}}.$$

To test the hypothesis about the slope, first calculate $U_i = Y_i - \beta_{10} X_i$, and find the Spearman coefficient of rank correlation $\hat{\rho}$ between the X_is and the U_is. For the case in which $\beta_{10} = 0$, this is no more than the standard Spearman correlation statistic. In any case, under the assumption of independence, $(\hat{\rho} - \rho)\sqrt{n-1} \sim \mathcal{N}(0,1)$, and the tests against alternatives H_1 are

Alternative	p-Value
$H_1 : \beta_1 \neq \beta_{10}$	$p = 2P(Z \geq |\hat{\rho}|\sqrt{n-1})$
$H_1 : \beta_1 < \beta_{10}$	$p = P(Z \leq \hat{\rho}\sqrt{n-1})$
$H_1 : \beta_1 > \beta_{10}$	$p = P(Z \geq \hat{\rho}\sqrt{n-1})$

where $Z \sim \mathcal{N}(0,1)$. The table represents a simple nonparametric regression test based only on Spearman's correlation statistic.

Example 12.1 Active Learning. Kvam (2000) examined the effect of active learning methods on student retention by examining students of an introductory statistics course eight months after the course finished. For a class taught using an emphasis on active learning techniques, scores were compared with the equivalent final exam scores:

Exam 1	14	15	18	16	17	12	17	15	17	14	17	13	15	18	14
Exam 2	14	10	11	8	17	9	11	13	12	13	14	11	11	15	9

Scores for the first (x-axis) and second (y-axis) exam are plotted in Figure 12.1a for 15 active-learning students. In Figure 12.1b, the solid line represents the

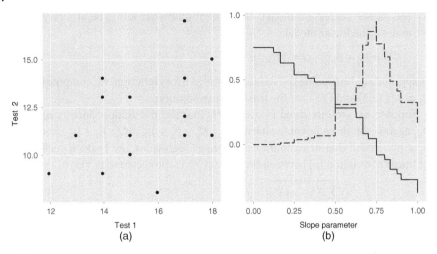

Figure 12.1 (a) Plot of test #1 scores (during term) and test #2 scores (eight months after). (b) Plot of Spearman correlation coefficient (*solid*) and corresponding *p*-value (*dashed*) for nonparametric test of slope for $-1 \le \beta_{10} \le 1$.

computed Spearman correlation coefficient for X_i and $U_i = Y_i - \beta_{10}X_i$ with β_{10} varying from -1 to 1. The dashed line is the *p*-value corresponding to the test $H_1 : \beta_1 \ne \beta_{10}$. For the hypothesis $H_0 : \beta_1 \ge 0$ versus $H_1 : \beta_1 < 0$, the *p*-value is about 0.12 (the *p*-value for the two-sided test, from the graph, is about 0.24).

Note that at $\beta_{10} = 0.498$, $\hat{\rho}$ is 0, and at $\beta_{10} = 0$, $\hat{\rho} = 0.387$. The *p*-value is highest at $\beta_{10} = 0.5$ and less than 0.05 for all values of β_{10} less than -0.332:

```
> trad1 <- c(18,14,14,18,18,15,18,18,18,9,15,12,17,18,15,13,17,18,14,13,
+ 16,14,15);
> trad2 <- c(11,13,6,16,14,12,17,16,13,1,10,6,14,6,14,7,14,12,7,6,11,8,13);
> act1 <- c(14,15,18,16,17,12,17,15,17,14,17,13,15,18,14);
> act2 <- c(14,10,11,8,17,9,11,13,12,13,14,11,11,15,9);
> trad <- cbind(trad1,trad2);
> act <- cbind(act1,act2);
> n0 <- 1000
> r <- rep(0,n0); p <- rep(0,n0); b <- rep(0,n0)
> for(i in 1:n0){
+ b[i] <- (i-(n0/2))/(n0/2)
+ ret <- cor.test(act1,act2-b[i]*act1,method="spearman")
+ r[i] <- ret$estimate
+ p[i] <- ret$p.value
+ }
>
> ggplot()+geom_point(aes(x=act1,y=act2),pch=16,size=5)
> ggplot()+geom_step(aes(x=seq(0,1,length=1000),y=r),lwd=0.8)+
+ geom_step(aes(x=seq(0,1,length=1000),y=p),lty=2,lwd=0.8)
```

12.2.1 Sen–Theil Estimator of Regression Slope

Among n bivariate observations, there are $\binom{n}{2}$ different pairs (X_i, Y_i) and (X_j, Y_j), $i \neq j$. For each pair (X_i, Y_i) and (X_j, Y_j), $1 \leq i < j \leq n$; we find the corresponding slope:

$$S_{ij} = \frac{Y_j - Y_i}{X_j - X_i}.$$

Compared with ordinary least-squares regression, a more robust estimator of the slope parameter β_1 is

$$\tilde{\beta}_1 = \text{median}\{S_{ij}, \ 1 \leq i < j \leq n\}.$$

Corresponding to the least-squares estimate, let

$$\tilde{\beta}_0 = \text{median}\{Y\} - \tilde{\beta}_1 \, \text{median}\{X\}.$$

Example 12.2 If we take the integers $\{1, \ldots, 20\}$ as our set of predictors $X_1, \ldots, X_{20}$, let Y be $2X + \epsilon$, where ϵ is a standard normal variable. Next, we change both Y_1 and Y_{20} to be outliers with value 20 and compare the ordinary least-squares regression with the more robust nonparametric method in Figure 12.2:

```
> x <- 1:20
> y <- 2*(1:20) + rnorm(20)
> y[c(1,20)]<-20
```

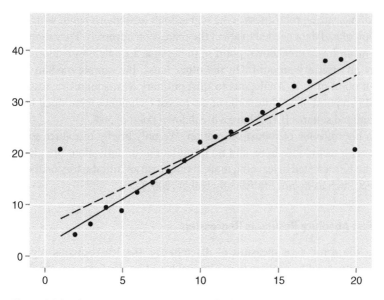

Figure 12.2 Regression: least squares (*dashed*) and nonparametric (*solid*).

```
> coef1 <- coef(lm(y~x))
> coef2 <- c(median(y)-median(x)*median(diff(y)/diff(x)),
+ median(diff(y)/diff(x)))
>
> dat <- data.frame(x=x,y=y)
> p <- ggplot(data=dat,aes(x=x,y=y)) + geom_point(data=dat,aes(x,y),col="black")
> p <- p + xlim(c(0,20)) + ylim(c(0,45)) + xlab("") + ylab("")
> p <- p + geom_line(aes(x=x,y=coef1[1]+coef1[2]*x),lty=2)
> p <- p + geom_line(aes(x=x,y=coef2[1]+coef2[2]*x),lty=1)
> p <- p + theme(axis.text.x=element_text(color="black"),axis.text.y=
+ element_text(color="black"))
> print(p)
```

12.3 Robust Regression

"Robust" estimators are ones that retain desired statistical properties even when the assumptions about the data are slightly off. Robust linear regression represents a modeling alternative to regular linear regression in the case the assumptions about the error distributions are potentially invalid. In the simple linear case, we observe n independent pairs (X_i, Y_i), where the linear regression of Y on X is the conditional expectation $\mathbb{E}(Y|X) = \beta_0 + \beta_1 X$.

For rank regression, the estimator of the regression slope is considered to be robust because no single observation (or small group of observations) will have an significant influence on estimated model; the regression slope picks out the median slope out of the $\binom{n}{2}$ different pairings of data points.

One way of measuring robustness is the regression's *breakdown point*, which is the proportion of bad data needed to affect the regression adversely. For example, the sample mean has a breakdown point of 0, because a single observation can change it by an arbitrary amount. On the other hand, the sample median has a breakdown point of 50%. Analogous to this, ordinary least-squares regression has a breakdown point of 0, while some of the robust techniques mentioned here (e.g. least trimmed squares [LTS]) have a breakdown point of 50%.

There is a big universe of robust estimation. We only briefly introduce some robust regression techniques here, and no formulations or derivations are given. A student who is interested in learning more should read an introductory textbook on the subject, such as *Robust Statistics* by Huber (2009).

12.3.1 Least Absolute Residuals Regression

By squaring the error as a measure of discrepancy, the least-squares regression is more influenced by outliers than a model based on, for example, absolute deviation errors: $\sum_i |Y_i - \hat{Y}_i|$, which is called least absolute residuals regression. By minimizing errors with a loss function that is more "forgiving" to large deviations, this method is less influenced by these outliers. In place of least-squares techniques, regression coefficients are found from linear programming.

12.3.2 Huber Estimate

The concept of robust regression is based on a more general class of estimates $(\hat{\beta}_0, \hat{\beta}_1)$ that minimize the function

$$\sum_{i=1}^{n} \frac{\psi(Y_i - \hat{Y}_i)}{\sigma},$$

where ψ is a loss function and σ is a scale factor. If $\psi(x) = x^2$, we have regular least-squares regression, and if $\psi(x) = |x|$, we have least absolute residuals regression. A general loss function introduced by Huber (1973) is

$$\psi(x) = \begin{cases} x^2, & |x| < c, \\ 2c|x| - c^2, & |x| > c. \end{cases}$$

Depending on the chosen value of $c > 0$, $\psi(x)$ uses squared-error loss for small errors, but the loss function flattens out for larger errors.

12.3.3 Least Trimmed Squares Regression

LTS is another robust regression technique proposed by Rousseeuw (1985) as a robust alternative to ordinary least-squares regression. Within the context of the linear model $y_i = \beta' x_i$, $i = 1, \ldots, n$, the LTS estimator is represented by the value of β that minimizes $\sum_{i=1}^{h} r_{i:n}$. Here, x_i is a $p \times 1$ vector, $r_{i:n}$ is the ith order statistic from the squared residuals $r_i = (y_i - \beta' x_i)^2$, and h is a trimming constant ($n/2 \le h \le n$) chosen so that the largest $n - h$ residuals do not affect the model estimate. Rousseeuw and Leroy (1987) showed that the LTS estimator has its highest level of robustness when $h = \lfloor n/2 \rfloor + \lfloor (p+1)/2 \rfloor$. While choosing h to be low leads to a more robust estimator, there is a tradeoff of robustness for efficiency.

12.3.4 Weighted Least-Squares Regression

For some data, one can improve model fit by including a scale factor (weight) in the deviation function. Weighted least squares minimizes

$$\sum_{i=1}^{n} w_i (y_i - \hat{y}_i)^2,$$

where w_i are weights that determine how much influence each response will have on the final regression. With the weights in the model, we estimate β in the linear model with

$$\hat{\beta} = \left(X'WX\right)^{-1}X'Wy,$$

where X is the design matrix made up of the vectors x_i, y is the response vector, and W is a diagonal matrix of the weights $w_1, \ldots, w_n$. This can be especially helpful if the responses seem not to have constant variances. Weights that counter the effect of heteroskedasticity such as

$$w_i = m\left(\sum_{i=1}^{m}(y_i - \bar{y})^2\right)^{-1},$$

work well if your data contain a lot of replicates; here m is the number of replicates at y_i. To compute this in R, the function lm computes least-squares estimates with known covariance; for example, the output of

```
lm(y ~ x, weights=w)
```

returns the weighted least-squares solution to the simple linear model $y = \beta_0 + \beta_1 x$ with weight vector w.

12.3.5 Least Median Squares Regression

The least median squares (LMS) regression finds the line through the data that minimizes the median (rather than the mean) of the squares of the errors. While the LMS method is proven to be robust, it cannot be easily solved like a weighted least-squares problem. The solution must be solved by searching in the space of possible estimates generated from the data, which is usually too large to do analytically. Instead, randomly chosen subsets of the data are chosen so that an approximate solution can be computed without too much trouble. The R function in MASS package

```
lmsreg()
```

computes the LMS for small- or medium-sized data sets.

Example 12.3 Star Data. Data from Rousseeuw and Leroy (1987), p. 27, Table 3, are given in all panels of Figure 12.3 as a scatterplot of temperature versus light intensity for 47 stars. The first variable is the logarithm of the effective temperature at the surface of the star (Te), and the second one is the logarithm of its light intensity ($L/L0$). In sequence, the four panels in Figure 12.3 show plots

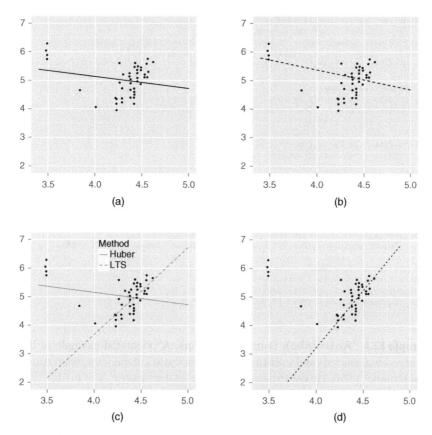

Figure 12.3 Star data with (a) Ordinary least squares (OLS) regression, (b) least absolute deviation, (c) Huber estimation and least trimmed squares, and (d) least median squares.

of the bivariate data with fitted regressions based on (a) least squares, (b) least absolute residuals, (c) Huber loss and least trimmed squares, and (d) least median squares. Observations far away from most of the other observations are called *leverage points*, effect of the leverage points:

```
> library(robustbase)
> library(MASS)
> library(quantreg)
>
> star <- data.frame(read.table("star.txt",col.names=c("no","x","y")))
> bols <- as.numeric(coef(lm(y~x,data=star)))
> blad <- as.numeric(coef(rq(y~x,data=star)))
> bhuber <- as.numeric(coef(rlm(y~x,data=star,scale.est="Huber",psi=psi.huber)))
```

```
> blts <- as.numeric(coef(ltsReg(y~x,data=star)))
> blms <- as.numeric(coef(lmsreg(y~x,data=star)))
> star2 <- data.frame(x=rep(x,5),y=c(bols[1]+bols[2]*x,blad[1]+blad[2]*x,
+ bhuber[1]+bhuber[2]*x,blts[1]+blts[2]*x,blms[1]+blms[2]*x),
+     method=c(rep("OLS",1),rep("LAD",1),rep("Huber",1),rep("LTS",1),rep("LMS",1)))
>
> # Ordinary Least Squares
> ggplot(data=subset(star2,method=="OLS"),aes(x=x,y=y)) +
+ geom_point(aes(x=x,y=y),data=star) + geom_line(aes(x=x,y=y),lty=1) +
+ xlim(c(3.4,5)) + ylim(c(2,7))
>
> # Least Absolute Deviation
> ggplot(data=subset(star2,method=="LAD"),aes(x=x,y=y)) +
+ geom_point(aes(x=x,y=y),data=star) + geom_line(aes(x=x,y=y),lty=2) +
+ xlim(c(3.4,5)) + ylim(c(2,7)) + xlab("") + ylab("")
>
> # Huber estimation and Least Trimmed Squares
> ggplot(data=subset(star2,method=="Huber"|method=="LTS"),aes(x=x,y=y)) +
+ geom_point(aes(x=x,y=y),data=star) + xlim(c(3.4,5)) +
+ geom_line(aes(x=x,y=y,lty=method,col=method,pch=method)) +
+ theme(legend.position=c(0.5,0.85),legend.background=element_rect(fill=NA))
>
> # Least Median Squares
> ggplot(data=subset(star2,method=="LMS"),aes(x=x,y=y)) +
+ geom_point(aes(x=x,y=y),data=star)+xlim(c(3.4,5)) +
+ geom_line(aes(x=x,y=y),lty=3) + ylim(c(2,7))
```

Example 12.4 Anscombe's Four Regressions. A celebrated example of the role of residual analysis and statistical graphics in statistical modeling was created by Anscombe (1973). He constructed four different data sets (X_i, Y_i), $i = 1, \ldots, 11$ that share the same descriptive statistics $(\bar{X}, \bar{Y}, \hat{\beta}_0, \hat{\beta}_1, MSE, R^2, F)$ necessary to establish linear regression fit $\hat{Y} = \hat{\beta}_0 + \hat{\beta}_1 X$. The following statistics are common for the four data sets:

Sample size N	11
Mean of X $(\bar{X})$	9
Mean of Y $(\bar{Y})$	7.5
Intercept $(\hat{\beta}_0)$	3
Slope $(\hat{\beta}_1)$	0.5
Estimator of σ, (s)	1.2366
Correlation $r_{X,Y}$	0.816

From inspection, one can ascertain that a linear model is appropriate for data set 1, but the scatter plots and residual analysis suggest that the data sets 2–4 are not amenable to linear modeling. Plotted in Figure 12.4 with the data are the lines for least-squares fit (*dotted*) and rank regression (*solid line*). See Exercise 12.1 for further examination of the three regression archetypes.

					Data set 1					
X	10	8	13	9	11	14	6	4	12	7	5
Y	8.04	6.95	7.58	8.81	8.33	9.96	7.24	4.26	10.84	4.82	5.68

					Data set 2					
X	10	8	13	9	11	14	6	4	12	7	5
Y	9.14	8.14	8.74	8.77	9.26	8.10	6.13	3.10	9.13	7.26	4.74

					Data set 3					
X	10	8	13	9	11	14	6	4	12	7	5
Y	7.46	6.77	12.74	7.11	7.81	8.84	6.08	5.39	8.15	6.42	5.73

					Data set 4					
X	8	8	8	8	8	8	8	19	8	8	8
Y	6.58	5.76	7.71	8.84	8.47	7.04	5.25	12.50	5.56	7.91	6.89

```
> library(nlme)
> library(Rfit)
>
> anscombe <- read.csv("./anscombe.csv",header=T)
> coef1 <- as.matrix(coef(lmList(y~x|set,data=anscombe)))
> coef2 <- matrix(0,nrow=4,ncol=2)
> for(i in 1:4){
+ x <- anscombe$x[which(anscombe$set==i)]
+ y <- anscombe$y[which(anscombe$set==i)]
+ coef2[i,] <- as.numeric(coef(rfit(y~x)))
+ }
>
> anscombe.plot <- function(setnum){
+ dat <- data.frame(anscombe[which(anscombe$set==setnum),1:2])
+ dat2 <- data.frame(x=4:20,y=coef1[setnum,1]+coef1[setnum,2]*4:20)
+ dat3 <- data.frame(x=4:20,y=coef2[setnum,1]+coef2[setnum,2]*4:20)
+ p <- ggplot() + geom_point(aes(x=x,y=y),data=dat)
+ p <- p + geom_line(aes(x=x,y=y),data=dat2,lty=2)
+ p <- p + geom_line(aes(x=x,y=y),data=dat3,lty=1)
+ p <- p + xlim(c(4,20)) + ylim(c(4,14)) + xlab("") + ylab("")
+ print(p)
```

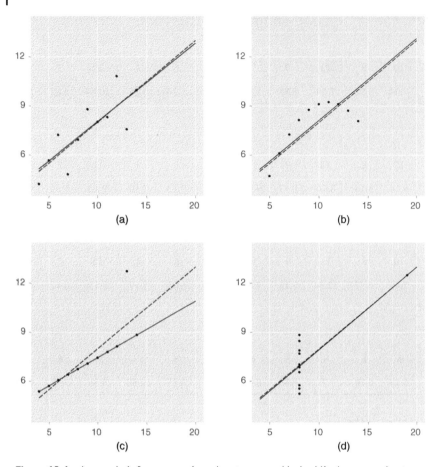

Figure 12.4 Anscombe's four regressions: least squares (dashed line) versus robust (solid line).

```
+ }
>
> anscombe.plot(1)
> anscombe.plot(2)
> anscombe.plot(3)
> anscombe.plot(4)
```

12.4   Isotonic Regression

In this section we consider bivariate data that satisfy an order or restriction in functional form. For example, if Y is known to be a decreasing function of X, a simple linear regression need only consider values of the slope parameter $\beta_1 < 0$. If we have no linear model, however, there is nothing in the empirical bivariate model to ensure such a constraint is satisfied. Isotonic regression considers a restricted class of estimators without the use of an explicit regression model.

Consider the dental study data in Table 12.1, which was used to illustrate isotonic regression by Robertson, Wright, and Dykstra (1988). The data are originally from a study of dental growth measurements of the distance (mm) from the center of the pituitary gland to the pterygomaxillary fissure (referring to the bone in the lower jaw) for 11 girls between the age of 8 and 14. It is assumed that PF increases with age, so the regression of PF on age is nondecreasing. However, it is also assumed that the relationship between PF and age is not necessarily linear. The means (or medians, for that matter) are *not* strictly increasing in the PF data. Least-squares regression does yield an increasing function for PF: $\hat{Y} = 0.065X + 21.89$, but the function is nearly flat and not altogether well suited to the data.

For an isotonic regression, we impose some order of the response as a function of the regressors.

Definition 12.1 If the regressors have a simple order $x_1 \leq \cdots \leq x_n$, a function f is *isotonic* with respect to x if $f(x_1) \leq \cdots \leq f(x_n)$. For our purposes, *isotonic* will be synonymous with *monotonic*. For some function g of X, we call the function g^* an *isotonic regression* of g with weights w if and only if g^* is isotonic (i.e. retains the necessary order) and minimizes

$$\sum_{i=1}^{n} w(x_i)\big(g(x_i) - f(x_i)\big)^2 \tag{12.3}$$

in the class of all isotonic functions f.

12.4.1 Graphical Solution to Regression

We can create a simple graph to show how the isotonic regression can be solved. Let $W_k = \sum_{i=1}^{k} w(x_i)$ and $G_k = \sum_{i=1}^{k} g(x_i)w(x_i)$. In the example, the means are

Table 12.1 Size of pituitary fissure for subjects of various ages.

Age	8	10	12	14
PF	21, 23.5, 23	24, 21, 25	21.5, 22, 19	23.5, 25
Mean	22.50	23.33	20.83	24.25
PAVA	22.22	22.22	22.22	24.25

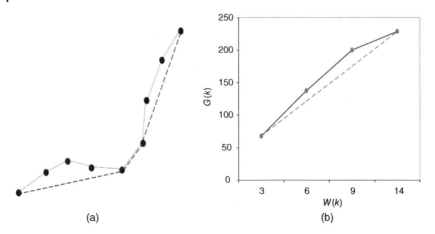

Figure 12.5 (a) Greatest convex minorant based on nine observations. (b) Greatest convex minorant for dental data.

ordered, so $f(x_i) = \mu_i$ and $w_i = n_i$, the number of observations at each age group. We let g be the set of PF means, and the plot of W_k versus G_k, called the *cumulative sum diagram* (CSD), shows that the empirical relationship between PF and age is not isotonic.

Define G^* to be the *greatest convex minorant* (GCM) that represents the largest convex function that lies below the CSD. You can envision G^* as a taut string tied to the left most observation (W_1, G_1) and pulled up and under the CSD, ending at the last observation. The example in Figure 12.5a shows that the GCM for the nine observations touches only four of them in forming a tight convex bowl around the data.

The GCM represents the isotonic regression. The reasoning follows below (and in the theorem that follows). Because G^* is convex, it is left differentiable at W_i. Let $g^*(x_i)$ be the left derivative of G^* at W_i. If the graph of the GCM is under the graph of CSD at W_i, the slopes of the GCM to the left and right of W_i remain the same, i.e. if $G^*(W_i) < G_i$, then $g^*(x_{i+1}) = g^*(x_i)$. This illustrates part of the intuition of the following theorem, which is not proven here (see Chapter 1 of Robertson, Wright, and Dykstra (1988)).

Theorem 12.1 *For function f in (12.3), the left-hand derivative g^* of the GCM is the unique isotonic regression of g on f. That is, if f is isotonic on X, then*

$$\sum_{i=1}^{n} w(x_i)\big(g(x_i) - f(x_i)\big)^2 \geq \sum_{i=1}^{n} w(x_i)\big(g(x_i) - g^*(x_i)\big)^2$$
$$+ \sum_{i=1}^{n} w(x_i)\big(g^*(x_i) - f(x_i)\big)^2.$$

Obviously, this graphing technique is going to be impractical for problems of any substantial size. The following algorithm provides an iterative way of solving for the isotonic regression using the idea of the GCM.

12.4.2 Pool Adjacent Violators Algorithm

In the CSD, we see that if $g(x_{i-1}) > g(x_i)$ for some i, then g is not isotonic. To construct an isotonic g^*, take the first such pair, and replace them with the weighted average

$$\bar{g}_i = \bar{g}_{i-1} = \frac{w(x_{i-1})g(x_{i-1}) + w(x_i)g(x_i)}{w(x_{i-1}) + w(x_i)}.$$

Replace the weights $w(x_i)$ and $w(x_{i-1})$ with $w(x_i) + w(x_{i-1})$. If this correction (replacing g with $\bar{g}$) makes the regression isotonic, we are finished. Otherwise, we repeat this process with until an isotonic is set. This is called the *pool adjacent violators algorithm* (PAVA).

Example 12.5 In Table 12.1, there is a decrease in PF between ages 10 and 12, which violates the assumption that pituitary fissure increases in age. Once we replace the PF averages by the average over both age groups (22.083), we still lack monotonicity because the PF average for girls of age 8 was 22.5. Consequently, these two categories, which now comprise three age groups, are averaged. The final averages are listed in the bottom row of Table 12.1, are plotted in Figure 12.5b.

12.5 Generalized Linear Models

Assume that $n(p + 1)$-tuples $(y_i, x_{1i}, x_{2i}, \dots, x_{pi})$, $i = 1, \dots, n$ are observed. The values y_i are responses, and components of vectors $x_i = (x_{1i}, x_{2i}, \dots, x_{pi})'$ are predictors. As we discussed at the beginning of this chapter, the standard theory of linear regression considers the model

$$Y = X\beta + \epsilon, \tag{12.4}$$

where $Y = (Y_1, \dots, Y_n)$ is the response vector, $X = (1_n\ x_1\ x_2\ \dots\ x_p)$ is the design matrix (1_n is a column vector of n 1's), and ϵ is vector of errors consisting of n i.i.d normal $\mathcal{N}(0, \sigma^2)$ random variables. The variance σ^2 is common for all Y_is and independent of predictors or the order of observation. The parameter β is a vector of $(p + 1)$ parameters in the linear relationship:

$$\mathbb{E}Y_i = x_i'\beta = \beta_0 + \beta_1 x_{1i} + \cdots + \beta_p x_{pi}.$$

The term *generalized linear model* (GLM) refers to a large class of models, introduced by Nelder and Wedderburn (1972) and popularized by McCullagh and Nelder (1994). In a canonical GLM, the response variable Y_i is assumed to follow an exponential family distribution with mean μ_i, which is assumed to be a function of $x_i'\beta$. This dependence can be nonlinear, but the distribution of Y_i depends on covariates only through their linear combination, $\eta_i = x_i'\beta$, called a *linear predictor*. As in the linear regression, the epithet *linear* refers to being linear in parameters, not in the explanatory variables. Thus, for example, the linear combination

$$\beta_0 + \beta_1 x_1 + \beta_2 x_2^2 + \beta_3 \log(x_1 + x_2) + \beta_4 x_1 \cdot x_2,$$

is a perfectly linear predictor. What, exactly, is *generalized* in the GLM model given in (12.4)? The three main generalizations concern the distributions of responses, the dependence of response on linear predictor, and variance of the error:

1. Although Y_is remain independent, their (common) distribution is generalized. Instead of normal, their distribution is selected from the exponential family of distributions (see Chapter 2). This family is quite versatile and includes normal, binomial, Poisson, negative binomial, and gamma as special cases.
2. In the linear model (12.4), the mean of Y_i, $\mu_i = \mathbb{E}Y_i$ was equal to $x_i'\beta$. In the GLM, the mean μ_i depends on the predictor $\eta_i = x_i'\beta$ as

$$g(\mu_i) = \eta_i \quad (= x_i'\beta). \tag{12.5}$$

   The function g is called the *link* function. For the model (12.4), the link is the identity function.
3. In the linear model, the variance of Y_i was constant. In GLM, it need not be constant and may depend on the mean μ_i.

Models and inference for categorical data, traditionally a nonparametric topic, are unified by a larger class of models that are parametric in nature and that are special cases of GLM. For example, in contingency tables, the cell counts N_{ij} could be modeled by multinomial $\mathcal{M}n(n, \{p_{ij}\})$ distribution. The standard hypothesis in contingency tables is concerning the independence of row/column factors. This is equivalent to testing $H_0 : p_{ij} = \alpha_i\beta_j$ for some unknown α_i and β_j such that $\sum_i \alpha_i = \sum_j \beta_j = 1$.

The expected cell count $\mathbb{E}N_{ij} = np_{ij}$ that under H_0 becomes $\mathbb{E}N_{ij} = n\alpha_i\beta_j$ by taking the logarithm of both sides obtains

$$\log \mathbb{E}N_{ij} = \log n + \log \alpha_i + \log \beta_j$$
$$= \text{const} + a_i + b_j,$$

for some parameters a_i and b_j. Thus, the test of goodness of fit for this model linear and additive in parameters a and b is equivalent to the test of the original independence hypothesis H_0 in the contingency table. More of such examples will be discussed in Chapter 18.

12.5.1 GLM Algorithm

The algorithms for fitting GLMs are robust and well established (see Nelder and Wedderburn (1972) and McCullagh and Nelder (1994)). The maximum likelihood estimates of β can be obtained using iterative weighted least squares (IWLS):

(i) Given vector $\hat{\mu}^{(k)}$, the initial value of the linear predictor $\hat{\eta}^{(k)}$ is formed using the link function, and components of adjusted dependent variate (working response), $z_i^{(k)}$, can be formed as

$$z_i^{(k)} = \hat{\eta}_i^{(k)} + \left(y_i - \hat{\mu}_i^{(k)}\right)\left(\frac{d\eta}{d\mu}\right)_i^{(k)},$$

where the derivative is evaluated at the available kth value.

(ii) The quadratic (working) weights, $W_i^{(k)}$, are defined so that

$$\frac{1}{W_i^{(k)}} = \left(\frac{d\eta}{d\mu}\right)_i^2 V_i^{(k)},$$

where V is the variance function evaluated at the initial values.

(iii) The working response $z^{(k)}$ is then regressed onto the covariates x_i, with weights $W_i^{(k)}$ to produce new parameter estimates, $\hat{\beta}^{(k+1)}$. This vector is then used to form new estimates

$$\eta^{(k+1)} = X'\hat{\beta}^{(k+1)} \quad \text{and} \quad \hat{\mu}^{(k+1)} = g^{-1}(\hat{\eta}^{(k+1)}).$$

We repeat iterations until changes become sufficiently small. Starting values are obtained directly from the data, using $\hat{\mu}^{(0)} = y$, with occasional refinements in some cases (for example, to avoid evaluating $\log 0$ when fitting a log-linear model with zero counts).

By default, the scale parameter should be estimated by the *mean deviance*

$$n^{-1}\sum_{i=1}^{n} D(y_i, \mu), \tag{12.6}$$

from p. 47 in Chapter 3, in the case of the normal and gamma distributions.

12.5.2 Link Functions

In the GLM, the predictors for Y_i are summarized as the linear predictor $\eta_i = x_i'\beta$. The link function is a monotone differentiable function g such that $\eta_i = g(\mu_i)$, where $\mu_i = \mathbb{E}Y_i$. We already mentioned that in the normal case $\mu = \eta$ and the link is identity, $g(\mu) = \mu$.

Example 12.6 For analyzing count data (e.g. contingency tables), the Poisson model is standardly assumed. As $\mu > 0$, the identity link is not appropriate because it allows for η to be negative. Instead, if $\mu = e^\eta$, the mean is always positive, and $\eta = \log(\mu)$ serves as an adequate link.

A link is called *natural* if it is connecting θ (the natural parameter in the exponential family of distributions) and μ. In the Poisson case,

$$f(y|\lambda) = \exp\left\{ y \log \lambda - (\lambda + \log y!) \right\},$$

$\mu = \lambda$ and $\theta = \log \mu$. Accordingly, the log is the natural link for the Poisson distribution.

Example 12.7   For the binomial distribution, the probability mass function

$$f(y|\pi) = \binom{n}{y} \pi^y (1 - \pi)^{n-y}$$

can be represented as

$$f(y|\pi) = \exp\left\{ y \log \frac{\pi}{1 - \pi} + n \log(1 - \pi) + \log \binom{n}{y} \right\}.$$

The natural link $\eta = \log(\pi/(1 - \pi))$ is called *logit* link. With the binomial distribution, several more links are commonly used. Examples are the *probit* link $\eta = \Phi^{-1}(\pi)$, where Φ is a standard normal CDF, and the *complementary log–log* link with $\eta = \log\{-\log(1 - \pi)\}$. For these three links, the probability π of interest is expressed as $\pi = e^\eta/(1 + e^\eta)$, $\pi = \Phi(\eta)$, and $\pi = 1 - \exp\{-e^\eta\}$, respectively:

Distribution	$\theta(\mu)$	$b(\theta)$	ϕ
$\mathcal{N}(\mu, \sigma^2)$	μ	$\theta^2/2$	σ^2
$Bin(1, \pi)$	$\log(\pi/(1 - \pi))$	$\log(1 + \exp(\theta))$	1
$P(\lambda)$	$\log \lambda$	$\exp(\theta)$	1
$Gamma(\mu, v/\mu)$	$-1/\mu$	$-\log(-\theta)$	$1/v$

When data y_i from the exponential family are expressed in grouped form (from which an average is considered as the group response), then the distribution for Y_i takes the form

$$f(y_i|\theta_i, \phi, \omega_i) = \exp\left\{ \frac{y_i\theta_i - b(\theta_i)}{\phi}\omega_i + c(y_i, \phi, \omega_i) \right\}. \tag{12.7}$$

The weights ω_i are equal to 1 if individual responses are considered, $\omega_i = n_i$ if response y_i is an average of n_i responses, and $\omega_i = 1/n_i$ if the sum of n_i individual responses is considered.

The variance of Y_i then takes the form

$$\mathbb{V}\mathrm{ar}\, Y_i = \frac{b''(\theta_i)\phi}{\omega_i} = \frac{\phi V(\mu_i)}{\omega_i}.$$

12.5.3 Deviance Analysis in GLM

In GLM, a proposed model's goodness of fit of can be assessed in several ways. The customary measure is *deviance* statistics. For a data set with n observations, assume the dispersion ϕ is known and equal to 1, and consider the two extreme models, the single parameter model stating $\mathbb{E}Y_i = \hat{\mu}$ and the n parameter *saturated* model setting $\mathbb{E}Y_i = \hat{\mu}_i = Y_i$. Most likely, the interesting model is between the two extremes. Suppose $\mathcal{M}$ is the interesting model with $1 < p < n$ parameters.

If $\hat{\theta}_i^{\mathcal{M}} = \hat{\theta}_i^{\mathcal{M}}(\hat{\mu}_i)$ are predictions of the model $\mathcal{M}$ and $\hat{\theta}_i^S = \hat{\theta}_i^S(y_i) = y_i$ are the predictions of the saturated model, then the deviance of the model $\mathcal{M}$ is

$$D_{\mathcal{M}} = 2\sum_{i=1}^{n}\left[(y_i\hat{\theta}_i^S(y_i) - b(\hat{\theta}_i^S)) - (y_i\hat{\theta}_i^{\mathcal{M}} - b(\hat{\theta}_i^{\mathcal{M}}))\right].$$

When the dispersion ϕ is estimated and different than 1, the *scaled deviance* of the model $\mathcal{M}$ is defined as $D_{\mathcal{M}}^* = D_{\mathcal{M}}/\phi$.

Example 12.8 For $y_i \in \{0,1\}$ in the binomial family,

$$D = 2\sum_{i=1}^{n}\left\{y_i\log\left(\frac{y_i}{\hat{y}_i}\right) + (n_i - y_i)\log\left(\frac{n_i - y_i}{n_i - \hat{y}_i}\right)\right\}.$$

- Deviance is minimized at saturated model S. Equivalently, the log-likelihood $\ell^S = \ell(y|y)$ is the maximal log likelihood with the data y.
- The scaled deviance $D_{\mathcal{M}}^*$ is asymptotically distributed as χ_{n-p}^2. Significant deviance represents the deviation from a good model fit.
- If a model $\mathcal{K}$ (with q parameters) is a subset of model $\mathcal{M}$ (with p parameters, $q < p$), then

$$\frac{D_{\mathcal{K}}^* - D_{\mathcal{M}}^*}{\phi} \sim \chi_{p-q}^2.$$

Residuals are critical for assessing the model (recall four Anscombe's regressions on p. 244). In standard normal regression models, residuals are calculated simply as $y_i - \hat{\mu}_i$, but in the context of GLMs, both predicted values and residuals are more ambiguous. For predictions, it is important to distinguish the scale: (i) predictions on the scale of $\eta = x_i'\beta$ and (ii) predictions on the scale of the observed responses y_i for which $\mathbb{E}Y_i = g^{-1}(\eta_i)$.

Regarding residuals, there are several approaches. *Response residuals* are defined as $r_i = y_i - g^{-1}(\eta_i) = y_i - \theta_i$. Also, the deviance residuals are defined as

$$r_i^D = \text{sign}(y_i - \mu_i)\sqrt{d_i},$$

where d_i are observation specific contributions to the deviance D.

Deviance residuals are ANOVA-like decompositions,

$$\sum_i (r_i^D)^2 = D,$$

thus testably assessing the contribution of each observation to the model deviance. In addition, the deviance residuals increase with $y_i - \hat{\mu}_i$ and are distributed approximately as standard normals, irrespective of the type of GLM.

Example 12.9 For $y_i \in \{0,1\}$ in the binomial family,

$$r_i^D = \mathrm{sign}(y_i - \hat{y}_i)\sqrt{2\left\{ y_i \log\left(\frac{y_i}{\hat{y}_i}\right) + (n_i - y_i)\log\left(\frac{n_i - y_i}{n_i - \hat{y}_i}\right)\right\}}.$$

Another popular measure of goodness of fit of GLM is Pearson statistic

$$X^2 = \sum_{i=1}^{n} \frac{(y_i - \hat{\mu}_i)^2}{V(\hat{\mu}_i)}.$$

The statistic X^2 also has a χ^2_{n-p} distribution.

Example 12.10 Cæsarean Birth Study. The data in this example come from Münich hospital (Fahrmeir and Tutz, 2001) and concern infection cases in births by Cæsarean section. The response of interest is occurrence of infection. Three covariates each at two levels were considered as important for the occurrence of infection:

- noplan – whether the Cæsarean section birth planned (0) or not (1).
- riskfac – the presence of Risk factors for the mother, such as diabetes, overweight, previous Cæsarean section birth, etc., where present = 1, not present = 0.
- antibio – whether antibiotics were given (1) or not given (0) as a prophylaxis.

Table 12.2 provides the counts.

The R function glm is instrumental in computing the solution in the example that follows:

```
> birth <- data.frame(
+ infection=c(1,11,0,0,28,23,8,0),
+ total=c(18,98,2,0,58,26,40,9),
+ noplan=c(0,1,0,1,0,1,0,1),
+ riskfac=c(1,1,0,0,1,1,0,0),
+ antibio=c(1,1,1,1,0,0,0,0));
> birth$prop <- birth$infection/birth$total
>
> fitglm <- glm(cbind(infection,total-infection)~noplan+riskfac+antibio,
+ data=birth,family=binomial(logit))
> pred <- predict(fitglm,birth[,3:5],type="response")
```

Table 12.2 Cæsarean section birth data.

	Planned		Not planned	
	Infection	No infection	Infection	No infection
Antibiotics				
Risk fact, yes	1	17	11	87
Risk fact, no	0	2	0	0
No antibiotics				
Risk fact, yes	28	30	23	3
Risk fact, no	8	32	0	9

```
>
> birth2 <- data.frame(x=c(1:8,1:8),y=c(birth$prop,pred),
+ type=c(rep("obs",8),rep("pred",8)))
> p <- ggplot() + geom_line(aes(x=1:8,y=pred),lty=2,col=4)
> p <- p + geom_point(aes(x=1:8,y=pred),pch=1,size=3,col=4)
> p <- p + geom_point(aes(x=1:8,y=birth$prop),pch=0,col=2,size=3)
> p <- p + xlim(c(1,8)) + ylim(c(0,1)) + xlab("") + ylab("")
> print(p)
```

The scaled deviance of this model is distributed as χ^2_3. The number of degrees of freedom is equal to 8 (*n*) vector `infection` minus 5 for the five estimated parameters, $\beta_0, \beta_1, \beta_2, \beta_3, \phi$. The deviance `deviance(fitglm)=10.997` is significant, yielding a *p*-value of `1 - pchisq(10.997,3)=0.0117`. The additive model (with no interactions) in R yields

$$\log \frac{P(\text{infection})}{P(\text{no infection})} = \beta_0 + \beta_1 \, \text{noplan} + \beta_2 \, \text{riskfac}$$
$$+ \beta_3 \, \text{antibio}.$$

The estimators of $(\beta_0, \beta_1, \beta_2, \beta_3)$ are, respectively, $(-1.89, 1.07, 2.03, -3.25)$. The interpretation of the estimators is made more clear if we look at the odds ratio

$$\frac{P(\text{infection})}{P(\text{no infection})} = e^{\beta_0} \cdot e^{\beta_1 \text{noplan}} \cdot e^{\beta_2 \text{riskfac}} \cdot e^{\beta_3 \text{antibio}}.$$

At the value `antibio = 1`, the antibiotics have the odds ratio of infection/no infection. This increases by the factor $\exp(-3.25) = 0.0376$, which is a decrease of more than 25 times. Figure 12.6 shows the observed proportions of infections for eight combinations of covariates (`noplan`, `riskfac`, `antibio`) marked by squares and model-predicted probabilities for the same combinations marked by circles. We will revisit this example in Chapter 18; see Example 18.5.

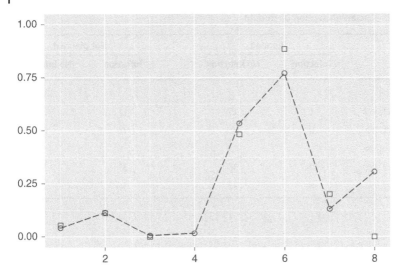

Figure 12.6 Cæsarean birth infection observed proportions (*squares*) and model predictions (*circles*). The numbers 1–8 on the *x*-axis correspond to following combinations of covariates (`noplan, riskfac, antibio`): $(0, 1, 1)$, $(1, 1, 1)$, $(0, 0, 1)$, $(1, 0, 1)$, $(0, 1, 0)$, $(1, 1, 0)$, $(0, 0, 0)$, and $(1, 0, 0)$.

12.6 Exercises

12.1 Using robust regression, find the intercept and slope $\tilde{\beta}_0$ and $\tilde{\beta}_1$ for each of the four data sets of Anscombe (1973) from p. 244. Plot the ordinary least-squares regression along with the rank regression estimator of slope. Contrast these with one of the other robust regression techniques. For which set does $\tilde{\beta}_1$ differ the most from its LS counterpart $\hat{\beta}_1 = 0.5$? Note that in the fourth set, 10 out of 11 Xs are equal, so one should use $S_{ij} = (Y_j - Y_i)/(X_j - X_i + \epsilon)$ to avoid dividing by 0. After finding $\tilde{\beta}_0$ and $\tilde{\beta}_1$, are they different than $\hat{\beta}_0$ and $\hat{\beta}_1$? Is the hypothesis $H_0 : \beta_1 = 1/2$ rejected in a robust test against the alternative $H_1 : \beta_1 < 1/2$, for data set 3? Note here $\beta_{10} = 1/2$.

12.2 Using the PF data in Table 12.1, compute a median squares regression, and compare it with the simple linear regression curve.

12.3 Using the PF data in Table 12.1, compute a nonparametric regression, and test to see if $\beta_{10} = 0$.

12.4 Consider the $Gamma(\alpha, \alpha/\mu)$ distribution. This parametrization was selected so that $\mathbb{E}y = \mu$. Identify θ and ϕ as functions of α and μ. Identify functions a, b, and c.

Hint: the density can be represented as

$$\exp\left\{ -\alpha \log \mu - \frac{\alpha y}{\mu} + \alpha \log(\alpha) + (\alpha - 1)\log y - \log(\Gamma(\alpha)) \right\}.$$

12.5 The zero-truncated Poisson distribution is given by

$$f(y|\lambda) = \frac{\lambda^j}{j!(e^\lambda - 1)}, \quad j = 1, 2, \ldots$$

Show that f is a member of exponential family with canonical parameter $\log \lambda$.

12.6 Dalziel, Lagen, and Thurston (1941) conducted an experiment to assess the effect of small electrical currents on farm animals, with the eventual goal of understanding the effects of high-voltage power lines on livestock. The experiment was carried out with seven cows and six shock intensities: 0, 1, 2, 3, 4, and 5 milliamps (note that shocks on the order of 15 milliamps are painful for many humans). Each cow was given 30 shocks, five at each intensity, in random order. The entire experiment was then repeated, so each cow received a total of 60 shocks. For each shock the response, mouth movement, was either present or absent. The data as quoted give the total number of responses, out of 70 trials, at each shock level. We ignore cow differences and differences between blocks (experiments):

Current (milliamps)	Number of responses	Number of trials	Proportion of responses
0	0	70	0.000
1	9	70	0.129
2	21	70	0.300
3	47	70	0.671
4	60	70	0.857
5	63	70	0.900

Propose a GLM in which the probability of a response is modeled with a value of current (in milliamps) as a covariate.

Table 12.3 Bliss beetle data.

Dose ($\log_{10}CS_2$ mg l^{-1})	Number of beetles	Number of killed
1.6907	59	6
1.7242	60	13
1.7552	62	18
1.7842	56	28
1.8113	63	52
1.8369	59	53
1.8610	62	61
1.8839	60	60

12.7 Bliss (1935) provides data showing the number of flour beetles killed after five-hour exposure to gaseous carbon disulfide at various concentrations. See Table 12.3. Propose a logistic regression model with a dose as a covariate. According to your model, what is the probability that a beetle will be killed if a dose of gaseous carbon disulfide is set to 1.8?

RELEVANT R FUNCTIONS AND DATA SETS IN THIS CHAPTER

R functions: cor.test, lm, lmList, rq, rlm, ltsReg, glm
R package: MASS, robustbase, quantreg, nlme

anscombe.csv, exer12.6.csv, exer12.7.csv

References

Anscombe, F. (1973), "Graphs in Statistical Analysis," *American Statistician*, 27, 17–21.

Bliss, C. I. (1935), "The Calculation of the Dose-Mortality Curve," *Annals of Applied Biology*, 22, 134–167.

Dalziel, C. F., Lagen, J. B., and Thurston, J. L. (1941), "Electric Shocks," *Transactions of IEEE*, 60, 1073–1079.

Fahrmeir, L., and Tutz, G. (2001), *Multivariate Statistical Modeling Based on Generalized Linear Models*, Second Edition, New York: Springer-Verlag.

Huber, P. J. (1973), "Robust Regression: Asymptotics, Conjectures, and Monte Carlo," *Annals of Statistics*, 1, 799–821.

Huber, P. J. (2009), *Robust Statistics*, Second Edition, New York: Wiley.

Kvam, P. H. (2000), "The Effect of Active Learning Methods on Student Retention in Engineering Statistics," *American Statistician*, 54 (2), 136–140.

McCullagh, P., and Nelder, J. A. (1994), *Generalized Linear Models*, Second Edition, London: Chapman & Hall.

Nelder, J. A., and Wedderburn, R. W. M. (1972), "Generalized Linear Models," *Journal of the Royal Statistical Society: Series A*, 135, 370–384.

Robertson, T., Wright, T. F., and Dykstra, R. L. (1988), *Order Restricted Statistical Inference*, New York: Wiley.

Rousseeuw, P. J. (1985), "Multivariate Estimation with High Breakdown Point," in *Mathematical Statistics and Applications B*, Eds. W. Grossmann et al., pp. 283–297, Dordrecht: Reidel Publishing Co.

Rousseeuw, P. J., and Leroy, A. M. (1987), *Robust Regression and Outlier Detection*, New York: Wiley.

13

Curve Fitting Techniques

"Now, my own suspicion is that the universe is not only queerer than we suppose, but queerer than we can suppose."

J.B.S. Haldane (Haldane's Law)

In this chapter, we will learn about a general class of nonparametric regression techniques that fit a response curve to input predictors without making strong assumptions about error distributions. The estimators, called *smoothing functions*, actually can be smooth or bumpy as the user sees fit. The final regression function can be made to bring out from the data what is deemed to be important to the analyst. Plots of a smooth estimator will give the user a good sense of the overall trend between the input X and the response Y. However, interesting nuances of the data might be lost to the eye. Such details will be more apparent with less smoothing, but a potentially noisy and jagged curve plotted made to catch such details might hide the overall trend of the data. Because no linear form is assumed in the model, this nonparametric regression approach is also an important component of *nonlinear regression*, which can also be parametric.

Let $(X_1, Y_1), \ldots, (X_n, Y_n)$ be a set of n independent pairs of observations from the bivariate random variable (X, Y). Define the regression function $m(x)$ as $\mathbb{E}(Y|X = x)$. Let $Y_i = m(X_i) + \varepsilon_i$, $i = 1, \ldots, n$ when ε_i's are errors with zero mean and constant variance. The estimators here are *locally weighted* with the form

$$\hat{m}(x) = \sum_{i=1}^{n} a_i Y_i.$$

The local weights a_i can be assigned to Y_i in a variety of ways. The straight line in Figure 13.1 is a linear regression of Y on X that represents an extremely smooth response curve. The curved line fit in Figure 13.1 represents an estimator that uses more local observations to fit the data at any X_i value. These two response curves

Nonparametric Statistics with Applications to Science and Engineering with R, Second Edition.
Paul Kvam, Brani Vidakovic, and Seong-joon Kim.
© 2023 John Wiley & Sons, Inc. Published 2023 by John Wiley & Sons, Inc.

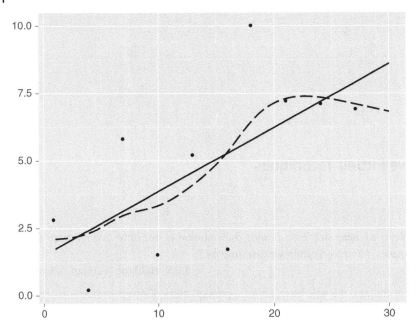

Figure 13.1 Linear Regression (solid line) and local estimator (dashed line) fit to data.

represent the tradeoff we make when making a curve more or less smooth. The tradeoff is between *bias* and *variance* of the estimated curve.

In the case of linear regression, the variance is estimated globally because it is assumed the unknown variance is constant over the range of the response. This makes for an optimal variance estimate. However, the linear model is often considered to be overly simplistic, so the true expected value of $\hat{m}(x)$ might be far from the estimated regression, making the estimator biased. The local (jagged) fit, on the other hand, uses only responses at the value X_i to estimate $\hat{m}(X_i)$, minimizing any potential bias. However, by estimating $m(x)$ locally, one does not pool the variance estimates, so the variance estimate at X is constructed using only responses at or close to X.

This illustrates the general difference between smoothing functions; those that estimate $m(x)$ using points only at x or close to it have less bias and high variance. Estimators that use data from a large neighborhood of x will produce a good estimate of variance but risk greater bias. In Sections 13.1 and 13.2, we feature two different ways of defining the local region (or neighborhood) of a design point. At an estimation point x, *kernel estimators* use fixed intervals around x such as $x \pm c_0$ for some $c_0 > 0$. *Nearest neighbor estimators* use the span produced by a fixed number of design points that are closest to x.

13.1 Kernel Estimators

Let $K(x)$ be a real-valued function for assigning local weights to the linear estimator, that is,

$$y(x) = \sum K\left(\frac{x - x_i}{h}\right) y_i.$$

If $K(u) \propto \mathbf{1}(|u| \le 1)$, then a fitted curve based on $K\left(\frac{x-x_i}{h}\right)$ will estimate $m(x)$ using only design points within h units of x. Usually it is assumed that $\int_R K(x)dx = 1$, so any bounded probability density could serve as a kernel. Unlike kernel functions used in density estimation, now $K(x)$ also can take negative values, and in fact such unrestricted kernels are needed to achieve optimal estimators in the asymptotic sense. An example is the *beta kernel* defined as

$$K(x) = \frac{1}{B(1/2, \gamma + 1)}\left(1 - x^2\right)^{\gamma}\mathbf{1}(|x| \le 1), \quad \gamma = 0, 1, 2\ldots \tag{13.1}$$

With the added parameter γ, the beta kernel is remarkably flexible. For $\gamma = 0$, the beta kernel becomes uniform. If $\gamma = 1$, we get the Epanechnikov kernel, $\gamma = 2$ produces the biweight kernel, $\gamma = 3$ the triweight, and so on (see Figure 11.4 on p. 269). For γ large enough, the beta kernel is close the Gaussian kernel

$$K(x) = \frac{1}{\sqrt{2\pi\sigma^2}}e^{-\frac{x}{2\sigma^2}},$$

with $\sigma^2 = 1/(2\gamma + 3)$, which is the variance of densities from (13.1). For example, if $\gamma = 10$, then $\int_{-1}^{1}\left(K(x) - \sigma^{-1}\phi(x/\sigma)\right)^2 dx \approx 0.00114$, where $\sigma = 1/\sqrt{2\gamma + 3}$. Define a scaling coefficient h so that

$$K_h(x) = \frac{1}{h}K\left(\frac{x}{h}\right), \tag{13.2}$$

where h is the associated *bandwidth*. By increasing h, the kernel function spreads weight away from its center, thus giving less weight to those data points close to x and sharing the weight more equally with a larger group of design points. A family of beta kernels and the Epanechnikov kernel ($\gamma = 1$) are given in Figure 13.2a. The Silverman kernel (Silverman, 1985) is given in Figure 13.2b.

13.1.1 Nadaraya–Watson Estimator

Nadaraya (1964) and Watson (1964) independently published the earliest results on for smoothing functions (but this is debatable), and the Nadaraya–Watson estimator (NWE) of $m(x)$ is defined as

$$\hat{m}(x) = \frac{\sum_{i=1}^{n} K_h(X_i - x)Y_i}{\sum_{i=1}^{n} K_h(X_i - x)}. \tag{13.3}$$

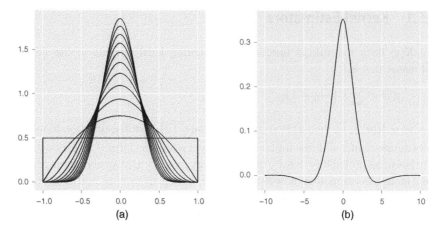

Figure 13.2 (a) A family of symmetric beta kernels. (b) $K(x) = \frac{1}{2} \exp\left\{-|x|/\sqrt{2}\right\}$ $\sin\left(|x|/\sqrt{2} + \pi/4\right)$.

For x fixed, the value $\hat{\theta}$ that minimizes

$$\sum_{i=1}^{n} (Y_i - \theta)^2 K_h(X_i - x), \tag{13.4}$$

is of the form $\sum_{i=1}^{n} a_i Y_i$. The NWE is the minimizer of (13.4) with $a_i = K_h(X_i - x)/\sum_{i=1}^{n} K_h(X_i - x)$.

Although several competing kernel-based estimators have been derived since, the NWE provided the basic framework for kernel estimators, including local polynomial (LP) fitting that is described later in this section. The R function

```
ksmooth(x,y,kernel,bandwidth)
```

computes the Nadaraya–Watson kernel estimate. Here, (X,Y) are input data, kernel is the kernel function, and bandwidth is the bandwidth.

Example 13.1 Noisy pairs (X_i, Y_i), $i = 1, \ldots, 200$ are generated in the following way:

```
> x <- sort(runif(200));
> y <- sort(4*pi*sort(runif(200))+0.3*rnorm(200));
```

Three bandwidths are selected $h = 0.07, 0.14$, and 0.21. The three NWE are shown in Figure 13.3. As expected, the estimators constructed with the larger bandwidths appear smoother than those with smaller bandwidths.

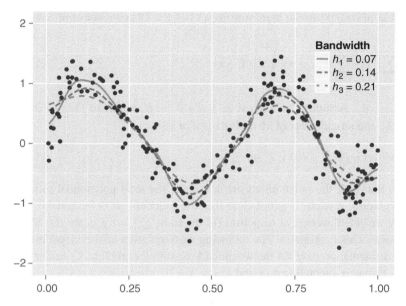

Figure 13.3 Nadaraya–Watson estimators for different values of bandwidth.

13.1.2 Gasser–Müller Estimator

The Gasser–Müller estimator proposed in 1979 uses areas of the kernel for the weights. Suppose X_i are ordered, $X_1 \leq X_2 \cdots \leq X_n$. Let $X_0 = -\infty$ and $X_{n+1} = \infty$, and define midpoints $s_i = (X_i + X_{i+1})/2$. Then

$$\hat{m}(x) = \sum_{i=1}^{n} Y_i \int_{s_{i-1}}^{s_i} K_h(u - x)du. \tag{13.5}$$

The Gasser–Müller estimator is the minimizer of (13.4) with the weights $a_i = \int_{s_{i-1}}^{s_i} K_h(u - x)du$.

13.1.3 Local Polynomial Estimator

Both NWE and Gasser–Müller estimator are *local constant fit* estimators, that is, they minimize weighted squared error $\sum_{i=1}^{n} (Y_i - \theta)^2 \omega_i$ for different values of weights ω_i. Assume that for z in a small neighborhood of x, the function $m(z)$ can well be approximated by a polynomial of order p:

$$m(z) \approx \sum_{j=0}^{p} \beta_j (z - x)^j,$$

where $\beta_j = m^{(j)}(x)/j!$ Instead of minimizing (13.4), the LP estimator minimizes

$$\sum_{i=1}^{n}\left(Y_i - \sum_{j=0}^{p}\beta_j(X_i - x)^j\right)^2 K_h(X_i - x) \tag{13.6}$$

over $\beta_1, \ldots, \beta_p$. Assume, for a fixed x, $\hat{\beta}_j, j = 0, \ldots, p$ minimize (13.6). Then, $\hat{m}(x) = \hat{\beta}_0$, and an estimator of jth derivative of m is

$$\hat{m}^{(j)}(x) = j!\hat{\beta}_j, \quad j = 0,1,\ldots,p. \tag{13.7}$$

If $p = 0$, that is, if the polynomials are constants, the local polynomial estimator is Nadaraya–Watson. It is not clear that the estimator $\hat{m}(x)$ for general p is a locally weighted average of responses (of the form $\sum_{i=1}^{n} a_i Y_i$) as are the NWE and Gasser–Müller estimator. The following representation of the LP estimator makes its calculation easy via the weighted least-squares problem. Consider the $n \times (p + 1)$ matrix depending on x and $X_i - x$, $i = 1, \ldots, n$:

$$X = \begin{pmatrix} 1 & X_1 - x & (X_1 - x)^2 & \ldots & (X_1 - x)^p \\ 1 & X_2 - x & (X_2 - x)^2 & \ldots & (X_2 - x)^p \\ \ldots & \ldots & \ldots & & \ldots \\ 1 & X_n - x & (X_n - x)^2 & \ldots & (X_n - x)^p \end{pmatrix}.$$

Define also the diagonal weight matrix W and response vector Y:

$$W = \begin{pmatrix} K_h(X_1 - x) & & & \\ & K_h(X_2 - x) & & \\ & & \ddots & \\ & & & K_h(X_n - x) \end{pmatrix}, \quad Y = \begin{pmatrix} Y_1 \\ Y_2 \\ \vdots \\ Y_n \end{pmatrix}.$$

Then the minimization problem can be written as $(Y - X\beta)'W(Y - X\beta)$. The solution is well known: $\hat{\beta} = (X'WX)^{-1}X'W\,Y$. Thus, if $(a_1\ a_2\ \ldots\ a_n)$ is the first row of matrix $(X'WX)^{-1}X'W$, $\hat{m}(x) = a \cdot Y = \sum_i a_i Y_i$. This representation (in matrix form) provides an efficient and elegant way to calculate the LP regression estimator. Although LP regression can be performed by the standard R functions, such as locPolSmootherC in locpol package and locfit in locfit package, we implemented the procedures from the estimator explained above. In R, use the custom function

```
lpfit(x, y, p, h),
```

where (x,y) is the input data, p is the order, and h is the bandwidth.

For general p, the first row $(a_1 \; a_2 \; \ldots \; a_n)$ of $(X'WX)^{-1}X'W$ is quite complicated. Yet, for $p = 1$ (the local linear estimator), the expression for $\hat{m}(x)$ simplifies to

$$\hat{m}(x) = \frac{1}{n}\sum_{i=1}^{n} \frac{(S_2(x) - S_1(x)(X_i - x))K_h(X_i - x)}{S_2(x)S_0(x) - S_1(x)^2}Y_i,$$

where $S_j = \frac{1}{n}\sum_{i=1}^{n}(X_i - x)^j K_h(X_i - x)$, $j = 0,1$, and 2. This estimator is implemented in R by the custom function

```
loc.lin.r.
```

13.2 Nearest Neighbor Methods

As an alternative to kernel estimators, nearest neighbor estimators define points local to X_i not through a kernel bandwidth, which is a fixed strip along the x-axis, but instead on a set of points closest to X_i. For example, a neighborhood for x might be defined to be the closest k design points on either side of x, where k is a positive integer such that $k \leq n/2$. Nearest neighbor methods make sense if we have spaces with clustered design points followed by intervals with sparse design points. The nearest neighbor estimator will increase its span if the design points are spread out. There is added complexity, however, if the data includes repeated design points. For illustration purposes, we will assume this is not the case in our examples.

Nearest neighbor and kernel estimators produce similar results, in general. In terms of bias and variance, the nearest neighbor estimator described in this section performs well if the variance decreases more than the squared bias increases (see Altman, 1992).

13.2.1 LOESS

William Cleveland (1979) introduced a curve fitting regression technique named "LOWESS," which stands for *locally weighted regression scatter plot smoothing*. Its derivative, LOESS,[1] stands more generally for a local regression, but many researchers consider LOWESS and LOESS as synonyms.

Consider a multiple linear regression set up with a set of regressors $\mathcal{X}_i = X_{i1}, \ldots, X_{ik}$ to predict Y_i, $i = 1, \ldots, n$. If $Y = f(x_1, \ldots, x_k) + \varepsilon$, where $\varepsilon \sim \mathcal{N}(0, \sigma^2)$. Adjacency of the regressors is defined by a distance function

---

1 Term actually defined by geologists as deposits of fine soil that are highly susceptible to wind erosion. We will stick with our less silty mathematical definition in this chapter.

$d(\mathcal{X}, \mathcal{X}^*)$. For $k = 2$, if we are fitting a curve at (X_{r1}, X_{r2}) with $1 \leq r \leq n$, then for $i = 1, \ldots, n$,

$$d_i = \sqrt{(X_{i1} - X_{r1})^2 + (X_{i2} - X_{r2})^2}.$$

Each data point influences the regression at (X_{r1}, X_{r2}) according to its distance to that point. In the LOESS method, this is done with a tri-cube weight function:

$$w_i = \begin{cases} \left(1 - \left(\frac{d_i}{d_q}\right)^3\right)^3, & d_i \leq d_q, \\ 0, & d_i > d_q, \end{cases}$$

where only q of n points closest to $\mathcal{X}_i$ is considered to be "in the neighborhood" of $\mathcal{X}_i$ and d_q is the distance of the furthest $\mathcal{X}_i$ that is in the neighborhood. Actually, many other weight functions can serve just as well as the triweight function; requirements for w_i are discussed in Cleveland (1979).

If q is large, the LOESS curve will be smoother but less sensitive to nuances in the data. As q decreases, the fit looks more like an interpolation of the data, and the curve is zigzaggy. Usually, q is chosen so that $0.10 \leq q/n \leq 0.25$. Within the window of observations in the neighborhood of $\mathcal{X}$, we construct the LOESS curve ($Y\mathcal{X}$) using either linear regression (called first order) or quadratic (second order).

There are great advantages to this curve estimation scheme. LOESS does not require a specific function to fit the model to the data; only a smoothing parameter ($\alpha = q/n$) and LP (first or second order) are required. Given that complex functions can be modeled with such a simple precept, the LOESS procedure is popular for constructing a regression equation with cloudy, multidimensional data.

On the other hand, LOESS requires a large data set in order for the curve fitting to work well. Unlike least-squares regression (and, for that matter, many non-linear regression techniques), the LOESS curve does not give the user a simple math formula to relate the regressors to the response. Because of this, one of the most valuable uses of LOESS is as an exploratory tool. It allows the practitioner to visually check the relationship between a regressor and response no matter how complex or convoluted the data appear to be.

In R, use the function

```
loess(formula,span,degree)
```

where `formula` is a symbolic description of the model specifying the response and one to four predictors. For example, if the formula is expressed as `y ~ x`, the variable on the left-hand side of the tilde operator is the response variable, and `x` on the right-hand side is the explanatory variable. The `span` argument is the smoothing parameter (usually 0.10 or 0.25), and `degree` is the order of polynomial (1 or 2). The fitted values can be obtained through the `fitted` function.

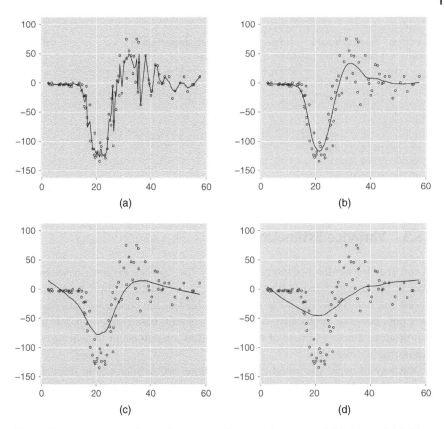

Figure 13.4 Loess curve fitting for motorcycle data using (a) $\alpha = 0.05$, (b) $\alpha = 0.20$, (c) $\alpha = 0.50$, and (d) $\alpha = 0.80$.

Example 13.2 Consider the motorcycle accident data found in Schmidt, Matter, and Schüler (1981). The first column is time, measured in milliseconds, after a simulated impact of a motorcycle. The second column is the acceleration factor of the driver's head (accel), measured in g (9.8 m s^{-2}). Time versus accel is graphed in Figure 13.4. The R code below creates an LOESS curve to model acceleration as a function of time (also in the figure). Note how the smoothing parameter influences the fit of the curve:

```
> motor <- read.table("./motorcycle.dat");
> time <- motor[,1];
> accel <- motor[,2];
> fit<-loess(accel ~ time,span=0.2,degree=1);
> plot(time,accel,ylim=c(-150,100))
> lines(time,fitted(fit),type="l")
```

```
>
> motor.plot <- function(alpha){
+ fit <- loess(accel ~ time,span=alpha,degree=1);
+ dat <- data.frame(x=time,y=fitted(fit));
+     p <- ggplot() + geom_point(aes(x=time,y=accel),shape=1)
+     p <- p + geom_line(aes(x=x,y=y),data=dat)
+     p <- p + xlab("") + ylab("") + ylim(c(-150,100))
+     print(p);
+ }
>
> motor.plot(0.05)
> motor.plot(0.2)
> motor.plot(0.5)
> motor.plot(0.8)
```

13.3   Variance Estimation

In constructing confidence intervals for $m(x)$, the variance estimate based on the smooth linear regression (with pooled-variance estimate) will produce the narrowest interval. However, if the estimate is biased, the confidence interval will have poor coverage probability. An estimator of $m(x)$ based only on points near x will produce a poor estimate of variance and as a result is apt to generate wide, uninformative intervals.

One way to avoid the worst pitfalls of these two extremes is to detrend the data locally and use the estimated variance from the detrended data. Altman and Paulson (1993) use pseudo-residuals $\tilde{e}_i = y_i - (y_{i+1} + y_{i-1})/2$ to form a variance estimator

$$\tilde{\sigma}^2 = \frac{2}{3(n-2)} \sum_{i=1}^{n-1} \tilde{e}_i^2,$$

where $\tilde{\sigma}^2/\sigma^2$ is distributed χ^2 with $(n-2)/2$ degrees of freedom. Because both the kernel and nearest neighbor estimators have linear form in y_i, a $100(1-\alpha)\%$ confidence interval for $m(t)$ can be approximated with

$$\hat{m}(t) \pm t_r(\alpha)\sqrt{\tilde{\sigma}^2 \sum a^2_i},$$

where $r = (n-2)/2$.

13.4   Splines

> **spline** (splīne) *n.* **1.** A flexible piece of wood, hard rubber, or metal used in drawing curves. **2.** A wooden or metal strip; a slat.
>
> The American Heritage Dictionary

Splines, in the mathematical sense, are concatenated piecewise polynomial functions that either interpolate or approximate the scatterplot generated by n observed pairs, $(X_1, Y_1), \ldots, (X_n, Y_n)$. Isaac J. Schoenberg, the "father of splines," was born in Galatz, Romania, on 21 April 1903 and died in Madison, Wisconsin, USA, on 21 February 1990. The more than 40 papers on splines written by Schoenberg after 1960 gave much impetus to the rapid development of the field. He wrote the first several in 1963, during a year's leave in Princeton at the Institute for Advanced Study; the others are part of his prolific output as a member of the Mathematics Research Center at the University of Wisconsin-Madison, in which he joined in 1965.

13.4.1 Interpolating Splines

There are many varieties of splines. Although piecewise constant, linear, and quadratic splines are easy to construct, cubic splines are most commonly used because they have a desirable extreme property.

Denote the cubic spline function by $m(x)$. Assume $X_1, X_2, \ldots, X_n$ are ordered and belong to a finite interval $[a, b]$. We will call $X_1, X_2, \ldots, X_n$ *knots*. On each interval $[X_{i-1}, X_i]$, $i = 1, 2, \ldots, n + 1$, $X_0 = a$, $X_{n+1} = b$, the spline $m(x)$ is a polynomial of degree less than or equal to 3. In addition, these polynomial pieces are connected in such a way that the second derivatives are continuous. That means that at the knot points X_i, $i = 1, \ldots, n$ where the two polynomials from the neighboring intervals meet, the polynomials have common tangent and curvature. We say that such functions belong to $C^2[a, b]$, the space of all functions on $[a, b]$ with continuous second derivative.

The cubic spline is called *natural* if the polynomial pieces on the intervals $[a, X_1]$ and $[X_n, b]$ are of degree 1, that is, linear. The following two properties distinguish natural cubic splines from other functions in $C^2[a, b]$.

Unique interpolation: Given the n pairs, $(X_1, Y_1), \ldots, (X_n, Y_n)$, with distinct knots X_i, there is a *unique* natural cubic spline m that interpolates the points, that is, $m(X_i) = Y_i$.

Extremal property: Given n pairs, $(X_1, Y_1), \ldots, (X_n, Y_n)$, with distinct and ordered knots X_i, the natural cubic spline $m(x)$ that interpolates the points also minimizes the curvature on the interval $[a, b]$, where $a < X_1$ and $X_n < b$. In other words, for any other function $g \in C^2[a, b]$,

$$\int_a^b (m''(t))^2 \, dt \leq \int_a^b (g''(t))^2 dt.$$

Example 13.3 One can "draw" the letter $\mathcal{V}$ using a simple spline. The bivariate set of points (X_i, Y_i) below lead the cubic spline to trace a shape reminiscent of the script letter $\mathcal{V}$. The result of R program is given in Figure 13.5:

Figure 13.5 A cubic spline drawing of letter $\mathcal{V}$.

```
> x <- c(10, 40, 40, 20,  60, 50, 25, 16, 30, 60, 80, 75, 65, 100);
> y <- c(85, 90, 65, 55, 100, 70, 35, 10, 10, 36, 60, 65, 55,  50);
> t <- 1:length(x);
> tt <- seq(1,length(t),length=250)
> fit1 <- splinefun(t,x); xx <- fit1(tt);
> fit2 <- splinefun(t,y); yy <- fit2(tt);
> plot(x,y,axes=FALSE,xlab="",ylab="",ylim=c(0,100),xlim=c(0,100));
> lines(xx,yy,type="l");
```

Example 13.4 In R, the functions splinefun and bicubic (in akima package) compute the cubic spline interpolant, and for the following x and y:

```
> x <- 4*pi*c(0,1,runif(20));
> y <- sin(x);
> fit <- splinefun(x,y);
> xx <- seq(0,max(x),length=100);yy<-fit(xx);
> p <- ggplot() + geom_point(aes(x=x,y=y),size=3)
> p <- p + geom_line(aes(x=xx,y=yy)) + xlab("") + ylab("")
> print(p)
```

The interpolation is plotted in Figure 13.6a, along with the data. A surface interpolation by 2D splines is demonstrated by the following R code and Figure 13.6b,c:

```
> x <- seq(-1,1,by=0.2);
> y <- seq(-1,1,by=0.25);
> z <- outer(x,y,function(x,y){sin(10*(x^2+y^2));});
> xy<-expand.grid(seq(-1,1,length=100),seq(-1,1,length=80));
> fit<-bicubic(x,y,z,xy[,1],xy[,2]);
>
> xx <- seq(-1,1,length=100); yy <- seq(-1,1,length=80);
> zz <- matrix(fit$z,nrow=100);
```

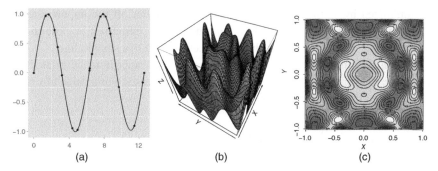

Figure 13.6 (a) Interpolating sine function. (b) Interpolating a surface. (c) Interpolating a contour.

```
> persp(x=xx,y=yy,z=zz,col="lightblue",phi=45,theta=-60,xlab="X",
+     ylab="Y",zlab="Z");
>
> image(x=seq(-1,1,length=100),y=seq(-1,1,length=80),
+     z=matrix(fit$z,nrow=100),xlab="X",ylab="Y");
> contour(x=seq(-1,1,length=100),y=seq(-1,1,length=80),
+     z=matrix(fit$z,nrow=100),add=TRUE);
```

There are important distinctions between spline regressions and regular polynomial regressions. The latter technique is applied to regression curves where the practitioner can see an interpolating quadratic or cubic equation that locally matches the relationship between the two variables being plotted. The Stone–Weierstrass theorem (Weierstrass, 1885) tells us that any continuous function in a closed interval can be approximated well by some polynomial. While a higher-order polynomial will provide a closer fit at any particular point, the loss of parsimony is not the only potential problem of over fitting; unwanted oscillations can appear between data points. Spline functions avoid this pitfall.

13.4.2 Smoothing Splines

Smoothing splines, unlike interpolating splines, may not contain the points of a scatterplot, but are rather a form of nonparametric regression. Suppose we are given bivariate observations (X_i, Y_i), $i = 1, \dots, n$. The continuously differentiable function $\hat{m}$ on $[a, b]$ that minimizes the functional

$$\sum_{i=1}^{n} (Y_i - m(X_i))^2 + \lambda \int_a^b (m''(t))^2 \, dt \tag{13.8}$$

is exactly a natural cubic spline. The cost functional in (13.8) has two parts: $\sum_{i=1}^{n} (Y_i - m(X_i))^2$ is minimized by an interpolating spline, and $\int_a^b (m''(t))^2 \, dt$

is minimized by a straight line. The parameter λ trades off the importance of these two competing costs in (13.8). For small λ, the minimizer is close to an interpolating spline. For λ large, the minimizer is closer to a straight line.

Although natural cubic smoothing splines do not appear to be related to kernel-type estimators, they can be similar in certain cases. For a value of x that is away from the boundary, if n is large and λ small, let

$$\hat{m}(x) = \frac{1}{n} \sum_{i=1}^{n} \frac{K_{h_i}(X - i - x)}{f(X_i)} Y_i,$$

where f is the density of the X's, $h_i = [\lambda/(nf(X_i))]^{1/4}$ and the kernel K is

$$K(x) = \frac{1}{2} \exp\left\{ -|x|/\sqrt{2} \right\} \sin\left(|x|/\sqrt{2} + \pi/4 \right). \tag{13.9}$$

As an alternative to minimizing (13.8), the following version is often used:

$$p \sum_{i=1}^{n} (Y_i - m(X_i))^2 + (1 - p) \int_a^b (m''(t))^2 \, dt. \tag{13.10}$$

In this case, $\lambda = (1 - p)/p$. Assume that h is an average spacing between the neighboring X's. An automatic choice for p is $6(6 + h^3)$ or $\lambda = h^3/6$.

13.4.2.1 Smoothing Splines as Linear Estimators

The spline estimator is linear in the observations, $\hat{m} = S(\lambda)Y$, for a smoothing matrix $S(\lambda)$. The Reinsch algorithm (Reinsch, 1967) efficiently calculates S as

$$S(\lambda) = \left(I + \lambda Q R^{-1} Q' \right)^{-1}, \tag{13.11}$$

where Q and R are structured matrices of dimensions $n \times (n - 2)$ and $(n - 2) \times (n - 2)$, respectively:

$$Q = \begin{pmatrix} q_{12} & & & \\ q_{22} & q_{23} & & \\ q_{32} & q_{33} & & \\ & q_{43} & & \\ & & \ddots & \\ & & q_{n-2,n-1} & \\ & & q_{n-1,n-1} & \\ & & q_{n,n-1} & \end{pmatrix}, \quad R = \begin{pmatrix} r_{22} & r_{23} & & \\ r_{32} & r_{33} & & \\ & r_{43} & & \\ & & \ddots & \\ & & q_{n-2,n-1} & \\ & & q_{n-1,n-1} & \end{pmatrix},$$

with entries

$$q_{ij} = \begin{cases} \dfrac{1}{h_{j-1}}, & i = j - 1 \\[2mm] -\left(\dfrac{1}{h_{j-1}} + \dfrac{1}{h_j}\right), & i = j \\[2mm] \dfrac{1}{h_j}, & i = j + 1 \end{cases}$$

and

$$r_{ij} = \begin{cases} \dfrac{1}{6} h_{j-1}, & i = j - 1 \\[2mm] \dfrac{1}{3}\left(h_{j-1} + h_j\right), & i = j \\[2mm] \dfrac{1}{6} h_j, & i = j + 1. \end{cases}$$

The values h_i are spacings between the X_i's, i.e. $h_i = X_{i+1} - X_i$, $i = 1, \ldots, n - 1$. For details about the Reinsch algorithm, see Green and Silverman (1994).

13.4.3   Selecting and Assessing the Regression Estimator

Let $\hat{m}_h(x)$ be the regression estimator of $m(x)$, obtained by using the set of n observations $(X_1, Y_1), \ldots, (X_n, Y_n)$ and parameter h. Note that for kernel-type estimators, h is the bandwidth, but for splines, h is λ in (13.8). Define the average mean square error of the estimator $\hat{m}_h$ as

$$\text{AMSE}(h) = \frac{1}{n} \sum_{i=1}^{n} \mathbb{E}\left[\hat{m}(X_i) - m(X_i)\right]^2.$$

Let $\hat{m}_{(i)h}(x)$ be the estimator of $m(x)$, based on bandwidth parameter h, obtained by using all the observation pairs except the pair (X_i, Y_i). Define the cross-validation score $\text{CV}(h)$ depending on the bandwidth/tradeoff parameter h as

$$\text{CV}(h) = \frac{1}{n} \sum_{i=1}^{n} \left[Y_i - \hat{m}_{(i)h}(x)\right]^2. \tag{13.12}$$

Because the expected $\text{CV}(h)$ score is proportional to the $\text{AMSE}(h)$ or, more precisely,

$$\mathbb{E}\left[\text{CV}(h)\right] \approx \text{AMSE}(h) + \sigma^2,$$

where σ^2 is constant variance of errors ε_i, the value of h that minimizes $\text{CV}(h)$ is likely, on average, to produce the best estimators.

For smoothing splines, and more generally, for linear smoothers $\hat{\mathbf{m}} = S(h)\mathbf{y}$, the computationally demanding procedure in (13.12) can be simplified by

$$CV(h) = \frac{1}{n} \sum_{i=1}^{n} \left[\frac{Y_i - \hat{m}_h(x)}{1 - S_{ii}(h)} \right]^2, \tag{13.13}$$

where $S_{ii}(h)$ is the diagonal element in the smoother (13.11). When n is large, constructing the smoothing matrix $S(h)$ is computationally difficult. There are efficient algorithms (Hutchinson and de Hoog, 1985) that calculate only needed diagonal elements $S_{ii}(h)$, for smoothing splines, with computational cost of $O(n)$.

Another simplification in finding the best smoother is the generalized cross-validation (GCV) criterion. The denominator in (13.13) $1 - S_{ii}(h)$ is replaced by overall average $1 - n^{-1} \sum_{i=1}^{n} S_{ii}(h)$ or, in terms of its trace, $1 - n^{-1} trS(h)$. Thus

$$GCV(h) = \frac{1}{n} \sum_{i=1}^{n} \left[\frac{Y_i - \hat{m}_h(x)}{1 - trS(h)/n} \right]^2. \tag{13.14}$$

Example 13.5 Assume that $\hat{m}$ is a spline estimator and that $\lambda_1, \ldots, \lambda_n$ are eigenvalues of matrix $QR^{-1}Q'$ from (13.11). Then, $trS(h) = \sum_{i=1}^{n} (1 + h\lambda_i)^{-1}$. The GCV criterion becomes

$$GCV(h) = \frac{nRSS(h)^2}{\left[n - \sum_{i=1}^{n} \frac{1}{1 + h\lambda_i} \right]^2}.$$

13.4.4 Spline Inference

Suppose that the estimator $\hat{m}$ is a linear combination of the Y_is:

$$\hat{m}(x) = \sum_{i=1}^{n} a_i(x)Y_i.$$

Then

$$\mathbb{E}(\hat{m}(x)) = \sum_{i=1}^{n} a_i(x)m(X_i), \quad \text{and} \quad \mathbb{V}ar(\hat{m}(x)) = \left(\sum_{i=1}^{n} a_i(x)^2 \right) \sigma^2.$$

Given $x = X_j$ we see that $\hat{m}$ is unbiased, that is, $\mathbb{E}\hat{m}(X_j) = m(X_j)$ only if all $a_i = 0$, $i \neq j$.

On the other hand, variance is minimized if all a_i are equal. This illustrates, once again, the tradeoff between the estimator's bias and variance. The variance of the errors is supposed to be constant. In linear regression we estimated the variance as

$$\hat{\sigma}_0^2 = \frac{RSS}{n - p},$$

where p is the number of free parameters in the model. Here we have an analogous estimator:

$$\hat{\sigma}^2 = \frac{\text{RSS}}{n - tr(S)},$$

where $\text{RSS} = \sum_{i=1}^{n} \left[Y_i - \hat{m}(X_i) \right]^2$.

13.5 Summary

This chapter has given a brief overview of both kernel estimators and local smoothers. An example from Gasser and Müller (1979) shows that choosing a smoothing method over a parametric regression model can make a crucial difference in the conclusions of a data analysis. A parametric model by Preece and Baines (1978) was constructed for predicting the future height of a human based on measuring children's heights at different stages of development. The parametric regression model they derived for was particularly complicated but provided a great improvement in estimating the human growth curve. Published six years later, the nonparametric regression by Gasser et al. (1984) brought out an important nuance of the growth data that could not be modeled with the Preece and Baines model (or any model that came before it). An example is a subtle growth spurt that seems to occur in children around seven years in age. Altman (1992) notes that such a growth spurt was discussed in past medical papers but had "disappeared from the literature following the development of the parametric models which did not allow for it."

13.6 Exercises

13.1 Describe how the LOESS curve can be equivalent to least-squares regression.

13.2 Data set `oj287.dat` is the light curve of the blazar OJ287. Blazars, also known as *BL Lac Objects* or *BL Lacertaes*, are bright, extragalactic, starlike objects that can vary rapidly in their luminosity. Rapid fluctuations of blazar brightness indicate that the energy-producing region is small. Blazars emit polarized light that is featureless on a light plot. These are interpreted to be active galaxy nuclei, not so different from quasars. From this interpretation, it follows that blazars are in the center of an otherwise normal galaxy and are probably powered by a supermassive black hole. Use a LP estimator to analyze the data in `oj287.dat` where column 1 is the Julian time and column 2 is the brightness. How does the fit compare for the three values of p in $\{0, 1, 2\}$?

13.3 Consider the function

$$s(x) = \begin{cases} 1 - x + x^2 - x^3, & 0 < x < 1, \\ -2(x - 1) - 2(x - 1)^2, & 1 < x < 2, \\ -4 - 6(x - 2) - 2(x - 2)^2, & 2 < x < 3. \end{cases}$$

Does $s(x)$ define a smooth cubic spline on $[0, 3]$ with knots 1 and 2? If so, plot the three polynomials on $[0, 3]$.

13.4 In R, open the data file `earthquake.dat` that contains water level records for a set of six wells in California. The measurements are made across time. Construct a LOESS smoother to examine trends in the data. Where does LOESS succeed? Where does it fail to capture the trends in the data?

13.5 Simulate a data set as follows:

```
x <- sort(runif(100));
y <- x^2 + 0.1*rnorm(100);
```

Fit an interpolating spline to the simulated data as shown in Figure 13.7a. The dashed line is $y = x^2$.

13.6 Refer to the motorcycle data from Figure 13.4. Fit a spline to the data. Variable `time` is the time in milliseconds, and `accel` is the acceleration of a head measured in (g). See Figure 13.7b as an example.

13.7 Star S in the Big Dipper constellation (Ursa Major) has a regular variation in its apparent magnitude[2]:

θ	−100	−60	−20	20	60	100	140
Magnitude	8.37	9.40	11.39	10.84	8.53	7.89	8.37

---

2 Campbell and Jacchia (1941).

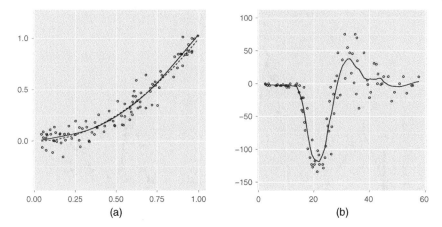

Figure 13.7 (a) Square plus noise. (b) Motorcycle data: time (X_i) and acceleration (Y_i), $i = 1, \ldots, 82$.

The magnitude is known to be periodic with period 240, so that the magnitude at $\theta = -100$ is the same as at $\theta = 140$. The spline `yy <- splinefun(x,y,'periodic')` constructs a cubic spline whose first and second derivatives are the same at the ends of the interval. Use it to interpolate the data. Plot the data and the interpolating curve in the same figure. Estimate the magnitude at $\theta = 0$.

13.8 Use the smoothing splines to analyze the data in `oj287.dat` that was described in Exercise 13.2. For your reference, the data and implementation of spline smoothing are given in Figure 13.8.

RELEVANT R FUNCTIONS AND DATA SETS IN THIS CHAPTER

R codes: `lpfit.r, loc.lin.r`
R functions: `bicubic, ksmooth, locfit, locPolSmootherC, loess, smooth.spline, spline, splinefunc`
R package: `akima, locfit, locpol`

`earthquake.dat, motorcycle.dat, oj287.dat`

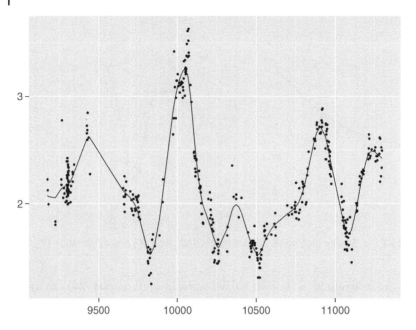

Figure 13.8 Blazar OJ287 luminosity.

References

Altman, N. S. (1992), "An Introduction to Kernel and Nearest Neighbor Nonparametric Regression," *American Statistician*, 46, 175–185.

Altman, N. S., and Paulson, C. P. (1993), "Some Remarks about the Gasser-Sroka-Jennen-Steinmetz Variance Estimator," *Communications in Statistics - Theory and Methods*, 22, 1045–1051.

Campbell, L., and Jacchia, L. (1941), *The Story of Variable Stars*, Philadelphia, PA: The Blackiston Co.

Cleveland, W. S. (1979), "Robust Locally Weighted Regression and Smoothing Scatterplots," *Journal of the American Statistical Association*, 74, 829–836.

Gasser, T., and Müller, H. G. (1979), "Kernel Estimation of Regression Functions," in *Smoothing Techniques for Curve Estimation*, Eds. Gasser and Rosenblatt, Eds. T. Gasser and M. Rosenblatt, Heidelberg: Springer-Verlag.

Gasser, T., Müller, H. G., Köhler, W., Molinari, L., and Prader, A. (1984), "Nonparametric Regression Analysis of Growth Curves," *Annals of Statistics*, 12, 210–229.

Green, P. J., and Silverman, B. W. (1994), *Nonparametric Regression and Generalized Linear Models: A Roughness Penalty Approach*, London: Chapman & Hall.

Hutchinson, M. F., and de Hoog, F. R. (1985), "Smoothing Noisy Data with Spline Functions," *Numerical Mathematics*, 1, 99–106.

Nadaraya, E. A. (1964), "On Estimating Regression," *Theory of Probability and Its Applications*, 10, 186–190.

Preece, M. A., and Baines, M. J. (1978), "A New Family of Mathematical Models Describing the Human Growth Curve," *Annals of Human Biology*, 5, 1–24.

Reinsch, C. H. (1967), "Smoothing by Spline Functions," *Numerical Mathematics*, 10, 177–183.

Schmidt, G., Mattern, R., and Schüler, F. (1981), "Biomechanical Investigation to Determine Physical and Traumatological Differentiation Criteria for the Maximum Load Capacity of Head and Vertebral Column with and without Helmet under Effects of Impact," *EEC Research Program on Biomechanics of Impacts. Final Report Phase III, 65*, Heidelberg, Germany: Institut fur Rechtsmedizin.

Silverman, B. W. (1985), "Some Aspects of the Spline Smoothing Approach to Non-parametric Curve Fitting," *Journal of the Royal Statistical Society. Series B*, 47, 1–52.

Watson, G. S. (1964), "Smooth Regression Analysis," *Sankhya, Series A*, 26, 359–372.

Weierstrass, K. (1885), "Über die analytische Darstellbarkeit sogenannter willkürlicher Functionen einer reellen Vernderlichen," Sitzungsberichte der Königlich Preußischen Akademie der Wissenschaften zu Berlin, 1885 (II). Erste Mitteilung (Part 1) 633–639; Zweite Mitteilung (Part 2) 789–805.

14

Wavelets

It is error only, and not truth, that shrinks from inquiry.

Thomas Paine (1737–1809)

14.1 Introduction to Wavelets

Wavelet-based procedures are now indispensable in many areas of modern statistics, for example, in regression, density and function estimation, factor analysis, modeling and forecasting of time series, functional data analysis, and data mining and classification, with ranges of application areas in science and engineering. Wavelets owe their initial popularity in statistics to *shrinkage*, a simple and yet powerful procedure efficient for many nonparametric statistical models.

Wavelets are functions that satisfy certain requirements. The name *wavelet* comes from the requirement that they integrate to zero, "waving" above and below the *x*-axis. The diminutive in *wavelet* suggests its good localization. Other requirements are technical and needed mostly to ensure quick and easy calculation of the direct and inverse wavelet transform.

There are many kinds of wavelets. One may choose between smooth wavelets, compactly supported wavelets, wavelets with simple mathematical expressions, or wavelets with short associated filters. The simplest is the *Haar wavelet*, and we discuss it as an introductory example in Section 14.2.1. Examples of some wavelets (from Daubechies' family) are given in Figure 14.1. Note that scaling and wavelet functions in panels (a, b) in Figure 14.1 (Daubechies 4) are supported on a short interval (of length 3) but are not smooth; the other family member, Daubechies 16 (panels (e, f) in Figure 14.1), is smooth, but its support is much larger.

Nonparametric Statistics with Applications to Science and Engineering with R, Second Edition.
Paul Kvam, Brani Vidakovic, and Seong-joon Kim.
© 2023 John Wiley & Sons, Inc. Published 2023 by John Wiley & Sons, Inc.

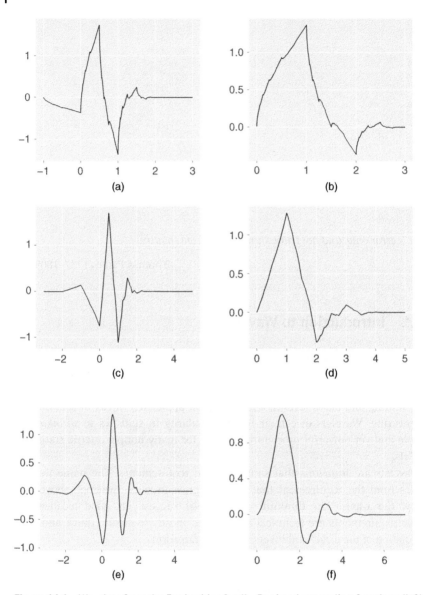

Figure 14.1 Wavelets from the Daubechies family. Depicted are scaling functions (*left*) and wavelets (*right*) corresponding to (a, b) 4, (c, d) 8, and (e, f) 16 tap filters.

Like sines and cosines in Fourier analysis, wavelets are used as atoms in representing other functions. Once the wavelet (sometimes informally called *the mother wavelet*) $\psi(x)$ is fixed, one can generate a family by its translations and dilations, $\{\psi(\frac{x-b}{a}), (a, b) \in \mathbb{R}^+ \times \mathbb{R}\}$. It is convenient to take special values for a and b in defining the wavelet basis: $a = 2^{-j}$ and $b = k \cdot 2^{-j}$, where k and j are integers. This choice of a and b is called *critical sampling* and generates a sparse basis. In addition, this choice naturally connects multiresolution analysis in discrete signal processing with the mathematics of wavelets.

Wavelets, as building blocks in modeling, are localized well in both time and scale (frequency). Functions with rapid local changes (functions with discontinuities, cusps, sharp spikes, etc.) can be well represented with a minimal number of wavelet coefficients. This parsimony does not, in general, hold for other standard orthonormal bases that may require many "compensating" coefficients to describe discontinuity artifacts or local bursts.

Heisenberg's principle states that time–frequency models cannot be precise in the time and frequency domains simultaneously. Wavelets, of course, are subject to Heisenberg's limitation but can adaptively distribute the time–frequency precision depending on the nature of function they are approximating. The economy of wavelet transforms can be attributed to this ability.

The above already hints at how the wavelets can be used in statistics. Large and noisy data sets can be easily and quickly transformed by a discrete wavelet transform (the counterpart of discrete Fourier transform). The data are coded by their wavelet coefficients. In addition, the descriptor "fast" in fast Fourier transforms can, in most cases, be replaced by "faster" for the wavelets. It is well known that the computational complexity of the fast Fourier transformation is $O(n \cdot \log_2(n))$. For the fast wavelet transform, the computational complexity goes down to $O(n)$. This means that the complexity of algorithm (in terms either of number of operations, time, or memory) is proportional to the input size, n.

Various data-processing procedures can now be done by processing the corresponding wavelet coefficients. For instance, one can do function smoothing by shrinking the corresponding wavelet coefficients and then back-transforming the shrunken coefficients to the original domain (Figure 14.2). A simple shrinkage method, thresholding, and some thresholding policies are discussed later.

Figure 14.2 Wavelet-based data processing.

An important feature of wavelet transforms is their *whitening* property. There is ample theoretical and empirical evidence that wavelet transforms reduce the dependence in the original signal. For example, it is possible, for any given stationary dependence in the input signal, to construct a biorthogonal wavelet basis such that the corresponding in the transform are uncorrelated (a wavelet counterpart of the so called Karhunen–Loève transform). For a discussion and examples, see Walter and Shen (2001).

We conclude this incomplete inventory of wavelet transform features by pointing out their sensitivity to self-similarity in data. The scaling regularities are distinctive features of self-similar data. Such regularities are clearly visible in the wavelet domain in the wavelet spectra, a wavelet counterpart of the Fourier spectra. More arguments can be provided: computational speed of the wavelet transform, easy incorporation of prior information about some features of the signal (smoothness, distribution of energy across scales), etc.

Basics on wavelets can be found in many texts, monographs, and papers at many different levels of exposition. Student interested in the exposition that is beyond this chapter's coverage should consult monographs by Daubechies (1992), Ogden (1997), Vidakovic (1999), and Walter and Shen (2001), among others.

14.2 How Do the Wavelets Work?

14.2.1 The Haar Wavelet

To explain how wavelets work, we start with an example. We choose the simplest and the oldest of all wavelets (we are tempted to say: grandmother of all wavelets!), the Haar wavelet, $\psi(x)$. It is a step function taking values 1 and -1 on intervals $[0, \frac{1}{2})$ and $[\frac{1}{2}, 1)$, respectively. The graphs of the Haar wavelet and some of its dilations/translations are given in Figure 14.3.

The Haar wavelet has been known for almost 100 years and is used in various mathematical fields. Any continuous function can be approximated uniformly by Haar functions, even though the "decomposing atom" is discontinuous.

Dilations and translations of the function ψ,

$$\psi_{jk}(x) = \text{const} \cdot \psi(2^j x - k), \quad j, k \in \mathbb{Z},$$

where $\mathbb{Z} = \{\dots, -2, -1, 0, 1, 2, \dots\}$ is set of all integers, define an orthogonal basis of $L^2(\mathbb{R})$ (the space of all square integrable functions). This means that any function from $L^2(\mathbb{R})$ may be represented as a (possibly infinite) linear combination of these basis functions.

The orthogonality of ψ_{jk}'s is easy to check. It is apparent that

$$\int \psi_{jk} \cdot \psi_{j'k'} = 0, \tag{14.1}$$

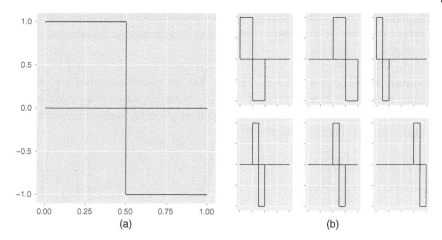

(a)   (b)

Figure 14.3 (a) Haar wavelet $\psi(x) = \mathbf{1}(0 \leq x < \frac{1}{2}) - \mathbf{1}\left(\frac{1}{2} < x \leq 1\right)$. (b) Some dilations and translations of Haar wavelet on [0,1].

whenever $j = j'$ and $k = k'$ are not satisfied simultaneously. If $j \neq j'$ (say $j' < j$), then nonzero values of the wavelet $\psi_{j'k'}$ are contained in the set where the wavelet ψ_{jk} is constant. That makes integral in (14.1) equal to zero: if $j = j'$, but $k \neq k'$, then at least one factor in the product $\psi_{j'k'} \cdot \psi_{jk}$ is zero. Thus the functions ψ_{ij} are mutually orthogonal. The constant that makes this orthogonal system orthonormal is $2^{j/2}$. The functions $\psi_{10}, \psi_{11}, \psi_{20}, \psi_{21}, \psi_{22}, \psi_{23}$ are depicted in Figure 14.3b.

The family $\{\psi_{jk}, j \in \mathbb{Z}, k \in \mathbb{Z}\}$ defines an orthonormal basis for $\mathbb{L}^2$. Alternatively we will consider orthonormal bases of the form $\{\phi_{L,k}, \psi_{jk}, j \geq L, k \in \mathbb{Z}\}$, where ϕ is called the *scaling function* associated with the wavelet basis ψ_{jk}, and $\phi_{jk}(x) = 2^{j/2}\phi(2^j x - k)$. The set of functions $\{\phi_{L,k}, k \in \mathbb{Z}\}$ spans the same subspace as $\{\psi_{jk}, j < L, k \in \mathbb{Z}\}$. For the Haar wavelet basis, the scaling function is simple. It is an indicator of the interval [0,1), that is,

$$\phi(x) = \mathbf{1}(0 \leq x < 1).$$

The data analyst is mainly interested in wavelet representations of functions generated by data sets. Discrete wavelet transforms map the data from the time domain (the original or input data, signal vector) to the wavelet domain. The result is a vector of the same size. Wavelet transforms are linear, and they can be defined by matrices of dimension $n \times n$ when they are applied to inputs of size n. Depending on a boundary condition, such matrices can be either orthogonal or "close" to orthogonal. A wavelet matrix W is close to orthogonal when the orthogonality is violated by nonperiodic handling of boundaries resulting in a small but nonzero value of the norm $\| WW' - I \|$, where I is the identity matrix. When the matrix is orthogonal, the corresponding transform can be thought as a rotation in $\mathbb{R}^n$ space where the data vectors represent coordinates of points. For a fixed point, the coordinates in the new, rotated space comprise the discrete wavelet transformation of the original coordinates.

Example 14.1 Let $y = (1, 0, -3, 2, 1, 0, 1, 2)$. The associated function f is given in Figure 14.4. The values $f(k) = y_k$, $k = 0, 1, \ldots, 7$ are interpolated by a piecewise constant function. The following matrix equation gives the connection between y and the wavelet coefficients d, $y = W'd$:

$$
\begin{bmatrix} 1 \\ 0 \\ -3 \\ 2 \\ 1 \\ 0 \\ 1 \\ 2 \end{bmatrix}
=
\begin{bmatrix}
\frac{1}{2\sqrt{2}} & \frac{1}{2\sqrt{2}} & \frac{1}{2} & 0 & \frac{1}{\sqrt{2}} & 0 & 0 & 0 \\
\frac{1}{2\sqrt{2}} & \frac{1}{2\sqrt{2}} & \frac{1}{2} & 0 & -\frac{1}{\sqrt{2}} & 0 & 0 & 0 \\
\frac{1}{2\sqrt{2}} & \frac{1}{2\sqrt{2}} & -\frac{1}{2} & 0 & 0 & \frac{1}{\sqrt{2}} & 0 & 0 \\
\frac{1}{2\sqrt{2}} & \frac{1}{2\sqrt{2}} & -\frac{1}{2} & 0 & 0 & -\frac{1}{\sqrt{2}} & 0 & 0 \\
\frac{1}{2\sqrt{2}} & -\frac{1}{2\sqrt{2}} & 0 & \frac{1}{2} & 0 & 0 & \frac{1}{\sqrt{2}} & 0 \\
\frac{1}{2\sqrt{2}} & -\frac{1}{2\sqrt{2}} & 0 & \frac{1}{2} & 0 & 0 & -\frac{1}{\sqrt{2}} & 0 \\
\frac{1}{2\sqrt{2}} & -\frac{1}{2\sqrt{2}} & 0 & -\frac{1}{2} & 0 & 0 & 0 & \frac{1}{\sqrt{2}} \\
\frac{1}{2\sqrt{2}} & -\frac{1}{2\sqrt{2}} & 0 & -\frac{1}{2} & 0 & 0 & 0 & -\frac{1}{\sqrt{2}}
\end{bmatrix}
\cdot
\begin{bmatrix} c_{00} \\ d_{00} \\ d_{10} \\ d_{11} \\ d_{20} \\ d_{21} \\ d_{22} \\ d_{23} \end{bmatrix} .
$$

(14.2)

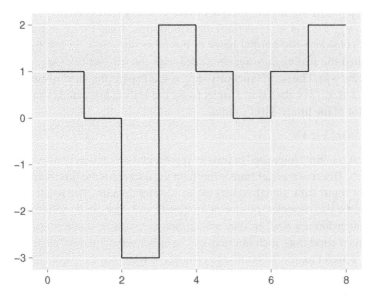

Figure 14.4 A function interpolating y on $[0,8)$.

The solution is $d = Wy$:

$$
\begin{bmatrix} c_{00} \\ d_{00} \\ d_{10} \\ d_{11} \\ d_{20} \\ d_{21} \\ d_{22} \\ d_{23} \end{bmatrix} = \begin{bmatrix} \sqrt{2} \\ -\sqrt{2} \\ 1 \\ -1 \\ \frac{1}{\sqrt{2}} \\ -\frac{5}{\sqrt{2}} \\ \frac{1}{\sqrt{2}} \\ -\frac{1}{\sqrt{2}} \end{bmatrix}.
$$

Accordingly

$$
f(x) = \sqrt{2}\phi_{0,0}(x) - \sqrt{2}\psi_{0,0}(x) + \psi_{1,0}(x) - \psi_{1,1}(x) \tag{14.3}
$$
$$
+ \frac{1}{\sqrt{2}}\psi_{2,0}(x) - \frac{5}{\sqrt{2}}\psi_{2,1}(x) + \frac{1}{\sqrt{2}}\psi_{2,2}(x) - \frac{1}{\sqrt{2}}\psi_{2,3}(x).
$$

The solution is easy to verify. For example, when $x \in [0,1)$,

$$
f(x) = \sqrt{2} \cdot \frac{1}{2\sqrt{2}} - \sqrt{2} \cdot \frac{1}{2\sqrt{2}} + 1 \cdot \frac{1}{2} + \frac{1}{\sqrt{2}} \cdot \frac{1}{\sqrt{2}} = 1/2 + 1/2 = 1 \ (= y_0).
$$

The R script Wavmat.r forms the wavelet matrix W, for a given wavelet base and dimension that is a power of 2. For example, W <- Wavmat(h, n, k0, shift) will calculate $n \times n$ wavelet matrix, corresponding to the filter h (connections between wavelets and filtering will be discussed in the following section), and k0 and shift are given parameters. We will see that Haar wavelet corresponds to a filter $h = \left\{ \sqrt{2}/2, \sqrt{2}/2 \right\}$. Here is the above example in R:

```
> source("Wavmat.r")
> W <- Wavmat(c(sqrt(2)/2,sqrt(2)/2),2^3,3,2);
> t(W)
          [,1]       [,2] [,3] [,4]       [,5]       [,6]       [,7]       [,8]
[1,] 0.3535534  0.3535534  0.5  0.0  0.7071068  0.0000000  0.0000000  0.0000000
[2,] 0.3535534  0.3535534  0.5  0.0 -0.7071068  0.0000000  0.0000000  0.0000000
[3,] 0.3535534  0.3535534 -0.5  0.0  0.0000000  0.7071068  0.0000000  0.0000000
[4,] 0.3535534  0.3535534 -0.5  0.0  0.0000000 -0.7071068  0.0000000  0.0000000
[5,] 0.3535534 -0.3535534  0.0  0.5  0.0000000  0.0000000  0.7071068  0.0000000
[6,] 0.3535534 -0.3535534  0.0  0.5  0.0000000  0.0000000 -0.7071068  0.0000000
[7,] 0.3535534 -0.3535534  0.0 -0.5  0.0000000  0.0000000  0.0000000  0.7071068
[8,] 0.3535534 -0.3535534  0.0 -0.5  0.0000000  0.0000000  0.0000000 -0.7071068
> dat <- c(1, 0, -3, 2, 1, 0, 1, 2);
> wt <- W %*% dat
> t(wt)
```

```
        [,1]       [,2]  [,3] [,4]      [,5]      [,6]      [,7]        [,8]
[1,]  1.414214 -1.414214    1   -1 0.7071068 -3.535534 0.7071068 -0.7071068
>
> data <- t(W) %*% wt
> t(data)
       [,1]         [,2] [,3] [,4] [,5]        [,6] [,7] [,8]
[1,]     1 1.110223e-16   -3    2    1 1.110223e-16    1    2
```

Performing wavelet transformations via the product of wavelet matrix W and input vector y is conceptually straightforward, but of limited practical value. Storing and manipulating wavelet matrices for inputs exceeding tens of thousands in length is not feasible.

14.2.2 Wavelets in the Language of Signal Processing

Fast discrete wavelet transforms become feasible by implementing the so-called *cascade algorithm* introduced by Mallat (1989). Let $\{h(k), k \in Z\}$ and $\{g(k), k \in Z\}$ be the *quadrature mirror filters* in the terminology of signal processing. Two filters h and g form a quadrature mirror pair when

$$g(n) = (-1)^n h(1 - n).$$

The filter $h(k)$ is a *low-pass* or *smoothing* filter while $g(k)$ is the *high-pass* or *detail* filter. The following properties of $h(n), g(n)$ can be derived by using the so-called scaling relationship, Fourier transforms, and orthogonality: $\Sigma_k h(k) = \sqrt{2}$, $\Sigma_k g(k) = 0$, $\Sigma_k h(k)^2 = 1$, and $\Sigma_k h(k)k(k - 2m) = 1(m = 0)$.

The most compact way to describe the cascade algorithm and to give efficient recipe for determining discrete wavelet coefficients is by using *operator representation of filters*. For a sequence $a = \{a_n\}$, the operators H and G are defined by the following coordinate-wise relations:

$$(Ha)_n = \Sigma_k h(k - 2n)a_k,$$
$$(Ga)_n = \Sigma_k g(k - 2n)a_k.$$

The operators H and G perform filtering and down-sampling (omitting every second entry in the output of filtering) and correspond to a single step in the wavelet decomposition. The wavelet decomposition thus consists of subsequent application of operators H and G in the particular order on the input data.

Denote the original signal y by $c^{(J)}$. If the signal is of length $n = 2^J$, then $c^{(J)}$ can be understood as the vector of coefficients in a series

$$f(x) = \Sigma_{k=0}^{2^J-1} c_k^{(J)} \phi_{nk},$$

for some scaling function ϕ. At each step of the wavelet transform, we move to a coarser approximation $c^{(j-1)}$ with $c^{(j-1)} = Hc^{(j)}$ and $d^{(j-1)} = Gc^{(j)}$. Here, $d^{(j-1)}$ represent the "details" lost by degrading $c^{(j)}$ to $c^{(j-1)}$.

The filters H and G are Decimating; thus the length of $c^{(j-1)}$ or $d^{(j-1)}$ is half the length of $c^{(j)}$. The discrete wavelet transform of a sequence $y = c^{(J)}$ of length $n = 2^J$ can then be represented as another sequence of length 2^J (notice that the sequence $c^{(j-1)}$ has half the length of $c^{(j)}$):

$$(c^{(0)}, d^{(0)}, d^{(1)}, \ldots, d^{(J-2)}, d^{(J-1)}). \tag{14.4}$$

In fact, this decomposition may not be carried until the singletons $c^{(0)}$ and $d^{(0)}$ are obtained, but could be curtailed at $(J - L)$th step:

$$(c^{(L)}, d^{(L)}, d^{(L+1)}, \ldots, d^{(J-2)}, d^{(J-1)}), \tag{14.5}$$

for any $0 \leq L \leq J - 1$. The resulting vector is still a valid wavelet transform. See Exercise 14.4 for Haar wavelet transform "by hand":

```
dwtr <- function(data,L,filterh){
#   function dwtr = dwtr(data, L, filterh);
#   Calculates the DWT of periodic data set
#   with scaling filter  filterh  and  L  detail levels.
#
#   Example of Use:
#   data <- c(1, 0, -3, 2, 1, 0, 1, 2); filterh <- c(sqrt(2)/2, sqrt(2)/2);
#   dwtr(data, 3, filterh)
#-----------------------------------------------------------------------

n <- length(filterh); # Length of wavelet filter
C <- data;
dwtr <- c();

H <- filterh;
G <- rev(filterh);                       # Make quadrature mirror
G[seq(1,n,by=2)] <- -G[seq(1,n,by=2)];   # counterpart

for(j in 1:L){                           # Start cascade
nn <- length(C);                    # Length
C <- c(C[((-(n-1):-1) %% nn)+1], C); # make periodic
D <- convolve(G,rev(C),type="open"); # Convolve is equivalent to filter
                                     #       with high-pass
D <- D[c(seq(n,(n+nn-2),by=2))+1];   # keep periodic and decimate
C <- convolve(H,rev(C),type="open"); # Convolve (Filter with low-pass)
C <- C[c(seq(n,(n+nn-2),by=2))+1];   # keep periodic and decimate

dwtr <- c(D,dwtr); # Add detail level to dwtr
}
dwtr <- c(C,dwtr); # Add the last ''smooth'' part
return(dwtr);
}
```

As a result, the discrete wavelet transformation can be summarized as

$$y \rightarrow (H^{J-L}y, GH^{J-1-L}y, GH^{J-2-L}y, \ldots, GHy, Gy), \quad 0 \leq L \leq J - 1.$$

The R script `dwtr.r` performs discrete wavelet transform:

```
> source("dwtr.r")
> data <- c(1, 0, -3, 2, 1, 0, 1, 2);
> filter <- c(sqrt(2)/2, sqrt(2)/2);
>
> wt <- dwtr(data,3,filter)
> wt
[1]  1.4142136 -1.4142136  1.0000000 -1.0000000  0.7071068 -3.5355339
     0.7071068 -0.7071068
```

The reconstruction formula is also simple in terms of H and G; we first define adjoint operators $H^\star$ and $G^\star$ as follows:

$$(H^\star a)_k = \Sigma_n h(k - 2n)a_n,$$

$$(G^\star a)_k = \Sigma_n g(k - 2n)a_n.$$

Recursive application leads to

$$(\boldsymbol{c}^{(L)}, \boldsymbol{d}^{(L)}, \boldsymbol{d}^{(L+1)}, \dots, \boldsymbol{d}^{(J-2)}, \boldsymbol{d}^{(J-1)}) \to \boldsymbol{y} = (H^\star)^J \boldsymbol{c}^{(L)} + \Sigma_{j=L}^{J-1}(H^\star)^j G^\star \boldsymbol{d}^{(j)},$$

for some $0 \le L \le J - 1$:

```
idwtr <- function( wtr,L,filterh){
# idwt(wtr, L, filterh);
# Calculates the IDWT of wavelet
# transformation wtr using wavelet filter  "filterh"  and  L  scales.
# Use
#>> data <- c(1, 0, -3, 2, 1, 0, 1, 2); filterh <- c(sqrt(2)/2, sqrt(2)/2);
#>> max(abs(data - idwtr(dwtr(data,3,filterh), 3,filterh)))
#
#ans = 5.551115e-16
#----------------------------------------------------------------------

nn <- length(wtr); n <- length(filterh);   # Lengths
H <- rev(filterh);                          # Wavelet H filter
G <- filterh;                               # Wavelet G filter
G[seq(2,n,by=2)] <- -G[seq(2,n,by=2)];

LL <- nn/(2^L);                             # Number of scaling coeffs
C <- wtr[1:LL];                             # Scaling coeffs

Cu<-c();Du<-c();
for(j in 1:L){                              # Cascade algorithm
w <- ((0:(n/2-1)) %% LL)+1;                 # Make periodic
D <- wtr[(LL+1):(2*LL)];                     # Wavelet coeffs
Cu[seq(1,2*LL+n,by=2)] <- c(C,C[w]);        # Upsample & keep periodic
Du[seq(1,2*LL+n,by=2)] <- c(D,D[w]);        # Upsample & keep periodic
Cu[which(is.na(Cu))]<-0;
Du[which(is.na(Du))]<-0;
C <- convolve(H,rev(Cu),type="open")+convolve(G,rev(Du),type="open");
C <- C[seq(n,n+2*LL-1)-1];                   # Periodic part
LL <- 2*LL;                                 # Double the size of level
}
return(C);                                  # The inverse DWT
}
```

Table 14.1   Some common wavelet filters from the Daubechies, Coiflet, and Symmlet families.

Name	h_0	h_1	h_2	h_3	h_4	h_5
Haar	$1/\sqrt{2}$	$1/\sqrt{2}$				
Daub 4	0.4829629	0.8365163	0.2241439	−0.1294095		
Daub 6	0.3326706	0.8068915	0.4598775	−0.1350110	−0.0854413	0.0352263
Coif 6	0.0385808	−0.1269691	−0.0771616	0.6074916	0.7456876	0.2265843
Daub 8	0.2303778	0.7148466	0.6308808	−0.0279838	−0.1870348	0.0308414
Symm 8	−0.0757657	−0.0296355	0.4976187	0.8037388	0.2978578	−0.0992195
Daub 10	0.1601024	0.6038293	0.7243085	0.1384281	−0.2422949	−0.0322449
Symm 10	0.0273331	0.0295195	−0.0391342	0.1993975	0.7234077	0.6339789
Daub 12	0.1115407	0.4946239	0.7511339	0.3152504	−0.2262647	−0.1297669
Symm 12	0.0154041	0.0034907	−0.1179901	−0.0483117	0.4910559	0.7876411

Name	h_6	h_7	h_8	h_9	h_{10}	h_{11}
Daub 8	0.0328830	−0.0105974				
Symm 8	−0.0126034	0.0322231				
Daub 10	0.0775715	−0.0062415	−0.0125808	0.0033357		
Symm 10	0.0166021	−0.1753281	−0.0211018	0.0195389		
Daub 12	0.0975016	0.0275229	−0.0315820	0.0005538	0.0047773	−0.0010773
Symm 12	0.3379294	−0.0726375	−0.0210603	0.0447249	0.0017677	−0.0078007

Because wavelet filters uniquely correspond to selection of the wavelet orthonormal basis, we give a table a few (and short) filters commonly used. See Table 14.1 for filters from the Daubechies, Coiflet, and Symmlet families.[1] See Exercise 14.5 for some common properties of wavelet filters.

The careful reader might have already noticed that when the length of the filter is larger than two, boundary problems occur (there are no boundary problems with the Haar wavelet). There are several ways to handle the boundaries, two main are *symmetric* and *periodic*, that is, extending the original function or data set in a symmetric or periodic manner to accommodate filtering that goes outside of domain of function/data.

---

1 Filters are indexed by the number of taps and rounded at seven decimal places.

14.3 Wavelet Shrinkage

Wavelet shrinkage provides a simple tool for nonparametric function estimation. It is an active research area where the methodology is based on optimal shrinkage estimators for the location parameters. Some references are Donoho and Johnstone (1994, 1995), Donoho et al. (1996), Vidakovic (1999), Antoniadis, Bigot, and Sapatinas (2001). In this section we focus on the simplest yet most important shrinkage strategy – wavelet thresholding.

In discrete wavelet transform, the filter H is an "averaging" filter while its mirror counterpart G produces details. The wavelet coefficients correspond to details. When detail coefficients are small in magnitude, they may be omitted without substantially affecting the general picture. Thus the idea of thresholding wavelet coefficients is a way of cleaning out unimportant details that correspond to noise.

An important feature of wavelets is that they provide unconditional bases[2] for functions that are more regular and smooth and have fast decay of their wavelet coefficients. As a consequence, wavelet shrinkage acts as a smoothing operator. The same cannot be said about Fourier methods. Shrinkage of Fourier coefficients in a Fourier expansion of a function affects the result globally due to the non-local nature of sines and cosines. However, trigonometric bases can be localized by properly selected window functions, so that they provide local, wavelet-like decompositions.

Why does wavelet thresholding work? Wavelet transforms disbalanced data. Informally, the "energy" in data set (sum of squares of the data) is preserved (equal to sum of squares of wavelet coefficients), but this energy is packed in a few wavelet coefficients. This *disbalancing property* ensures that the function of interest can be well described by a relatively small number of wavelet coefficients. The normal i.i.d. noise, on the other hand, is invariant with respect to orthogonal transforms (e.g. wavelet transforms) and passes to the wavelet domain structurally unaffected. Small wavelet coefficients likely correspond to a noise because the signal part gets transformed to a few big-magnitude coefficients.

The process of thresholding wavelet coefficients can be divided into two steps. The first step is the policy choice, which is the choice of the threshold function T. Two standard choices are *hard* and *soft* thresholding with corresponding transformations given by

$$T^{\text{hard}}(d, \lambda) = d\ \mathbf{1}(|d| > \lambda),$$
$$T^{\text{soft}}(d, \lambda) = (d - sign(d)\lambda)\ \mathbf{1}(|d| > \lambda), \tag{14.6}$$

---

2 Informally, a family $\{\psi_i\}$ is an unconditional basis for a space of functions S if one can determine if the function $f = \Sigma_i a_i \psi_i$ belongs to S by inspecting only the magnitudes of coefficients, $|a_i|$s.

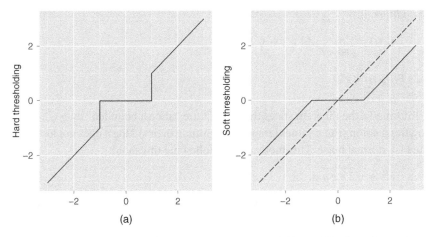

Figure 14.5 (a) Hard and (b) soft thresholding with $\lambda = 1$ (dashed line for reference).

where λ denotes the threshold and d generically denotes a wavelet coefficient. Figure 14.5 shows graphs of (a) hard- and (b) soft-thresholding rules when the input is wavelet coefficient d.

Another class of useful functions are general shrinkage functions. A function S from that class exhibits the following properties:

$$S(d) \approx 0, \quad \text{for } d \text{ small}; \quad S(d) \approx d, \quad \text{for } d \text{ large}.$$

Many state-of-the-art shrinkage strategies are in fact of type $S(d)$.

The second step is the choice of a threshold if the shrinkage rule is thresholding or appropriate parameters if the rule has S-functional form. In the following subsection we briefly discuss some of the standard methods of selecting a threshold.

14.3.1 Universal Threshold

In the early 1990s, Donoho and Johnstone proposed a threshold λ (Donoho and Johnstone, 1994, 1995) based on the result in theory of extrema of normal random variables.

Theorem 14.1 *Let* $Z_1, \dots, Z_n$ *be a sequence of i.i.d. standard normal random variables. Define*

$$A_n = \{ \max_{i=1,\dots,n} |Z_i| \leq \sqrt{2 \log n} \}.$$

Then

$$\pi_n = P(A_n) \to 0, \ n \to \infty.$$

In addition, if

$$B_n(t) = \{ \max_{i=1,\dots,n} |Z_i| > t + \sqrt{2 \log n} \},$$

then $P(B_n(t)) < e^{-\frac{t^2}{2}}$.

Informally, the theorem states that the Z_is are "almost bounded" by $\pm\sqrt{2 \log n}$. Anything among the n values larger in magnitude than $\sqrt{2 \log n}$ does not look like the i.i.d. normal noise. This motivates the following threshold:

$$\lambda^U = \sqrt{2 \log n}\, \hat{\sigma}, \tag{14.7}$$

which Donoho and Johnstone call *universal*. This threshold is one of the first proposed and provides an easy and automatic thresholding.

In the real-life problems, the level of noise σ is not known; however wavelet domains are suitable for its assessment. Almost all methods for estimating the variance of noise involve the wavelet coefficients at the scale of finest detail. The signal-to-noise ratio is smallest at this level for almost all reasonably behaved signals, and the level coefficients correspond mainly to the noise.

Some standard estimators of σ are

(i)
$$\hat{\sigma} = \sqrt{\frac{1}{N/2-1} \Sigma_{k=1}^{N/2}(d_{n-1,k} - {}^-d)^2}, \text{ with } d = \frac{1}{N/2}; \Sigma d_{n-1,k} \tag{14.8}$$

or a more robust MAD estimator.

(ii)
$$\hat{\sigma} = 1/0.6745 \; median_k |d_{n-1,k} - median_m(d_{n-1,m})|, \tag{14.9}$$

where $d_{n-1,k}$ are coefficients in the level of finest detail. In some situations, for instance, when data sets are large or when σ is overestimated, the universal thresholding oversmooths.

Example 14.2 The following R script demonstrates how the wavelets smooth the functions. A Doppler signal of size 1024 is generated, and random normal noise of size $\sigma = 0.1$ is added. By using the Symmlet wavelet 8-tap filter the noisy signal is transformed. After thresholding in the wavelet domain the signal is back-transformed to the original domain (See Figure 14.6):

```
# Demo of wavelet-based function estimation
library(ggplot2)
source("dwtr.r");source("idwtr.r")
# (i) Make "Doppler" signal on [0,1]
t <- seq(0,1,length=1024);
```

Figure 14.6 Demo output: (a) original `doppler` signal, (b) noisy `doppler`, (c) wavelet coefficients that "survived" thresholding, and (d) inverse-transformed thresholded coefficients.

```
sig <- sqrt(t*(1-t))*sin(2*pi*1.05/(t+0.05))
# and plot it
ggplot() + geom_line(aes(x=t,y=sig)) + xlab("") + ylab("")

# (ii) Add noise of size 0.1. We are fixing
# the seed of random number generator for repeatability
# of example. We add the random noise to the signal
# and make a plot.

set.seed(1)
sign <- sig + 0.1*rnorm(length(sig));
ggplot() + geom_line(aes(x=t,y=sign)) + xlab("") + ylab("")

# (iii) Take the filter H, in this case this is SYMMLET 8

filt <- c(-0.07576571478934,   -0.02963552764595,
           0.49761866763246,    0.80373875180522,
           0.29785779560554,   -0.09921954357694,
          -0.01260396726226,    0.03222310060407);

# (iv) Transform the noisy signal in the wavelet domain.
# Choose L=8, eight detail levels in the decomposition.

sw <- dwtr(sign,8,filt)

# At this point you may view the sw. Is it disbalanced?
```

```
# Is it decorrelated?

# (v) Let's now threshold the small coefficients.
# The universal threshold is determined as
# lambda = sqrt(2 * log(1024)) * 0.1 = 0.3723
#
# Here we assumed $sigma=0.1$ is known. In real life
# this is not the case and we estimate sigma.
# A robust estimator is 'MAD' from the finest level of detail
# believed to be mostly transformed noise.

finest <- sw[513:1024];
sigma_est = 1/0.6745 *median(abs(finest-median(finest)));
lambda <- sqrt(2*log(1024))*sigma_est;
# Hard threshold in the wavelet domain

swt <- sw*(abs(sw)>lambda);
ggplot() + geom_line(aes(x=1:1024,y=swt)) + xlab("") + ylab("")

# (vi) Back-transform the thresholded object to the time
# domain. Of course, retain the same filter and value L.

sig.est <- idwtr(swt,8,filt);
ggplot() + geom_line(aes(x=t,y=sig.est)) + xlab("") + ylab("")
```

Example 14.3   A researcher was interested in predicting earthquakes by the level of water in nearby wells. She had a large ($8192 = 2^{13}$ measurements) data set of water levels taken every hour in a period of time of about one year in a California well. Here is the description of the problem:

> The ability of water wells to act as strain meters has been observed for centuries. Lab studies indicate that a seismic slip occurs along a fault prior to rupture. Recent work has attempted to quantify this response, in an effort to use water wells as sensitive indicators of volumetric strain. If this is possible, water wells could aid in earthquake prediction by sensing precursory earthquake strain.
>
> We obtained water level records from a well in southern California, collected over a year time span. Several moderate size earthquakes (magnitude 4.0–6.0) occurred in close proximity to the well during this time interval. There is a significant amount of noise in the water level record that must first be filtered out. Environmental factors such as earth tides and atmospheric pressure create noise with frequencies ranging from seasonal to semidiurnal. The amount of rainfall also affects the water level, as do surface loading, pumping, recharge (such as an increase in water level due to irrigation), and sonic booms, to name a few. Once the noise is

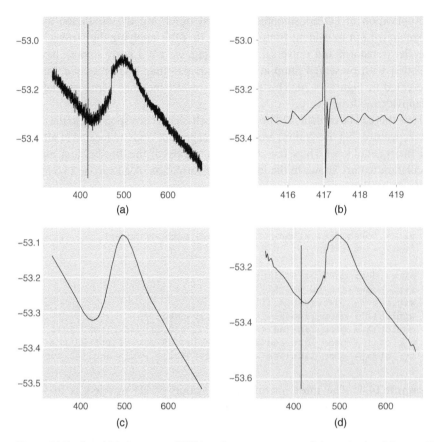

Figure 14.7 Panel (a) shows $n = 8192$ hourly measurements of the water level for a well in an earthquake zone. Notice the wide range of water levels at the time of an earthquake around $t = 417$. Panel (b) focuses on the data around the earthquake time. Panel (c) shows the result of LOESS. Panel (d) gives a wavelet-based reconstruction.

subtracted from the signal, the record can be analyzed for changes in water level, either an increase or a decrease depending upon whether the aquifer is experiencing a tensile or compressional volume strain, just prior to an earthquake.

This data set is given in earthquake.dat. A plot of the raw data for hourly measurements over one year ($8192 = 2^{13}$ observations) is given in Figure 14.7a. The detail showing the oscillation at the earthquake time is presented in Figure 14.7b.

Application of LOESS smoother captured trend, but the oscillation artifact is smoothed out as evident from Figure 14.7c. After applying the Daubechies 8 wavelet transform and universal thresholding, we got a fairly smooth baseline function with preserved jump at the earthquake time. The processed data are presented in Figure 14.7d. This feature of wavelet methods demonstrated data adaptivity and locality.

How this can be explained? The wavelet coefficients corresponding to the earthquake feature (big oscillation) are large in magnitude and are located at all even the finest detail level. These few coefficients "survived" the thresholding, and the oscillation feature shows in the inverse transformation. See Exercise 14.6 for the suggested follow-up:

```
> dat <- read.table("earthquake.dat",sep="\t")
>
> y2 <- loess(dat[,2]~dat[,1],span=0.3,method="loess",
+ family="gaussian")
>
> sw <- dwtr(dat[,2],8,filt);
> swt <- sw*(abs(sw)>0.15);
> sig.est <- idwtr(swt,8,filt);
> ggplot() + geom_line(aes(x=dat[,1],y=dat[,2]))
> ggplot() + geom_line(aes(x=dat[1930:2030,1],y=dat[1930:2030,2]))
> ggplot() + geom_line(aes(x=as.numeric(y2$x),y=y2$fitted))
> ggplot() + geom_line(aes(x=dat[100:7950,1],y=sig.est[100:7950]))
```

Example 14.4 The most important application of 2D wavelets is in image processing. Any gray-scale image can be represented by a matrix A in which the entries a_{ij} correspond to color intensities of the pixel at location (i,j). We assume as standardly done that A is a square matrix of dimension $2^n \times 2^n$, n integer.

The process of wavelet decomposition proceeds as follows. On the rows of the matrix A, the filters H and G are applied. Two resulting matrices H_rA and G_rA are obtained, both of dimension $2^n \times 2^{n-1}$ (subscript r suggest that the filters are applied on rows of the matrix A, 2^{n-1} is obtained in the dimension of H_rA and G_rA because wavelet filtering decimate). Now, the filters H and G are applied on the columns of H_rA and G_rA, and matrices $H_cH_rA, G_cH_rA, H_cG_rA$ and G_cG_rA of dimension $2^{n-1} \times 2^{n-1}$ are obtained. The matrix H_cH_rA is the average, while the matrices G_cH_rA, H_cG_rA, and G_cG_rA are details (see Figure 14.8).[3]

The process could be continued in the same fashion with the *smoothed* matrix H_cH_rA as an input and can be carried out until a single number is obtained as

---

3 This image of Lenna (Sjooblom) Soderberg, a Playboy centerfold from 1972, has become one of the most widely used standard test images in signal processing.

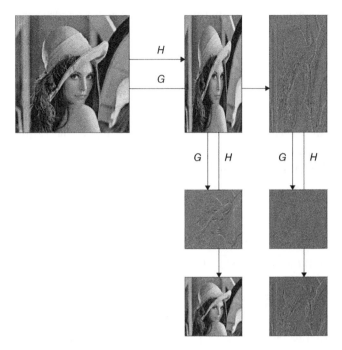

Figure 14.8   One step in wavelet transformation of 2-D data exemplified on celebrated Lenna image.

an overall "smooth" or can be stopped at any step. Notice that in decomposition exemplified in Figure 14.8, the matrix is decomposed to one smooth and three detail submatrices.

A powerful generalization of wavelet bases is the concept of wavelet packets. Wavelet packets result from applications of operators H and G, discussed on p. 290, in *any* order. This corresponds to an overcomplete system of functions from which the best basis for a particular data set can be selected.

14.4   Exercises

14.1   Show that the matrix W' in (14.2) is orthogonal.

14.2   In (14.1) we argued that ψ_{jk} and $\psi_{j'k'}$ are orthogonal functions whenever $j = j'$ and $k = k'$ are not satisfied simultaneously. Argue that ϕ_{jk} and $\psi_{j'k'}$

are orthogonal whenever $j' \geq j$. Find an example in which ϕ_{jk} and $\psi_{j'k'}$ are not orthogonal if $j' < j$.

14.3 In Example 14.1 it was verified that in (14.4) $f(x) = 1$ whenever $x \in [0,1)$. Show that $f(x) = 0$ whenever $x \in [1,2)$.

14.4 Verify that $(\sqrt{2}, -\sqrt{2}, 1, -1, \frac{1}{\sqrt{2}}, -\frac{5}{\sqrt{2}}, \frac{1}{\sqrt{2}}, -\frac{1}{\sqrt{2}})$ is a Haar wavelet transform of data set $\mathbf{y} = (1,0,-3,2,1,0,1,2)$ by using operators H and G from (14.4).

Hint: For the Haar wavelet, low- and high-pass filters are $h = \left(1/\sqrt{2} \; 1/\sqrt{2}\right)$ and $g = (1/\sqrt{2} \; -1/\sqrt{2})$, so

$$
\begin{aligned}
Hy &= H((1,0,-3,2,1,0,1,2)) \\
&= (1 \cdot 1/\sqrt{2} + 0 \cdot 1/\sqrt{2}, \; -3 \cdot 1/\sqrt{2} + 2 \cdot 1/\sqrt{2}, \\
&\quad 1 \cdot 1/\sqrt{2} + 0 \cdot 1/\sqrt{2}, \; 1 \cdot 1/\sqrt{2} + 2 \cdot 1/\sqrt{2}) \\
&= \left(\frac{1}{\sqrt{2}}, -\frac{1}{\sqrt{2}}, \frac{1}{\sqrt{2}}, \frac{3}{\sqrt{2}}\right), \text{ and} \\
Gy &= G((1,0,-3,2,1,0,1,2)) = \left(\frac{1}{\sqrt{2}}, -\frac{5}{\sqrt{2}}, \frac{1}{\sqrt{2}}, -\frac{1}{\sqrt{2}}\right).
\end{aligned}
$$

Repeat the G operator on Hy and $H(Hy)$. The final filtering is $H(H(Hy))$. Organize result as

$$(H(H(Hy)), G(H(Hy)), G(Hy), Gy).$$

14.5 Demonstrate that all filters in Table 14.1 satisfy the following properties (up to rounding error):

$$\Sigma_i h_i = \sqrt{2}, \quad \Sigma_i h_i^2 = 1, \quad \text{and} \quad \Sigma_i h_i h_{i+2} = 0.$$

14.6 Refer to Example 14.3 in which wavelet-based smoother exhibited notable difference from the standard smoother LOESS. Read the data `earthquake.dat` into R, select the wavelet filter, and apply the wavelet transform to the data.

(a) Estimate the size of the noise by estimating σ using MAD from page 296, and find the universal threshold λ_U.

(b) Show that finest level of detail contains coefficients exceeding the universal threshold.

(c) Threshold the wavelet coefficients using hard-thresholding rule with λ_U that you have obtained in (b), and apply inverse wavelet transform. Comment. How do you explain oscillations at boundaries?

RELEVANT R FUNCTIONS AND DATA SETS IN THIS CHAPTER

R codes: dwtr.r, idwtr.r, Wavmat.r
R package: wavethresh

earthquake.dat

References

Antoniadis, A., Bigot, J., and Sapatinas, T. (2001), "Wavelet Estimators in Nonparametric Regression: A Comparative Simulation Study," *Journal of Statistical Software*, 6, 1–83.

Daubechies, I. (1992), *Ten Lectures on Wavelets*, Philadelphia, PA: SIAM.

Donoho, D., and Johnstone, I. (1994), "Ideal Spatial Adaptation by Wavelet Shrinkage," *Biometrika*, 81, 425–455.

Donoho, D., and Johnstone, I. (1995), Adapting to Unknown Smoothness via Wavelet Shrinkage," *Journal of the American Statistical Association*, 90, 1200–1224.

Donoho, D., Johnstone, I., Kerkyacharian, G., and Pickard, D. (1996), "Density Estimation by Wavelet Thresholding," *Annals of Statistics*, 24, 508–539.

Mallat, S. (1989), "A Theory for Multiresolution Signal Decomposition: The Wavelet Representation," *IEEE Transactions on Pattern Analysis and Machine Intelligence*, 11, 674–693.

Ogden, T. (1997), *Essential Wavelets for Statistical Applications and Data Analysis*, Boston, MA: Birkhäuser.

Vidakovic, B. (1999), *Statistical Modeling by Wavelets*, New York: Wiley.

Walter, G. G., and Shen, X. (2001), *Wavelets and Others Orthogonal Systems*, Second Edition, Boca Raton, FL: Chapman & Hall/CRC.

15

Bootstrap

Confine! I'll confine myself no finer than I am:
these clothes are good enough to drink in; and so be these boots too:
an they be not, let them hang themselves in their own straps.

William Shakespeare (*Twelfth Night*, Act 1, Scene III)

15.1 Bootstrap Sampling

The idea of resampling is relatively recent. There is little that seems honest or intuitive about simulating or resampling observations from our original data set as a way of gaining improved information about the population, in general. However, despite the apparent controversy, resampling methods like the bootstrap, jackknife, or even permutation tests are powerful tools for modern statistical inference now that we can handle the computational costs they bear.

Bootstrapping might be the most frequently used sort of resampling, and to some, it is the most controversial. *Resampling* means we take a random sample *from the sample*, as if your sampled data $X_1, \dots, X_n$ represented a finite population of size n. This new sample (typically of the same size n) is taken by "sampling with replacement," so some of the n items from the original sample can appear more than once. This new collection is called a *bootstrap sample* and can be used to assess statistical properties such as an estimator's variability and bias, predictive performance of a rule, significance of a test, and so forth, when the exact analytic methods are impossible or intractable.

By simulating directly from the data, the bootstrap avoids making unnecessary assumptions about parameters and models – we are figuratively pulling ourselves up by our bootstraps rather than relying on the outside help of parametric assumptions. In that sense, the bootstrap is a nonparametric procedure. In fact,

Nonparametric Statistics with Applications to Science and Engineering with R, Second Edition.
Paul Kvam, Brani Vidakovic, and Seong-joon Kim.

this resampling technique includes both parametric and nonparametric forms, but it is essentially empirical.

The term *bootstrap* was coined by Bradley Efron at his 1977 Stanford University Reitz Lecture to describe a resampling method that can help us to understand characteristics of an estimator (e.g. uncertainty, bias) without the aid of additional probability modeling. The bootstrap described by Efron (1979) is not the first resampling method to help out this way (e.g. permutation methods of Fisher (1935) and Pitman (1937), spatial sampling methods of Mahalanobis (1946), or jackknife methods of Quenouille (1949)). However, it is the most popular resampling tool used in statistics today.

So what good is a bootstrap sample? For any direct inference on the underlying distribution, it is obviously inferior to the original sample. In Chapter 10, we learned that for estimating a parameter $\theta = \theta(F)$ from a distribution F, we obviously prefer to use $\theta_n = \theta(F_n)$. What the bootstrap sample *can* tell us is how θ_n might change from sample to sample. While we can only compute θ_n once (because we have just the one sample of n), we can resample (and form a bootstrap sample) an infinite amount of times, in theory. So a meta-estimator built from a bootstrap sample (say, $\tilde{\theta}$) tells us not about θ, but about θ_n. If we generate repeated bootstrap samples $\tilde{\theta}_1, \dots, \tilde{\theta}_B$, we can form an indirect picture of how θ_n is distributed, and from this we generate confidence statements for θ. B is not really limited – it is as large as you want as long as you have the patience for generating repeated bootstrap samples.

For example, $\bar{x} \pm z_{\alpha/2}\sigma_{\bar{x}}$ constitutes an exact $(1 - \alpha)100\%$ confidence interval for μ if we know $X_1, \dots, X_n \sim \mathcal{N}(\mu, \sigma^2)$ and $\sigma_{\bar{x}} = \sigma/\sqrt{n}$. We are essentially finding the appropriate quantiles from the sampling distribution of point estimate $\bar{x}$. Unlike this simple example, characteristics of the sample estimator often are much more difficult to ascertain, and even an interval based on a normal approximation seems out of reach or provides poor coverage probability. This is where resampling comes in most useful.

The idea of bootstrapping was met with initial trepidation. After all, it might seem to be promising something for nothing. The stories of Baron Von Munchausen (Raspe, 1785), based mostly on folk tales, include astounding feats such as riding cannonballs, traveling to the Moon, and being swallowed by a whale before escaping unharmed. In one adventure, the baron escapes from a swamp

Figure 15.1 Baron Von Munchausen: the first bootstrapper. Source: Gustave Dore / Wikipedia Commons / Public Domain.

by pulling himself up by his own hair. In later versions he was using his own bootstraps to pull himself out of the sea, and this fable gave rise to the term *bootstrapping* (Figure 15.1).

15.2 Nonparametric Bootstrap

The *percentile bootstrap* procedure provides a $1 - \alpha$ nonparametric confidence interval for θ directly. We examine the EDF from the bootstrap sample for

$$\tilde{\theta}_1 - \theta_n, \ldots, \tilde{\theta}_B - \theta_n.$$

If θ_n is a good estimate of θ, then we *know* $\tilde{\theta} - \theta_n$ is a good estimate of $\theta_n - \theta$. We do not know the distribution of $\theta_n - \theta$ because we do not know θ, so we cannot use the quantiles from $\theta_n - \theta$ to form a confidence interval. However, we do know the distribution of $\tilde{\theta} - \theta_n$, and the quantiles serve the same purpose. Order the outcomes of the bootstrap sample $(\tilde{\theta}_1 - \theta_n, \ldots, \tilde{\theta}_B - \theta_n)$.

To construct the $(1 - \alpha)$ interval, we first choose the $\alpha/2$ and $1 - \alpha/2$ sample quantiles from the bootstrap sample: $[\tilde{\theta}(\alpha/2) - \theta_n, \tilde{\theta}(1 - \alpha/2) - \theta_n]$. Then

$$P\left(\tilde{\theta}(\alpha/2) - \theta_n < \theta - \theta_n < \tilde{\theta}(1 - \alpha/2) - \theta_n\right)$$
$$= P\left(\tilde{\theta}(\alpha/2) < \theta < \tilde{\theta}(1 - \alpha/2)\right) \approx 1 - \alpha.$$

The quantiles of the bootstrap samples form an approximate confidence interval for θ that is computationally simple to construct.

15.2.1 Parametric Case

If the actual data are assumed to be generated from a distribution $F(x; \theta)$ (with unknown θ), we can improve over the nonparametric bootstrap. Instead of resampling from the data, we can generate a more efficient bootstrap sample by simulating data from $F(x; \theta_n)$.

Example 15.1 Hubble Telescope and Hubble Correlation. The Hubble constant (H) is one of the most important numbers in cosmology because it is instrumental in estimating the size and age of the universe. This long-sought number indicates the rate at which the universe is expanding, from the primordial "Big Bang." The Hubble constant can be used to determine the intrinsic brightness and masses of stars in nearby galaxies, examine those same properties in more distant galaxies and galaxy clusters, deduce the amount of dark matter present in the universe, obtain the scale size of faraway galaxy clusters, and serve as a test for theoretical cosmological models.

In 1929, Edwin Hubble investigated the relationship between the distance of a galaxy from the earth and the velocity with which it appears to be receding. Galaxies appear to be moving away from us no matter which direction we look. This is thought to be the result of the Big Bang. Hubble hoped to provide some knowledge about how the universe was formed and what might happen in the future. The data collected include distances (megaparsecs[1]) to $n = 24$ galaxies and their recessional velocities (km s^{-1}). The scatter plot of the pairs is given in Figure 15.2.

---

1 1 parsec = 3.26 light years.

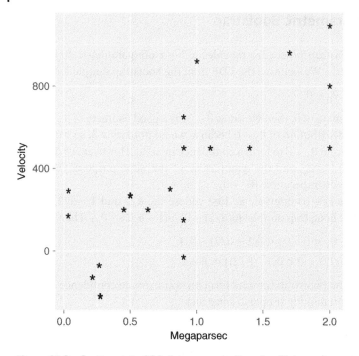

Figure 15.2 Scatter plot of 24 distance–velocity pairs. Distance is measured in parsecs and velocity in km h^{-1}.

Hubble's law claims that recessional velocity is directly proportional to the distance and the coefficient of proportionality is Hubble's constant, H. By working backward in time, the galaxies appear to meet in the same place. Thus $1/H$ can be used to estimate the time since the Big Bang – a measure of the age of the universe. Thus, because of this simple linear model, it is important to estimate correlation between distances and velocities and see if the no-intercept linear regression model is appropriate:

Distance in megaparsecs	0.032	0.034	0.214	0.263	0.275	0.275
(Mpc)	0.45	0.5	0.5	0.63	0.8	0.9
	0.9	0.9	0.9	1.0	1.1	1.1
	1.4	1.7	2.0	2.0	2.0	2.0
The recessional velocity	170	290	−130	−70	−185	−220
(km s^{-1})	200	290	270	200	300	−30
	650	150	500	920	450	500
	500	960	500	850	800	1090

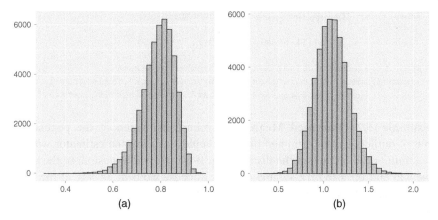

Figure 15.3 (a) Histogram of correlations from 50 000 bootstrap samples. (b) Histogram of correlations of Fisher's z transformations of the bootstrap correlations.

The correlation coefficient between mpc and v based on $n = 24$ pairs is 0.7896. How confident are we about this estimate? To answer this question we resample data and obtain $B = 50\,000$ subrogate samples, each consisting of 24 randomly selected (with repeating) pairs from the original set of 24 pairs. The histogram of all correlations r_i^*, $i = 1, \ldots, 50\,000$ among bootstrap samples is shown in Figure 15.3a. From the bootstrap samples we find that the standard deviation of r can be estimated by 0.0707. From the empirical density for r, we can generate various bootstrap summaries about r.

Figure 15.3b shows the Fisher z-transform of the r^*s, $z_i^* = 0.5 \log[(1 + r_i^*)/(1 - r_i^*)]$ that are bootstrap replicates of $z = 0.5 \log[(1 + r)/(1 - r)]$. Theoretically, when normality is assumed, the standard deviation of z is $(n - 3)^{-1/2}$. Here, we estimate standard deviation of z using bootstrap samples as 0.1893 that is close to $(24 - 3)^{-1/2} = 0.2182$. In the R script below, the bootstrap sample stored in `bootsamp` uses the `sample` function. The `replace=TRUE` argument ensures the samples are generated *with* replacement, so that elements from the original sample can be randomly sampled multiple times:

```
> mpc <- c(.032,.034,.214,.263,.275,.275,.45,.5,.5,.63,
+.8,.9,.9,.9,.9, 1.0, 1.1, 1.1, 1.4, 1.7, 2.0, 2.0, 2.0, 2.0);
>
> v <- c(170,    290,   -130,    -70, -185,   -220, 200,    290,
+            270,   200,   300,    -30, 650,   150,   500,   920,
+            450,   500,   500,   960,   500,   850,   800,   1090);
>
> n <- length(mpc);
> B <- 50000; bsam <- rep(0,B);
> for(i in 1:B){
+ bootsamp <- sample(n,n,replace=TRUE);
```

```
+   bsam[i]   <- cor(mpc[bootsamp],v[bootsamp]);
+   }
>   fisherZ<- 0.5*log((1+bsam)/(1-bsam));
>
>   ggplot()+geom_point(aes(x=mpc,y=v),pch="*",size=8)
>   ggplot()+geom_histogram(aes(x=bsam),col="black",fill="gray")
>   ggplot()+geom_histogram(aes(x=fisherZ),col="black",fill="gray")
```

Example 15.2 Trimmed Mean. For robust estimation of the population mean, outliers can be trimmed off the sample, ensuring the estimator will be less influenced by tails of the distribution. If we trim off almost all of the data, we will end up using the sample median. Suppose we trim off 50% of the data by excluding the smallest and largest 25% of the sample. Obviously, the standard error of this estimator is not easily tractable, so no exact confidence interval can be constructed. This is where the bootstrap technique can help out. In this example, we will focus on constructing a two-sided 95% confidence interval for μ, where

$$\mu = \frac{\int_{x_{1/4}}^{x_{3/4}} t\, dF(t)}{F(x_{3/4}) - F(x_{1/4})} = 2 \int_{x_{1/4}}^{x_{3/4}} t\, dF(t)$$

is an alternative measure of central tendency, the same as the population mean if the distribution is symmetric.

If we compute the trimmed mean from the sample as μ_n, it is easy to generate bootstrap samples and do the same. In this case, limiting B to 1000 or 2000 will make computing easier, because each repeated sample must be ranked and trimmed before $\tilde{\mu}$ can be computed. Let $\tilde{\mu}(0.025)$ and $\tilde{\mu}(0.975)$ be the lower and upper quantiles from the bootstrap sample $\tilde{\mu}_1, \ldots, \tilde{\mu}_B$.

The R function mean(x, P/2) trims $100P\%$ (so $0 < P < 1$) of the data or $P/2\%$ of the biggest and smallest observations. The R scripts

```
bs.fun  <-  function(x,i,P){mean(x[i],trim=P/2);}
    bs  <-  boot(x,bs.fun,R=2000,P=0.1);
bs.ci  <-  boot.ci(bs,conf=c(0.9,0.95),type="all")
```

acquires 2000 bootstrap samples from x and performs the *mean(x, P/2)* function (its additional argument, $P = 0.1$, is left on the end of boot() function call), and 90% and 95% (two-sided) confidence intervals are generated. Below, the vector x represents a skewed sample of test scores, and 90% and 95% confidence intervals for the trimmed mean are given. The third argument in the boot.ci function can take six options, and this input dictates the type of bootstrap to construct. The input options are as follows:

1. "norm": normal approximation.
2. "basic": basic bootstrap method.
3. "student": studentized bootstrap method.
4. "percent": bootstrap percentile method.

5. "bca": adjusted bootstrap percentile (BCa) method.
6. "all": compute all five types of intervals.

```
> x <- c(11,13,14,32,55,58,61,67,69,73,73,89,90,93,94,94,95,96,99,99);
> m <- mean(x,trim=0.1/2);m
[1] 70.27778
> m2 <- mean(x);m2
[1] 68.75
> library(boot);
> bs.fun <- function(x,i,P){mean(x[i],trim=P/2)}
> bs<-boot(x,bs.fun,R=2000,P=0.1);
> bs.ci<-boot.ci(bs,conf=c(0.9,0.95),type=c("norm","basic","perc","bca"));
> boot.ci$t0
[1] 70.27778
> bs.ci
BOOTSTRAP CONFIDENCE INTERVAL CALCULATIONS
Based on 2000 bootstrap replicates

CALL:
boot.ci(boot.out = boot, conf = c(0.9, 0.95), type = c("norm",
    "basic", "perc", "bca"))

Intervals:
Level      Normal                  Basic
90%    (58.79, 81.59)      (58.94, 82.06)
95%    (56.60, 83.77)      (57.17, 84.77)

Level      Percentile              BCa
90%    (58.50, 81.61)      (56.72, 80.22)
95%    (55.79, 83.39)      (54.20, 82.06)
Calculations and Intervals on Original Scale
```

15.2.2   Estimating Standard Error

The most common application of a simple bootstrap is to estimate the standard error of the estimator θ_n. The algorithm is similar to the general nonparametric bootstrap:

- Generate B bootstrap samples of size n.
- Evaluate the bootstrap estimators $\tilde{\theta}_1, \dots, \tilde{\theta}_B$.
- Estimate standard error of θ_n as

$$\hat{\sigma}_{\theta_n} = \sqrt{\frac{\sum_{i=1}^{B} \left(\tilde{\theta}_i - \tilde{\theta}^* \right)^2}{B - 1}},$$

where $\tilde{\theta}^* = B^{-1} \sum \tilde{\theta}_i$.

15.3   Bias Correction for Nonparametric Intervals

The percentile method described in the last section is simple and easy to use and has good large sample properties. However, the coverage probability is not accurate for many small sample problems. The *acceleration and bias correction* (or BC_a) method improves on the percentile method by adjusting the percentiles (e.g. $\tilde{\theta}(1 - \alpha/2, \tilde{\theta}(\alpha/2))$ chosen from the bootstrap sample. A detailed discussion is provided in Efron and Tibshirani (1993).

The BC_a interval is determined by the proportion of the bootstrap estimates $\tilde{\theta}$ less than θ_n, i.e. $p_0 = B^{-1} \sum I(\tilde{\theta}_i < \theta_n)$ define the *bias factor* as

$$z_0 = \Phi^{-1}(p_0)$$

express this bias, where Φ is the standard normal CDF, so that values of z_0 away from zero indicate a problem. Let

$$a_0 = \frac{\sum_{i=1}^{B} \left(\tilde{\theta}^* - \tilde{\theta}_i\right)^3}{6\left(\sum_{i=1}^{B} (\tilde{\theta}^* - \tilde{\theta}_i)^2\right)^{3/2}}$$

be the *acceleration factor*, where $\tilde{\theta}^*$ is the average of the bootstrap estimates $\tilde{\theta}_1, \ldots, \tilde{\theta}_B$. It gets this name because it measures the rate of change in σ_{θ_n} as a function of θ.

Finally, the $100(1 - \alpha)\%$ BC_a interval is computed as

$$\left[\tilde{\theta}(q_1), \ \tilde{\theta}(q_2)\right],$$

where

$$q_1 = \Phi\left(z_0 + \frac{z_0 + z_{\alpha/2}}{1 - a_0(z_0 + z_{\alpha/2})}\right),$$

$$q_2 = \Phi\left(z_0 + \frac{z_0 + z_{1-\alpha/2}}{1 - a_0(z_0 + z_{1-\alpha/2})}\right).$$

Note that if $z_0 = 0$ (no measured bias) and $a_0 = 0$, then (15.1) is the same as the percentile bootstrap interval. In the R function boot.ci, the BC_a is an option "bca" for the nonparametric interval. For the trimmed mean example, the bias-corrected interval is shifted upward.

Example 15.3 Recall the data from Crowder et al. (1991) that was discussed in Example 10.2. The data contain strength measurements (in coded units) for 48 pieces of weathered cord. Seven of the pieces of cord were damaged and yielded strength measurements that are considered right censored. The following R code uses a bias-corrected bootstrap to calculate a 95% confidence interval for the probability that the strength measure is equal to or less than 50, that is, $F(50)$:

```
> library(survival)
> source("kme.at.50.r")
> source("kme.all.x.r")
>
> data <- c(36.3, 41.7, 43.9, 49.9, 50.1, 50.8, 51.9, 52.1, 52.3, 52.3,
+           52.4, 52.6, 52.7, 53.1, 53.6, 53.6, 53.9, 53.9, 54.1, 54.6,
+           54.8, 54.8, 55.1, 55.4, 55.9, 56.0, 56.1, 56.5, 56.9, 57.1,
+           57.1, 57.3, 57.7, 57.8, 58.1, 58.9, 59.0, 59.1, 59.6, 60.4,
+           60.7, 26.8, 29.6, 33.4, 35.0, 40.0, 41.9, 42.5);
```

```
>
> censor <- c(rep(1,41), rep(0,7));
>
> kmest<-(1-survfit(Surv(data,event=censor,type="right")~1,
+ type="kaplan-meier")$surv)
> kmest[sum(50.0>=data)]
[1] 0.09491897
```

Using kme.at.50 and boot.ci functions, we obtain a confidence interval for $F(50)$ based on 1000 bootstrap replicates:

```
> bs <- boot(cbind(data,censor),kme.at.50,R=1000)
> bs.ci <- boot.ci(bs,conf=0.95,type="perc")
> bs.ci
BOOTSTRAP CONFIDENCE INTERVAL CALCULATIONS
Based on 1000 bootstrap replicates

CALL:
boot.ci(boot.out = bs, conf = 0.95, type = "perc")

Intervals:
Level      Percentile
95%    ( 0.0217,   0.1887)
Calculations and Intervals on Original Scale
```

The R functions boot.ci and kme.all.x are used to produce Figure 15.4:

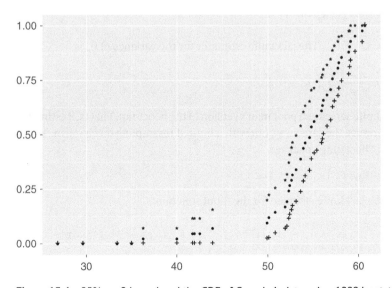

Figure 15.4 95% confidence band the CDF of Crowder's data using 1000 bootstrap samples. Lower boundary of the confidence band is plotted with marker "+," the estimate is plotted with the dots, and the upper boundary is plotted with marker "*."

```
> dat <- sort(unique(data));
> bs <- boot(cbind(data,censor),kme.all.x,R=1000)
> ci <- matrix(0,nrow=length(dat),ncol=3);
> for(i in 1:length(dat)){
+ tryCatch({rm(tmp);tmp<-boot.ci(bs,conf=0.95,type="perc",index=i)
            $percent[4:5]},
+ error=function(e){tmp<-NULL})
+ if(!is.null(tmp)){
+ ci[i,]<-c(bs$t0[i],tmp);
+ }else{
+ ci[i,]<-rep(bs$t0[i],3);
+ }}
> p <- ggplot() + geom_point(aes(x=dat,y=ci[,1]))
> p <- p + geom_point(aes(x=dat,y=ci[,2]),pch="+",size=5)
> p <- p + geom_point(aes(x=dat,y=ci[,3]),pch="*",size=5)
> print(p)
```

15.4 The Jackknife

The *jackknife* procedure, introduced by Quenouille (1949), is a resampling method for estimating bias and variance in θ_n. It predates the bootstrap and actually serves as a special case. The resample is based on the "leave one out" method, which was computationally easier when computing resources were limited.

The ith jackknife sample is $(x_1, \ldots, x_{i-1}, x_{i+1}, \ldots, x_n)$. Let $\hat{\theta}_{(i)}$ be the estimator of θ based only on the ith jackknife sample. The jackknife estimate of the *bias* is defined as

$$\hat{b}_J = (n-1)\left(\hat{\theta}_n - \hat{\theta}^*\right),$$

where $\hat{\theta}^* = n^{-1}\sum \hat{\theta}_{(i)}$. The jackknife estimator for the variance of θ_n is

$$\sigma_J^2 = \frac{n-1}{n}\sum_{i=1}^{n}\left(\hat{\theta}_{(i)} - \hat{\theta}^*\right)^2.$$

The jackknife serves as a poor man's version of the bootstrap. That is, it estimates bias and variance the same way, but with a limited resampling mechanism. The R function in "bootstrap" package

```
jackknife(x,function,...)
```

produces the jackknife estimate for the input function:

```
> library(bootstrap)
> x <- c(11,13,14,32,55,58,61,67,69,73,73,89,90,93,94,94,95,96,99,99);
> jackknife(x,mean,trim=0.1/2)
$jack.se
[1] 6.72422
```

```
$jack.bias
[1] -29.02778

$jack.values
 [1] 71.78947 71.68421 71.63158 ...
 [8] 68.84211 68.73684 68.52632 ...
[15] 67.42105 67.42105 67.36842 ...

$call
jackknife(x = x, theta = mean, trim = 0.1/2)
```

The jackknife performs well in most situations, but poorly in some. In case θ_n can change significantly with slight changes to the data, the jackknife can be temperamental. This is true with $\theta =$ median, for example. In such cases, it is recommended to augment the resampling by using a *delete-d jackknife*, which leaves out d observations for each jackknife sample. See Chapter of Efron and Tibshirani (1993) for details.

15.5 Bayesian Bootstrap

The Bayesian bootstrap (BB), a Bayesian analogue to the bootstrap, was introduced by Rubin (1981). In Efron's standard bootstrap, each observation X_i from the sample $X_1, \dots, X_n$ has a probability of $1/n$ to be selected, and after the selection process the relative frequency f_i of X_i in the bootstrap sample belongs to the set $\{0, 1/n, 2/n, \dots, (n-1)/n, 1\}$. Of course, $\sum_i f_i = 1$. Then, for example, if the statistic to be evaluated is the sample mean, its bootstrap replicate is $\bar{X}^* = \sum_i f_i X_i$.

In Bayesian bootstrapping, at each replication a discrete probability distribution $\mathbf{g} = \{g_1, \dots, g_n\}$ on $\{1, 2, \dots, n\}$ is generated and used to produce bootstrap statistics. Specifically, the distribution $\mathbf{g}$ is generated by generating $n - 1$ uniform random variables $U_i \sim U(0,1)$, $i = 1, \dots, n - 1$ and ordering them according to $\tilde{U}_j = U_{j:n-1}$ with $\tilde{U}_0 \equiv 0$ and $\tilde{U}_n \equiv 1$. Then the probability of X_i is defined as

$$g_i = \tilde{U}_i - \tilde{U}_{i-1}, i = 1, \dots, n.$$

If the sample mean is the statistic of interest, its Bayesian bootstrap replicate is a weighted average if the sample $\bar{X}^* = \sum_i g_i X_i$. The following example explains why this resampling technique is Bayesian.

Example 15.4 Suppose that $X_1, \dots, X_n$ are i.i.d. $Ber(p)$, and we seek a BB estimator of p. Let n_1 be the number of ones in the sample and $n - n_1$ the number of zeros. If the BB distribution $\mathbf{g}$ is generated, then let $P_1 = \sum g_i 1(X_i = 1)$ be the probability of 1 in the sample. The distribution for P_1 is simple, because the gaps in the $U_1, \dots, U_{n-1}$ follow the $(n - 1)$-variate Dirichlet distribution, $Dir(1, 1, \dots, 1)$. Consequently, P_1 is the sum of n_1 gaps and is distributed $Be(n_1, n - n_1)$. Note that

$Be(n_1, n - n_1)$ is, in fact, the posterior for P_1 if the prior is $\propto [P_1(1 - P_1)]^{-1}$. That is, for $x \in \{0,1\}$,

$$P(X = x|P_1) = P_1^x(1 - P_1)^{1-x}, \quad P_1 \propto [P_1(1 - P_1)]^{-1},$$

then the posterior is

$$[P_1|X_1, \dots, X_n] \sim Be(n_1, n - n_1).$$

For general case when X_i takes $d \le n$ different values, the Bayesian interpretation is still valid; see Rubin's (1981) article.

Example 15.5 We revisit Hubble's data and give a BB estimate of variability of observed coefficient of correlation r. For each BB distribution **g**, calculate

$$r^* = \frac{\sum_{i=1}^n g_i X_i Y_i - (\sum_{i-1}^n g_i X_i)(\sum_{i-1}^n g_i X_i)}{[\sum_{i=1}^n g_i X_i^2 - (\sum_{i=1}^n g_i X_i)^2]^{1/2} [\sum_{i=1}^n g_i Y_i^2 - (\sum_{i=1}^n g_i Y_i)^2]^{1/2}},$$

where $(X_i, Y_i), i = 1, \dots, 24$ are observed pairs of distances and velocities. The R script below performs the BB resampling:

```
> x <- mpc; y <- v  # from previous example
> n <- 24; B <- 50000; # Number of BB replicates
>
> bbcorr <- rep(0,B);
> for(i in 1:B){
+     all <- c(0, sort(runif(n-1)),1);
+     gis <- diff(all);
+     # gis is BB distribution, corrbb is correlation
+     # with gis as weights
+     ssx <- sum(gis*x); ssy <- sum(gis*y);
+     ssx2 <- sum(gis*x^2); ssy2 <- sum(gis*y^2);
+     ssxy <- sum(gis*x*y);
+     corrbb <- (ssxy-ssx*ssy)/(sqrt((ssx2-ssx^2)*(ssy2-ssy^2)));
+     # correlation replicate
+     bbcorr[i] <- corrbb;
+     }
> zs <- 0.5*log((1+bbcorr)/(1-bbcorr));
> ggplot() + geom_histogram(aes(x=bbcorr),col="black",fill="gray",
+     binwidth=0.01) + xlab("") + ylab("")
> ggplot() + geom_histogram(aes(x=zs),col="black",fill="gray",
+     binwidth=0.02) + xlab("") + ylab("")
```

The histograms of correlation bootstrap replicates and their z-transforms in Figure 15.5a,b look similar to those in Figure 15.3a,b. Numerically, $B = 50\,000$ replicates gave standard deviation of observed r as 0.0635 and standard deviation of $z = 1/2\log((1 + r)/(1 - r))$ as 0.1704 slightly smaller than theoretical $24 - 3^{-1/2} = 0.2182$.

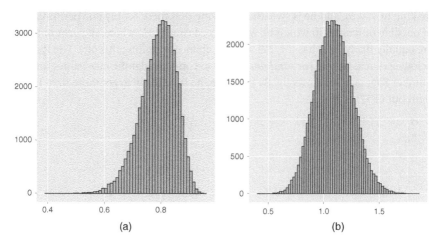

Figure 15.5 (a) The histogram of 50,000 BB resamples for the correlation between the distance and velocity in the Hubble data. (b) Fisher z-transform of the BB correlations.

15.6 Permutation Tests

The original idea of the permutation test was to exhaust all possible values of a test statistic (under the null hypothesis) by considering all possible rearrangements of labels for the observations.

Example 15.6 The Magician. Four dinner guests of Sagi the Magician randomly select an ace from his deck of cards. Sagi uses his mind-reading ability to figure out which card ($A\spadesuit$, $A\clubsuit$, $A\diamondsuit$, $A\heartsuit$) belongs to which dinner guest, and they are impressed when he is right. As a scientist, however, you might not be as impressed. If we attribute Sagi's selections to random guesses, we know there is one permutation out of $4! = 24$, so he has a 0.0417 chance of matching the cards by guessing.

For such a small experiment, it is pretty easy to consider all possible arrangements of cards and guesses, but in cases where the samples are large, we inventively resample from all the possible permutations available in hope of getting a sampling distribution that is representative of the unobtainable permutation sample.

Example 15.7 The Magician, Part II. Sagi is tasked to show off his telepathic abilities to a larger audience. Now an audience of 52 people randomly selects cards from a single deck, and Sagi must match each audience member with a unique playing card. Sagi's supernatural abilities can be called into question once he starts making mistakes, and he is sweating knowing that there is only one permutation of possible outcomes in which he will succeed (under H_0 where all permutations are equally likely, Sagi has a $1/52! \approx 10^{-68}$ chance at pulling this off). The R code below uses resampling (or *randomization*) to compute significance values for the number Sagi guesses correctly. By guessing, he is unlikely to get more than three right out of 52:

```
> n <- 100000; sagi <- rep(NULL,n)
> for (i in 1:n){
+    x <- 1:52
+    y <- sample(x)
+    sagi[i] <- sum(x==y)
     }
table(sagi)/n
   sagi
      0        1        2        3        4        5        6
 0.3676   0.3687   0.1831   0.0618   0.0152   0.0030   0.0006
```

For a general permutation (or randomization) test, we presume that for our statistical experiment the sample or samples are taken. and a statistic S is constructed for testing a particular hypothesis H_0. The values of S that seem extreme from the viewpoint of H_0 are critical for this hypothesis. The decision if the observed value of statistics S is extreme is made by looking at the distribution of S when H_0 is true. However, what if such distribution is unknown or too complex to find? What if the distribution for S is known only under stringent assumptions that we are not willing to make?

Resampling methods consisting of permuting the original data can be used to approximate the null distribution of S. Given the sample, one forms the permutations that are *consistent with experimental design and H_0* and then calculates the value of S. The values of S are used to estimate its density (often as a histogram), and using this empirical density, we find an approximate p-value, often called a *permutation p-value*.

What permutations are consistent with H_0? Suppose that in a two-sample problem we want to compare the means of two populations based on two independent samples $X_1, \ldots, X_m$ and $Y_1, \ldots, Y_n$. The null hypothesis H_0 is $\mu_X = \mu_Y$. The permutations consistent with H_0 would be all permutations of a combined (concatenated) sample $(X_1, \ldots, X_m, Y_1, \ldots, Y_n)$. Otherwise, suppose we have a repeated measures design in which observations are triplets corresponding to three treatments, i.e. (X_{11}, X_{12}, X_{13}), ..., (X_{n1}, X_{n2}, X_{n3}), and that H_0 states that the three treatment means are the same, $\mu_1 = \mu_2 = \mu_3$. Then permutations consistent with this experimental design are random reconfigurations of the triplets (X_{i1}, X_{i2}, X_{i3}), $i = 1, \ldots, n$, and a possible permutation might be

$$(X_{13}, X_{11}, X_{12}), (X_{21}, X_{23}, X_{22}), (X_{32}, X_{33}, X_{31}), \ldots, (X_{n2}, X_{n1}, X_{n3}).$$

Figure 15.6 A coin of Manuel I Comnenus (1143–1180).

Thus, depending on the design and H_0, consistent permutations can be quite different.

Example 15.8 Byzantine Coins. To illustrate the spirit of permutation tests, we use data from a paper by Hendy and Charles (1970) (see also Hand et al., 1994) that represent the silver content (%Ag) of a number of Byzantine coins discovered in Cyprus. The coins (Figure 15.6) are from the first and fourth coinages in the reign of King Manuel I, Comnenus (1143–1180):

First coinage	5.9	6.8	6.4	7.0	6.6	7.7	7.2	6.9	6.2
Fourth coinage	5.3	5.6	5.5	5.1	6.2	5.8	5.8		

The question of interest is whether or not there is statistical evidence to suggest that the silver content of the coins was significantly different in the later coinage.

Of course, the two-sample t-test or one of its nonparametric counterparts is possible to apply here, but we will use the permutation test for purposes of illustration. The following R scripts perform the test:

```
> coins <- c(5.9,  6.8,  6.4,  7.0,  6.6,  7.7,  7.2,  6.9,  6.2,
+            5.3,  5.6,  5.5,  5.1,  6.2,  5.8,  5.8);
> coins1 <- coins[1:9]; coins2 <- coins[10:16];
> S<- (mean(coins1)-mean(coins2))/sqrt(var(coins1)+var(coins2));
>
> N <- 10000
> Sps <- rep(0,N); asl <- 0;
> for(i in 1:N){
+     coinsp <- coins[sample(16,16)]
+     coinsp1 <- coinsp[1:9]; coinsp2 <- coinsp[10:16]
+     Sp <- (mean(coinsp1)-mean(coinsp2))/sqrt(var(coinsp1)+var(coinsp2))
+     Sps[i] <- Sp
+      asl <- asl + (abs(Sp)>S)
+     }
> asl <- asl / N;
> S
[1] 1.730115
> asl
[1] 4e-04
> ggplot() + geom_histogram(aes(x=Sps),col="black",fill="gray")
+   geom_line(aes(x=c(1.7301,1.7301),y=c(0,400)),lwd=1,lty=3)
```

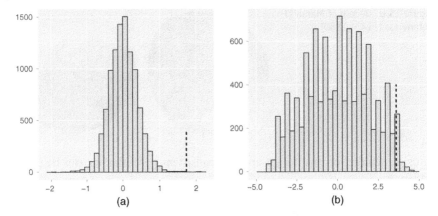

Figure 15.7 Panels (a) and (b) show permutation null distribution of statistics S and the observed value of S (marked by *dotted* line) for the cases of (a) Byzantine coins and (b) left-handed grippers.

The value for S is 1.7301, and the permutation p-value or the achieved significance level is `asl` = 0.0004. Panel (a) in Figure 15.7 shows the permutation null distribution of statistics S, and the observed value of S is indicated by the dotted vertical line. Note that there is nothing special about selecting

$$S = \frac{\overline{X}_1 - \overline{X}_2}{\sqrt{s_1^2/n_1 + s_2^2/n_2}}$$

and that any other statistics that sensibly measures deviation from $H_0 : \mu_1 = \mu_2$ could be used. For example, one could use $S = \text{median}(X_1)/s_1 - \text{median}(X_2)/s_2$ or simply $S = \overline{X}_1 - \overline{X}_2$.

To demonstrate how the choice what to permute depends on statistical design, we consider again the two-sample problem but with paired observations. In this case, the permutations are done within the pairs, independently from pair to pair.

Example 15.9 Left-Handed Grippers. Measurements of the left- and right-hand gripping strengths of 10 left-handed writers are recorded:

Person	1	2	3	4	5	6	7	8	9	10
Left hand (X)	140	90	125	130	95	121	85	97	131	110
Right hand (Y)	138	87	110	132	96	120	86	90	129	100

Do the data provide strong evidence that people who write with their left hand have greater gripping strength in the left hand than they do in the right hand? In the R solution provided below, `dataL` and `dataR` are paired measurements, and `pdataL` and `pdataR` are random permutations, either {1,2} or {2,1} of the 10 original pairs. The statistic S is the difference of the sample means. The permutation null distribution is shown as non-normalized histogram in Figure 15.7b. The position of S with respect to the histogram is marked by dotted line:

```
> dataL <- c(140, 90, 125, 130, 95, 121, 85, 97, 131, 110);
> dataR <- c(138, 87, 110, 132, 96, 120, 86, 90, 129, 100);
>
> S <- mean(dataL-dataR);
> data <- cbind(dataL,dataR);
> N <- 10000; asl <- 0;
> means <- rep(0,N);
> for(i in 1:N){
+ pdata <- c()
+ for( j in 1:10){
+ pairs <- data[j,sample(2,2)];
+ pdata <- rbind(pdata,pairs);
+ }
+ pdataL <- pdata[,1];
+ pdataR <- pdata[,2];
+ pmean <- mean(pdataL-pdataR);
+ means[i] <- pmean;
+ asl <- asl + (abs(pmean) > S);
+ }
> S
[1] 3.6
> asl/N
[1] 0.0395
> p <- ggplot() + geom_histogram(aes(x=means),col="black",fill="lightblue")
> p <- p + geom_line(aes(x=c(3.6,3.6),y=c(0,400)),lwd=1,lty=3)
> print(p)
```

15.7 More on the Bootstrap

There are several excellent resources for learning more about bootstrap techniques, and there are many different kinds of bootstraps that work on various problems. Besides Efron and Tibshirani (1993), books by Chernick (1999) and Davison and Hinkley (1997) provide excellent overviews with numerous helpful examples. In the case of dependent data, various bootstrapping strategies are proposed such as block bootstrap, stationary bootstrap, wavelet-based bootstrap (wavestrap), and so on. A monograph by Good (2000) gives a comprehensive coverage of permutation tests.

Bootstrapping is not infallible. Data sets that might lead to poor performance include those with missing values and excessive censoring. Choice of statistics is also critical; see Exercise 15.6. If there are few observations in the tail of the distribution, bootstrap statistics based on the EDF perform poorly because they are deduced using only a few of those extreme observations.

15.8 Exercises

15.1 Generate a sample of 20 from the gamma distribution with $\lambda = 0.1$ and $r = 3$. Compute a 90% confidence interval for the mean using (a) the standard normal approximation, (b) the percentile method, and (c) the bias-corrected method. Repeat this 1000 times, and report the actual *coverage probability* of the three intervals you constructed.

15.2 For the case of estimating the sample mean with $\overline{X}$, derive the expected value of the jackknife estimate of bias and variance.

15.3 Refer to insect waiting times for the *female* Western White Clematis in Table 10.1. Use the percentile method to find a 90% confidence interval for $F(30)$, the probability that the waiting time is less than or equal to 30 minutes.

15.4 In a data set of size n generated from a continuous F, how many *distinct* bootstrap samples are possible?

15.5 Refer to the dominance–submissiveness data in Exercise 7.4. Construct a 95% confidence interval for the correlation using the percentile bootstrap and the jackknife. Compare your results with the normal approximation described in that example.

15.6 Suppose we have three observations from $\mathcal{U}(0, \theta)$. If we are interested in estimating θ, the maximum likelihood estimate (MLE) for it is $\hat{\theta} = X_{3:3}$, the largest observation. If we obtain a bootstrap sampling procedure to estimate the variance of the MLE, what is the distribution of the bootstrap estimator for θ?

15.7 Seven patients each underwent three different methods of kidney dialysis. The following values were obtained for weight change in kilograms between dialysis sessions:

Patient	Treatment 1	Treatment 2	Treatment 3
1	2.90	2.97	2.67
2	2.56	2.45	2.62
3	2.88	2.76	1.84
4	2.73	2.20	2.33
5	2.50	2.16	1.27
6	3.18	2.89	2.39
7	2.83	2.87	2.39

Test the null hypothesis that there is no difference in mean weight change among treatments. Use properly designed permutation test.

15.8 In a controlled clinical trial *Physician's Health Study I* that began in 1982 and ended in 1987, more than 22 000 physicians participated. The participants were randomly assigned to two groups: (i) *aspirin* and (ii) *placebo*, where the aspirin group have been taking 325 mg aspirin every second day. At the end of trial, the number of participants who suffered from myocardial infarction was assessed. The counts are given in the following table:

	MyoInf	No MyoInf	Total
Aspirin	104	10 933	11 037
Placebo	189	10 845	11 034

The popular measure in assessing results in clinical trials is risk ratio (RR) that is the ratio of proportions of cases (risks) in the two groups/treatments. From the table,

$$RR = R_a/R_p = \frac{104/11\ 037}{189/11\ 034} = 0.55.$$

Interpretation of RR is that the risk of myocardial infarction for the placebo group is approximately $1/0.55 = 1.82$ times higher than that for the aspirin group. With R, construct a bootstrap estimate for the variability of RR. *Hint*:

```
aspi <- c(rep(0,10933),rep(1,104));
plac <- c(rep(0,10845),rep(1,189));
RR <- (sum(aspi)/11037)/(sum(plac)/11034);
B <- 10000; BRR <- rep(0,B);
for(b in 1:B){
    baspi <- aspi[sample(11037,11037,replace=TRUE)];
    bplac <- plac[sample(11034,11034,replace=TRUE)];
    BRR[b] <- (sum(baspi)/11037)/(sum(bplac)/11034);
}
```

(i) Find the variability of the difference of the risks $R_a - R_p$ and of logarithm of the odds ratio, $\log(R_a/(1 - R_a)) - \log(R_p/(1 - R_p))$.

(ii) Using the BB, estimate the variability of RR, $R_a - R_p$, and $\log(R_a/(1 - R_a)) - \log(R_p/(1 - R_p))$.

15.9 Let f_i and g_i be frequency/probability of the observation X_i in an ordinary/Bayesian bootstrap resample from $X_1, \ldots, X_n$. Prove that

$\mathbb{E}f_i = \mathbb{E}g_i = 1/n$, i.e. the expected probability distribution is discrete uniform, $\mathbb{V}\text{ar}\, f_i = (n+1)/n$, $\mathbb{V}\text{ar}\, g_i = (n-1)/n^2$, and for $i \neq j$, $\text{Corr}(f_i, f_j) = \text{Corr}(g_i, g_j) = -1/(n-1)$.

15.10 Return to Example 7.5 in which tread wear for tires is measured using weight loss and groove wear. Construct a randomization test to compute the significance of the Pearson correlation coefficient. How does this value compare with the p-value in the R function cor.test?

RELEVANT R FUNCTIONS AND DATA SETS IN THIS CHAPTER

R codes: kme.all.x.r, kme.at.50.r
R functions: boot, boot.ci, survfit, Surv, jackknife
R package: boot, survival, bootstrap

References

Chernick, M. R., (1999), *Bootstrap Methods – A Practitioner's Guide*, New York: Wiley.

Crowder, M. J., Kimber, A. C., Smith, R. L., and Sweeting, T. J., (1991), *Statistical Analysis of Reliability Data*, London: Chapman Hall.

Davison, A. C., and Hinkley, D. V. (1997), *Bootstrap Methods and Their Applications*, Boston, MA: Cambridge University Press.

Efron, B. (1979), "Bootstrap Methods: Another Look at the Jackknife," *Annals of Statistics*, 7, 1–26.

Efron, B., and Tibshirani, R. J. (1993), *An Introduction to the Bootstrap*, Boca Raton, FL: CRC Press.

Fisher, R. A. (1935), *The Design of Experiments*, New York: Hafner.

Good, P. I. (2000), *Permutation Tests: A Practical Guide to Resampling Methods for Testing Hypotheses*, Second Edition, New York: Springer-Verlag.

Hand, D. J., Daly, F., Lunn, A. D., McConway, K. J., and Ostrowski, E. (1994), *A Handbook of Small Datasets*, New York: Chapman & Hall.

Hendy, M. F., and Charles, J. A. (1970), "The Production Techniques, Silver Content, and Circulation History of the Twelfth-Century Byzantine Trachy," *Archaeometry*, 12, 13–21.

Mahalanobis, P. C. (1946), "On Large-Scale Sample Surveys," *Philosophical Transactions of the Royal Society of London Series B*, 231, 329–451.

Pitman, E. J. G., (1937), "Significance Tests Which May Be Applied to Samples from Any Population," *Royal Statistical Society Supplement*, 4, 119–130 and 225–232 (Parts I and II).

Quenouille, M. H. (1949), "Approximate Tests of Correlation in Time Series," *Journal of the Royal Statistical Society Series B*, 11, 18–84.

Raspe, R. E. (1785). *The Travels and Surprising Adventures of Baron Munchausen*, London: Trubner, 1859 [1st Ed. 1785].

Rubin, D. (1981), "The Bayesian Bootstrap," *Annals of Statistics*, 9, 130–134.

16

EM Algorithm

Insanity is doing the same thing over and over again and expecting different results.

Albert Einstein

The expectation–maximization (EM) algorithm is broadly applicable statistical technique for maximizing complex likelihoods while handling problems with incomplete data. Within each iteration of the algorithm, two steps are performed: (i) the *E*-step consisting of projecting an appropriate functional containing the augmented data on the space of the original, incomplete data and (ii) the *M*-step consisting of maximizing the functional.

The name EM algorithm was coined by Dempster, Laird, and Rubin (1977) in their fundamental paper, referred to here as the DLR paper. However, as is usually the case, if one comes to a smart idea, one may be sure that other smart guys in the history had already thought about it. Long before, McKendrick (1926) and Healy and Westmacott (1956) proposed iterative methods that are examples of the EM algorithm. In fact, before the DLR paper appeared in 1997, dozens of papers proposing various iterative solvers were essentially applying the EM Algorithm in some form.

However, the DLR paper was the first to formally recognize these separate algorithms as having the same fundamental underpinnings, so perhaps their 1977 paper prevented further reinventions of the same basic math tool. While the algorithm is not guaranteed to converge in every type of problem (as mistakenly claimed by DLR), Wu (1983) showed convergence is guaranteed if the densities making up the full data belong to the exponential family. This does not prevent the EM method from being helpful in nonparametric problems; Tsai and Crowley (1985) first applied it to a general nonparametric setting, and numerous applications have appeared since.

Nonparametric Statistics with Applications to Science and Engineering with R, Second Edition.
Paul Kvam, Brani Vidakovic, and Seong-joon Kim.
© 2023 John Wiley & Sons, Inc. Published 2023 by John Wiley & Sons, Inc.

Definition

Let Y be a random vector corresponding to the observed data y and having a postulated PDF $f(y, \psi)$, where $\psi = (\psi_1, \dots, \psi_d)$ is a vector of unknown parameters. Let x be a vector of augmented (so-called complete) data, and let z be the missing data that completes x, so that $x = [y, z]$.

Denote by $g_c(x, \psi)$ the PDF of the random vector corresponding to the complete data set x. The log likelihood for ψ, if x were fully observed, would be

$$\log L_c(\psi) = \log g_c(x, \psi).$$

The incomplete data vector y comes from the "incomplete" sample space $\mathcal{Y}$. There is a one-to-one correspondence between the complete sample space $\mathcal{X}$ and the incomplete sample space $\mathcal{Y}$. Thus, for $x \in \mathcal{X}$, one can uniquely find the "incomplete" $y = y(x) \in \mathcal{Y}$. Also, the incomplete pdf can be found by properly integrating out the complete pdf,

$$g(y, \psi) = \int_{\mathcal{X}(y)} g_c(x, \psi) dx,$$

where $\mathcal{X}(y)$ is the subset of $\mathcal{X}$ constrained by the relation $y = y(x)$.

Let $\psi^{(0)}$ be some initial value for ψ. At the kth step, the EM algorithm one performs the following two steps:

E-Step. Calculate

$$Q\left(\psi, \psi^{(k)}\right) = \mathbb{E}_{\psi^{(k)}} \{\log L_c(\psi) | y\}.$$

M-Step. Choose any value $\psi^{(k+1)}$ that maximizes $Q\left(\psi, \psi^{(k)}\right)$, that is,

$$(\forall \psi) Q\left(\psi^{(k+1)}, \psi^{(k)}\right) \geq Q\left(\psi, \psi^{(k)}\right).$$

The E- and M-steps are alternated until the difference

$$L\left(\psi^{(k+1)}\right) - L\left(\psi^{(k)}\right)$$

becomes small in absolute value.

Next we illustrate the EM algorithm with a famous example first considered by Fisher and Balmukand (1928). It is also discussed in Rao (1973) and later by McLachlan and Krishnan (1997) and Slatkin and Excoffier (1996).

16.1 Fisher's Example

The following genetics example was recognized by as an application of the EM algorithm by Dempster et al. (1979). The description provided here essentially follows a lecture by Terry Speed of UC at Berkeley. In basic genetics terminology, suppose there are two linked biallelic loci, A and B, with alleles A and a and B and b, respectively, where A is dominant over a and B is dominant over b. A double heterozygote $AaBb$ will produce gametes of four types: AB, Ab, aB, and ab. As the loci are linked, the types AB and ab will appear with a frequency different from that of Ab and aB, say, $1 - r$ and r, respectively, in males and $1 - r'$ and r', respectively, in females.

Here we suppose that the parental origin of these heterozygotes is from the mating $AABB \times aabb$, so that r and r' are the male and female recombination rates between the two loci. The problem is to estimate r and r', if possible, from the offspring of selfed double heterozygotes. Because gametes AB, Ab, aB, and ab are produced in proportions $(1 - r)/2, r/2, r/2$, and $(1 - r)/2$, respectively, by the male parent and $(1 - r')/2, r'/2, r'/2$, and $(1 - r')/2$, respectively, by the female parent, zygotes with genotypes $AABB$, $AaBB$, etc. are produced with frequencies $(1 - r)(1 - r')/4, (1 - r)r'/4$, etc.

The problem here is this: although there are 16 distinct offspring genotypes, taking parental origin into account, the dominance relations imply that we only observe four distinct phenotypes, which we denote by A^*B^*, A^*b^*, a^*B^*, and a^*b^*. Here A^* (B^*) denotes the dominant, while a^* (b^*) denotes the recessive phenotype determined by the alleles at A (B).

Thus individuals with genotypes $AABB$, $AaBB$, $AABb$, or $AaBb$ (which account for 9/16 of the gametic combinations) exhibit the phenotype A^*B^*, i.e. the dominant alternative in both characters, while those with genotypes $AAbb$ or $Aabb$ (3/16) exhibit the phenotype A^*b^*, those with genotypes $aaBB$ and $aaBb$ (3/16) exhibit the phenotype a^*B^*, and finally the double recessive $aabb$ (1/16) exhibits the phenotype a^*b^*. It is a slightly surprising fact that the probabilities of the four phenotypic classes are definable in terms of the parameter $\psi = (1 - r)(1 - r')$, as follows: a^*b^* has probability $\psi/4$ (easy to see), a^*B^* and A^*b^* both have probabilities $(1 - \psi)/4$, while A^*B^* has rest of the probability, which is $(2 + \psi)/4$. Now suppose we have a random sample of n offspring from the selfing of our double heterozygote. The four phenotypic classes will be represented roughly in proportion to their theoretical probabilities, their joint distribution being multinomial:

$$\mathcal{M}n\left(n; \frac{2 + \psi}{4}, \frac{1 - \psi}{4}, \frac{1 - \psi}{4}, \frac{\psi}{4}\right). \tag{16.1}$$

Note that here neither r nor r' will be separately estimable from these data, but only the product $(1 - r)(1 - r')$. Because we know that $r \leq 1/2$ and $r' \leq 1/2$, it follows that $\psi \geq 1/4$.

How do we estimate ψ? Fisher and Balmukand listed a variety of methods that were in the literature at the time and compare them with maximum likelihood, which is the method of choice in problems like this. We describe a variant on their approach to illustrate the EM algorithm.

Let $y = (125, 18, 20, 34)$ be a realization of vector $y = (y_1, y_2, y_3, y_4)$ believed to be coming from the multinomial distribution given in (16.1). The probability mass function, given the data, is

$$g(y, \psi) = \frac{n!}{y_1! y_2! y_3! y_4!} (1/2 + \psi/4)^{y_1} (1/4 - \psi/4)^{y_2 + y_3} (\psi/4)^{y_4}.$$

The log likelihood after omitting an additive term not containing ψ is

$$\log L(\psi) = y_1 \log(2 + \psi) + (y_2 + y_3) \log(1 - \psi) + y_4 \log(\psi).$$

By differentiating with respect to ψ, one gets

$$\partial \log L(\psi)/\partial \psi = \frac{y_1}{2 + \psi} - \frac{y_2 + y_3}{1 - \psi} + \frac{y_4}{\psi}.$$

The equation $\partial \log L(\psi)/\partial \psi = 0$ can be solved, and solution is $\psi = (15 + \sqrt{53\,809})/394 \approx 0.626821$.

Now assume that instead of original value y_1 the counts y_{11} and y_{12}, such that $y_{11} + y_{12} = y_1$, could be observed and that their probabilities are $1/2$ and $\psi/4$, respectively. The complete data can be defined as $x = (y_{11}, y_{12}, y_2, y_3, y_4)$. The probability mass function of incomplete data y is $g(y, \psi) = \sum g_c(x, \psi)$, where

$$g_c(x, \psi) = c(x)(1/2)^{y_{11}} (\psi/4)^{y_{12}} (1/4 - \psi/4)^{y_2 + y_3} (\psi/4)^{y_4},$$

$c(x)$ is free of ψ, and the summation is taken over all values of x for which $y_{11} + y_{12} = y_1$. The complete log likelihood is

$$\log L_c(\psi) = (y_{12} + y_4) \log(\psi) + (y_2 + y_3) \log(1 - \psi). \tag{16.2}$$

Our goal is to find the conditional expectation of $\log L_c(\psi)$ given y, using the starting point for $\psi^{(0)}$:

$$Q(\psi, \psi^{(0)}) = \mathbb{E}_{\psi^{(0)}} \{ \log L_c(\psi) | y \}.$$

As $\log L_c$ is linear function in y_{11} and y_{12}, the *E-step* is done by simply by replacing y_{11} and y_{12} by their conditional expectations, given y. If Y_{11} is the random variable corresponding to y_{11}, it is easy to see that

$$Y_{11} \sim Bin \left(y_1, \frac{1/2}{1/2 + \psi^{(0)}/4} \right)$$

so that the conditional expectation of Y_{11} given y_1 is

$$\mathbb{E}_{\psi^{(0)}}(Y_{11} | y_1) = \frac{\frac{y_1}{2}}{\frac{1}{2} + \frac{\psi^{(0)}}{4}} = y_{11}^{(0)}.$$

Of course, $y_{12}^{(0)} = y_1 - y_{11}^{(0)}$. This completes the *E-step* part.

In the *M-step*, one chooses $\psi^{(1)}$ so that $Q(\psi, \psi^{(0)})$ is maximized. After replacing y_{11} and y_{12} by their conditional expectations $y_{11}^{(0)}$ and $y_{12}^{(0)}$ in the Q-function, the maximum is obtained at

$$\psi^{(1)} = \frac{y_{12}^{(0)} + y_4}{y_{12}^{(0)} + y_2 + y_3 + y_4} = \frac{y_{12}^{(0)} + y_4}{n - y_{11}^{(0)}}.$$

The EM algorithm is composed of alternating these two steps. At the iteration k, we have

$$\psi^{(k+1)} = \frac{y_{12}^{(k)} + y_4}{n - y_{11}^{(k)}},$$

where $y_{11}^{(k)} = \frac{1}{2}y_1/(1/2 + \psi^{(k)}/4)$ and $y_{12}^{(k)} = y_1 - y_{11}^{(k)}$. To see how the EM algorithm computes the MLE for this problem, see the R script `emexample.r`.

16.2 Mixtures

Recall from Chapter 2 that mixtures are compound distributions of the form

$$F(x) = \int F(x|t)dG(t). \tag{16.3}$$

The CDF $G(t)$ in (16.3) serves as a mixing distribution on kernel distribution $F(x|t)$. Recognizing and estimating mixtures of distributions is an important task in data analysis. Pattern recognition, data mining, and other modern statistical tasks often call for mixture estimation.

For example, suppose an industrial process produces machine parts with lifetime distribution F_1, but a small proportion of the parts (say, ω) are defective and have CDF $F_2 \gg F_1$. If we cannot sort out the good ones from the defective ones, the lifetime of a randomly chosen part is

$$F(x) = (1 - \omega)F_1(x) + \omega F_2(x).$$

This is a simple two-point mixture where the mixing distribution has two discrete points of positive mass. With (finite) discrete mixtures like this, the probability points of G serve as weights for the kernel distribution. In the nonparametric likelihood, we see immediately how difficult it is to solve for the MLE in the presence of the weight ω, especially if ω is unknown.

Suppose we want to estimate the weights of a fixed number k of fully known distributions. We illustrate EM approach that introduces unobserved indicators with the goal of simplifying the likelihood. The weights are estimated by maximum likelihood. Assume that a sample $X_1, X_2, \ldots, X_n$ comes from the mixture

$$f(x, \omega) = \sum_{j=1}^{k} \omega_j f_j(x),$$

where $f_1, \ldots, f_k$ are continuous and the weights $0 \le \omega_j \le 1$ are unknown and constitute $(k-1)$-dimensional vector $\omega = (\omega_1, \ldots, \omega_{k-1})$ and $\omega_k = 1 - \omega_1 - \cdots - \omega_{k-1}$. The class densities $f_j(x)$ are fully specified.

Even in this simplest case when $f_1, \ldots, f_k$ are given and the only parameters are the weights ω, the log likelihood assumes a complicated form:

$$\sum_{i=1}^{n} \log f(x_i, \omega) = \sum_{i=1}^{n} \log \left(\sum_{j=1}^{k} \omega_j f_j(x_i) \right).$$

The derivatives with respect to ω_j lead to the system of equations, not solvable in a closed form.

Here is a situation where the EM algorithm can be applied with a little creative foresight. Augment the data $x = (x_1, \ldots, x_n)$ by an unobservable matrix $z = (z_{ij}, i = 1, \ldots, n; j = 1, \ldots, k)$. The values z_{ij} are indicators, defined as

$$z_{ij} = \begin{cases} 1, & x_i \text{ from } f_j \\ 0, & \text{otherwise} \end{cases}$$

The unobservable matrix z (our "missing value") tells us (in an oracular fashion) where the ith observation x_i comes from. Note that each row of z contains a single 1 and $k-1$ 0's. With augmented data, $x = (y, z)$, the (complete) likelihood takes quite a simple form:

$$\prod_{i=1}^{n} \prod_{j=1}^{k} \left(\omega_j f_j(x_i) \right)^{z_{ij}}.$$

The complete log-likelihood is simply

$$\log L_c(\omega) = \sum_{i=1}^{n} \sum_{j=1}^{k} z_{ij} \log \omega_j + C,$$

where $C = \sum_i \sum_j z_{ij} \log f_j(x_i)$ is free of ω. This is easily solved.

Assume that mth iteration of the weight estimate $\omega^{(m)}$ is already obtained. The mth *E-step* is

$$\mathbb{E}_{\omega^{(m)}}(z_{ij}|x) = \mathbb{P}_{\omega^{(m)}}(z_{ij} = 1|x) = z_{ij}^{(k)},$$

where $z_{ij}^{(m)}$ is the posterior probability of ith observation coming from the jth mixture component, f_j, in the iterative step m:

$$z_{ij}^{(m)} = \frac{\omega_j^{(m)} f_j(x_i)}{f\left(x_i, \omega^{(m)}\right)}.$$

Because $\log L_c(\omega)$ is linear in z_{ij}, $Q\left(\omega, \omega^{(m)}\right)$ is simply $\sum_{i=1}^{n} \sum_{j=1}^{k} z_{ij}^{(m)} \log \omega_j + C$. The subsequent M-step is simple: $Q\left(\omega, \omega^{(m)}\right)$ is maximized by

$$\omega_j^{(m+1)} = \frac{\sum_{i=1}^{n} z_{ij}^{(m)}}{n}.$$

The R script (`mixture_cla.r`) and below codes illustrate the algorithm above. A sample of size 150 is generated from the mixture $f(x) = 0.5\mathcal{N}(-5, 2^2) + 0.3\mathcal{N}(0,0.5^2) + 0.2\mathcal{N}(2,1)$. The mixing weights are estimated by the EM algorithm. $M = 20$ iterations of EM algorithm yielded $\hat{\omega} = (0.5033, 0.2822, 0.2145)$. Figure 16.1 gives histogram of data, theoretical mixture, and EM estimate:

```
> source("mixture_cla.r")
> omega.current  # The estimated mixing weights (omega hat)
[1] 0.5032592 0.2822136 0.2145272
>
> xx<- seq(-12,5,by=0.05)
> omega <- c(0.5,0.3,0.2);
> mixt <- 0; mixe <- 0;
>
> for(j in 1:3){
+ mixt <- mixt + omega[j]/(sqrt(2*pi*sig2s[j]))*
+          exp(-(xx-mus[j])^2/(2*sig2s[j]));
+ mixe <- mixe + omega.current[j]/(sqrt(2*pi*sig2s[j]))*
```

Figure 16.1 Observations from the $0.5\mathcal{N}(-5, 2^2) + 0.3\mathcal{N}(0,0.5^2) + 0.2\mathcal{N}(2,1)$ mixture (histogram), the mixture (*dotted line*), and EM estimated mixture (*solid line*).

```
+               exp(-(xx-mus[j])^2/(2*sig2s[j])));
+ }
>
> p <- ggplot() + geom_histogram(aes(x=x,y=..density..),col="black",
+ fill="lightblue",binwidth=0.8)
> p <- p + geom_density()+geom_line(aes(x=xx,y=mixt),lwd=1,lty=2)
> p <- p + geom_line(aes(x=xx,y=mixe),lwd=1,lty=1,col=2)
> print(p)
```

Example 16.1 As an example of a specific mixture of distributions, we consider application of EM algorithm in the so-called zero-inflated Poisson (ZIP) model. In ZIP models the observations come from two populations, one in which all values are identically equal to 0 and the other Poisson $P(\lambda)$. The "zero" population is selected with probability ξ and the Poisson population with complementary probability of $1 - \xi$. Given the data, both λ and ξ are to be estimated. To illustrate EM algorithm in fitting ZIP models, we consider data set (Thisted, 1988) on distribution of number of children in a sample of $n = 4075$ widows, given in Table 16.1.

At first glance the Poisson model for this data seems to be appropriate; however, the sample mean and variance are quite different (theoretically, in Poisson models they are the same):

```
> number <- 0:6;           # number of children
> freqs <- c(3062, 587, 284, 103, 33, 4, 2);
> n <- sum(freqs);
> sum(freqs*number/n)   # sample mean
[1] 0.3995092
> sum(freqs*(number-0.3995)^2/(n-1)) # sample variance
[1] 0.6626409
```

This indicates presence of *over-dispersion* and the ZIP model can account for the apparent excess of zeros. The ZIP model can be formalized as

$$P(X = 0) = \xi + (1 - \xi)\frac{\lambda^0}{0!}e^{-\lambda} = \xi + (1 - \xi)e^{-\lambda}$$

$$P(X = i) = (1 - \xi)\frac{\lambda^i}{i!}e^{-\lambda}, \quad i = 1, 2, \ldots,$$

and the estimation of ξ and λ is of interest. To apply the EM algorithm, we treat this problem as an *incomplete data* problem. The complete data would involve knowledge of frequencies of zeros from both populations, n_{00} and n_{01}, such that the observed frequency of zeros n_0 is split as $n_{00} + n_{01}$. Here n_{00} is number of cases coming from the point mass at 0-part, and n_{01} is number of cases coming from the

Table 16.1 Frequency distribution of the number of children among 4075 widows.

Number of children (`number`)	0	1	2	3	4	5	6
Number of widows (`freq`)	3062	587	284	103	33	4	2

Poisson part of the mixture. If values of n_{00} and n_{01} are available, the estimation of ξ and λ is straightforward. For example, the MLEs are

$$\hat{\xi} = \frac{n_{00}}{n} \quad \text{and} \quad \hat{\lambda} = \frac{\sum_i i n_i}{n - n_{00}},$$

where n_i is the observed frequency of i children. This will be a basis for M-step in the EM implementation, because the estimator of ξ comes from the fact that $n_{00} \sim Bin(n, \xi)$, while the estimator of λ is the sample mean of the Poisson part. The E-step involves finding $\mathbb{E}n_{00}$ if ξ and λ are known. With $n_{00} \sim Bin(n_0, p_{00}/(p_{00} + p_{01}))$, where $p_{00} = \xi$ and $p_{01} = (1 - \xi)e^{-\lambda}$, the expectation of n_{00} is

$$\mathbb{E}(n_{00}|\text{observed frequencies}, \xi, \lambda) = n_0 \times \frac{\xi}{\xi + (1 - \xi)e^{-\lambda}}.$$

From this expectation, the iterative procedure can be set with

$$n_{00}^{(t)} = n_0 \times \frac{\xi^{(t)}}{\xi^{(t)} + (1 - \xi^{(t)})\exp\{-\lambda^{(t)}\}}$$
$$\xi^{(t+1)} = n_{00}^{(t)}/n, \text{ and}$$
$$\lambda^{(t+1)} = \frac{1}{n - n_{00}^{(t)}} \sum_i i n_i,$$

where t is the iteration step. The following R code performs 20 iterations of the algorithm and collects the calculated values of n_{00}, ξ, and λ in three sequences newn00s, newxis, and newlambdas. The initial values are given for ξ and λ as $\xi_0 = 3/4$ and $\lambda_0 = 1/4$:

```
> newn00s <- rep(0,20);
> newxis <- rep(0,20);
> newlambdas <- rep(0,20);
> newxis[1] <- 3/4;  newlambdas[1] <- 1/4;     # initial values
>
> for(i in 1:19){
+ # collect the values in three sequences
+ newn00s[i] <- freqs[1]*newxis[i]/(newxis[i]+(1-newxis[i])*
+           exp(-newlambdas[i]));
+ newxis[i+1] <- newn00s[i]/n;
+ newlambdas[i+1] <- sum((1:6)*freqs[2:7])/(n-newn00s[i]);
+ }
> newn00s[20] <- freqs[1]*newxis[20]/(newxis[20]+(1-newxis[20])*
+           exp(-newlambdas[20]));
> head(cbind(newlambdas,newxis,newn00s))
      newlambdas      newxis   newn00s
[1,]   0.2500000   0.7500000   2430.930
[2,]   0.9902254   0.5965472   2447.161
[3,]   1.0000992   0.6005304   2460.056
[4,]   1.0080845   0.6036947   2470.239
[5,]   1.0144816   0.6061937   2478.244
[6,]   1.0195671   0.6081580   2484.512
```

Table 16.2 Some of the 20 steps in the EM implementation of ZIP modeling on widow data.

Step	newlambdas	newxis	newn00s
0	1/4	3/4	2430.9
1	0.5965	0.9902	2447.2
2	0.6005	1.0001	2460.1
3	0.6037	1.0081	2470.2
⋮			
18	0.6149	1.0372	2505.6
19	0.6149	1.0373	2505.8
20	0.6149	1.0374	2505.9

Table 16.2 gives the partial output of the R program. The values for `newxi`, `newlambda`, and `newn00` will stabilize after several iteration steps.

16.3 EM and Order Statistics

When applying nonparametric maximum likelihood to data that contain (independent) order statistics, the EM algorithm can be applied by assuming that with the observed order statistic $X_{i:k}$ (the ith smallest observation from an i.i.d. sample of k), there are associated with it $k-1$ missing values: $i-1$ values smaller than $X_{i:k}$ and $k-i$ values that are larger. Kvam and Samaniego (1994) exploited this opportunity to use the EM for finding the nonparametric MLE for i.i.d. component lifetimes based on observing only k-out-of-n system lifetimes. Recall a k-out-of-n system needs k or more working components to operate and fails after $n-k+1$ components fail; hence the system lifetime is equivalent to $X_{n-k+1:n}$.

Suppose we observe independent order statistics $X_{r_i:k_i}$, $i = 1, \ldots, n$ where the unordered values are independently generated from F. When F is absolutely continuous, the density for $X_{r_i:k_i}$ is expressed as

$$f_{r_i:k_i}(x) = r_i \binom{k_i}{r_i} F^{r_i-1}(x)(1 - F(x))^{k_i-r_i} f(x).$$

For simplicity, let $k_i = k$. In this application, we assign the complete data to be $X_i = \{X_{i1}, \ldots, X_{ik}, Z_i\}$, $i = 1, \ldots, n$ where Z_i is defined as the rank of the value observed from X_i. The observed data can be written as $Y_i = \{W_i, Z_i\}$, where W_i is the Z_i^{th} smallest observation from X_i.

With the complete data, the MLE for $F(x)$ is the EDF, which we will write as $N(x)/(nk)$ where $N(x) = \sum_i \sum_j \mathbf{1}(X_{ij} \leq x)$. This makes the M-step simple, but for the E-step, N is estimated through the log likelihood. For example, if $Z_i = z$, we observe W_i distributed as $X_{z:k}$. If $W_i \leq x$, out of the subgroup of size k from which W_i was measured,

$$z + (k - z) \frac{F(t) - F(W_i)}{1 - F(W_i)}$$

are expected to be less than or equal to x. On the other hand, if $W_i > x$, we know $k - z + 1$ elements from X_i are larger than x, and

$$(z - 1) \frac{F(W_i)}{F(x)}$$

are expected in $(-\infty, x]$.

The E-step is completed by summing all of these expected counts out of the complete sample of nk based on the most recent estimator of F from the M-step. Then, if $F^{(j)}$ represents our estimate of F after j iterations of the EM algorithm, it is updated as

$$
\begin{aligned}
F^{(j+1)}(x) = \frac{1}{nk} \sum_{i=1}^{n} \Bigg[& Z_i + (k - Z_i) \frac{F^{(j)}(x) - F^{(j)}(W_i)}{1 - F^{(j)}(x)} \mathbf{1}(W_i \leq x) \\
& + (Z_i - 1) \frac{F^{(j)}(x)}{F^{(j)}(W_i)} \mathbf{1}(W_i > x) \Bigg].
\end{aligned}
\tag{16.4}
$$

Equation (16.4) essentially joins the two steps of the EM algorithm together. All that is needed is an initial estimate $F^{(0)}$ to start it off. The observed sample EDF suffices. Because the full likelihood is essentially a multinomial distribution, convergence of $F^{(j)}$ is guaranteed. In general, the speed of convergence is dependent upon the amount of information. Compared with the mixture application, there is a great amount of missing data here, and convergence is expected to be relatively slow.

16.4 MAP via EM

The EM algorithm can be readily adapted to Bayesian context to maximize the posterior distribution. A maximum of the posterior distribution is the so-called MAP (maximum a posteriori) estimator, used widely in Bayesian inference. The benefit of MAP estimators over some other posterior parameters was pointed out on p. 57 of Chapter 4 in the context of Bayesian estimators. The maximum of the posterior $\pi(\psi|y)$, if it exists, coincides with the maximum of the product of the likelihood and prior $f(y|\psi)\pi(\psi)$. In terms of logarithms, finding the MAP estimator amounts to maximizing

$$\log \pi(\psi|y) = \log L(\psi) + \log \pi(\psi).$$

The EM algorithm can be readily implemented as follows:

E-Step. At $(k+1)^{\text{st}}$ iteration, calculate

$$\mathbb{E}_{\psi^{(k)}}\{\log \pi(\psi|x)|y\} = Q\left(\psi, \psi^{(k)}\right) + \log \pi(\psi).$$

The *E-step* coincides with the traditional EM algorithm, that is, $Q\left(\psi, \psi^{(k)}\right)$ has to be calculated.

M-Step. Choose $\psi^{(k+1)}$ to maximize $Q\left(\psi, \psi^{(k)}\right) + \log \pi(\psi)$. The *M-step* here differs from that in the EM, because the objective function to be maximized with respect to ψ's contains additional term, logarithm of the prior. However, the presence of this additional term contributes to the concavity of the objective function, thus improving the speed of convergence.

Example 16.2   MAP Solution to Fisher's Genomic Example. Assume that we elicit a $Be(\nu_1, \nu_2)$ prior on ψ:

$$\pi(\psi) = \frac{1}{B(\nu_1, \nu_2)}\psi^{\nu_1-1}(1-\psi)^{\nu_2-1}.$$

The beta distribution is a natural conjugate for the missing data distribution, because $y_{12} \sim Bin(y_1, (\psi/4)/(1/2 + \psi/4))$. Thus the log posterior (additive constants ignored) is

$$\log \pi(\psi|x) = \log L(\psi) + \log \pi(\psi)$$
$$= (y_{12} + y_4 + \nu_1 - 1)\log \psi + (y_2 + y_3 + \nu_2 - 1)\log(1-\psi).$$

The *E-step* is completed by replacing y_{12} by its conditional expectation $y_1 \times \left(\psi^{(k)}/4\right)/\left(1/2 + \psi^{(k)}/4\right)$. This step is the same as in the standard EM algorithm.

The *M-step*, at $(k+1)$st iteration, is

$$\psi^{(k+1)} = \frac{y_{12}^{(k)} + y_4 + \nu_1 - 1}{y_{12}^{(k)} + y_2 + y_3 + y_4 + \nu_1 + \nu_2 - 2}.$$

When the beta prior coincides with uniform distribution (that is, when $\nu_1 = \nu_2 = 1$), the MAP and MLE solutions coincide.

16.5 Infection Pattern Estimation

Reilly and Lawlor (1999) applied the EM algorithm to identify contaminated lots in blood samples. Here the observed data contain the disease exposure history of a person over k points in time. For the ith individual, let

$$X_i = \mathbf{1}(i\text{th person infected by end of trial}),$$

where $P_i = P(X_i = 1)$ is the probability that the ith person was infected at least once during k exposures to the disease. The exposure history is defined as a vector $y_i = \{y_{i1}, \dots, y_{ik}\}$, where

$$y_{ij} = \mathbf{1}(\text{th person exposed to disease at } j\text{th time point } k).$$

Let λ_j be the rate of infection at time point j. The probability of not being infected in time point j is $1 - y_{ij}\lambda_j$, so we have $P_i = 1 - \prod(1 - y_{ij}\lambda_j)$. The corresponding likelihood for $\lambda = \{\lambda_1, \dots, \lambda_k\}$ from observing n patients is a bit daunting:

$$
\begin{aligned}
L(\lambda) &= \prod_{i=1}^{n} p_i^{x_i}\left(1 - p_i\right)^{1-x_i} \\
&= \prod_{i=1}^{n}\left(1 - \prod_{j=1}^{k}(1 - y_{ij}\lambda_j)\right)^{x_i}\left(\prod_{j=1}^{k}(1 - y_{ij}\lambda_j)\right)^{1-x_i}.
\end{aligned}
$$

The EM algorithm helps if we assign the unobservable

$$Z_{ij} = \mathbf{1}(\text{person } i \text{ infected at time point } j),$$

where $P(Z_{ij} = 1) = \lambda_j$ if $y_{ij} = 1$ and $P(Z_{ij} = 1) = 0$ if $y_{ij} = 0$. Averaging over y_{ij}, $P(Z_{ij} = 1) = y_{ij}\lambda_j$. With z_{ij} in the complete likelihood ($1 \le i \le n$, $1 \le j \le k$), we have the observed data changing to $x_i = \max\{z_{i1}, \dots, z_{ik}\}$, and

$$
L(\lambda|Z) = \prod_{i=1}^{n}\prod_{j=1}^{k}\left(y_{ij}\lambda_j\right)^{z_{ij}}\left(1 - y_{ij}\lambda_j\right)^{1-z_{ij}},
$$

which has the simple binomial form.

For the *E-step*, we find $\mathbb{E}\left(Z_{ij}|x_i, \lambda^{(m)}\right)$, where $\lambda^{(m)}$ is the current estimate for $(\lambda_1, \dots, \lambda_k)$ after m iterations of the algorithm. We need only concern ourselves with the case $x_i = 1$, so that

$$
\mathbb{E}(Z_{ij}|x_i = 1) = P(y_{ij} = 1|x_i = 1) = \frac{y_{ij}\lambda_j}{1 - \prod_{j=1}^{k}\left(1 - y_{ij}\lambda_j\right)}.
$$

In the *M-step*, MLEs for $(\lambda_1, \ldots, \lambda_k)$ are updated in iteration $m + 1$ from $\lambda_1^{(m)}, \ldots, \lambda_k^{(m)}$ to

$$\lambda_j^{(m+1)} = \frac{\sum_{i=1}^{n} y_{ij} \left[\dfrac{y_{ij} \lambda_j^{(m+1)}}{1 - \prod_{j=1}^{k} \left(1 - y_{ij} \lambda_j^{(m+1)} \right)} \right]}{\sum_{i=1}^{n} y_{ij}}.$$

16.6 Exercises

16.1 Suppose we have data generated from a mixture of two normal distributions with a common known variance. Write an R script to determine the MLE of the unknown means from an i.i.d. sample from the mixture by using the EM algorithm. Test your program using a sample of 10 observations generated from an equal mixture of the two kernels $\mathcal{N}(0,1)$ and $\mathcal{N}(1,1)$.

16.2 The data in the following table come from the mixture of two Poisson random variables, $\mathcal{P}(\lambda_1)$ with probability ϵ and $\mathcal{P}(\lambda_2)$ with probability $1 - \epsilon$:

Value	0	1	2	3	4	5	6	7	8	9	10
Frequency	708	947	832	635	427	246	121	51	19	6	1

(i) Develop an EM algorithm for estimating ϵ, λ_1, and λ_2.
(ii) Write R program that uses (i) in estimating ϵ, λ_1, and λ_2 for data from the table.

16.3 The following data give the numbers of occupants in 1768 cars observed on a road junction in Jakarta, Indonesia, during a certain time period on a weekday morning:

Number of occupants	1	2	3	4	5	6	7
Number of cars	897	540	223	85	17	5	1

The proposed model for number of occupants X is truncated Poisson (TP), defined as

$$P(X = i) = \frac{\lambda^i \exp\{-\lambda\}}{(1 - \exp\{-\lambda\}) \, i!}, \quad i = 1, 2, \ldots$$

(i) Write down the likelihood (or the log-likelihood) function. Is it straightforward to find the MLE of λ by maximizing the likelihood or log likelihood directly?

(ii) Develop an EM algorithm for approximating the MLE of λ. *Hint*: Assume that missing data is i_0 – the number of cases when $X = 0$, so with the complete data the model is Poisson, $\mathcal{P}(\lambda)$. Estimate λ from the complete data. Update i_0 given the estimator of λ.

(iii) Write R program that will estimate the MLE of λ for Jakarta cars data using the EM procedure from (ii).

16.4 Consider the problem of right censoring in lifetime measurements in Chapter 10. Set up the EM algorithm for solving the nonparametric MLE for a sample of possibly right censored values $X_1, \ldots, X_n$.

16.5 Write R program that will approximate the MAP estimator in Fisher's problem (Example 16.2), if the prior on ψ is $Be(2,2)$. Compare the MAP and MLE solutions.

References

Dempster, A. P., Laird, N. M., and Rubin, D. B. (1977), "Maximum Likelihood from Incomplete Data via the EM Algorithm" (with discussion), *Journal of the Royal Statistical Society. Series B*, 39, 1–38.

Fisher, R. A., and Balmukand, B. (1928), "The Estimation of Linkage from the Offspring of Selfed Heterozygotes," *Journal of Genetics*, 20, 79–92.

Healy, M. J. R., and Westmacott, M. H. (1956), "Missing Values in Experiments Analysed on Automatic Computers," *Applied Statistics*, 5, 203–306.

Kvam, P. H., and Samaniego, F. J. (1994), "Nonparametric Maximum Likelihood Estimation Based on Ranked Set Samples," *Journal of the American Statistical Association*, 89, 526–537.

McKendrick, A. G. (1926), "Applications of Mathematics to Medical Problems," *Proceedings of the Edinburgh Mathematical Society*, 44, 98–130.

McLachlan, G. J., and Krishnan, T. (1997), *The EM Algorithm and Extensions*, New York: Wiley.

Rao, C. R. (1973), *Linear Statistical Inference and its Applications*, Second Edition, New York: Wiley.

Reilly, M., and Lawlor, E. (1999), "A Likelihood Method of Identifying Contaminated Lots of Blood Product," *International Journal of Epidemiology*, 28, 787–792.

Slatkin, M., and Excoffier, L. (1996), "Testing for Linkage Disequilibrium in Genotypic Data Using the Expectation–Maximization Algorithm," *Heredity*, 76, 377–383.

Tsai, W. Y., and Crowley, J. (1985), "A Large Sample Study of Generalized Maximum Likelihood Estimators from Incomplete Data via Self-Consistency," *Annals of Statistics*, 13, 1317–1334.

Thisted, R. A. (1988), *Elements of Statistical Computing: Numerical Computation*, New York: Chapman & Hall.

Wu, C. F. J. (1983), "On the Convergence Properties of the EM Algorithm," *Annals of Statistics*, 11, 95–103.

17

Statistical Learning

Learning is not compulsory …neither is survival.

W. Edwards Deming (1900–1993)

A general type of artificial intelligence, called *machine learning,* refers to techniques that sift through data and find patterns that lead to optimal decision rules, such as classification rules. In a way, these techniques allow computers to "learn" from the data, adapting as trends in the data become more clearly understood with the computer algorithms. Statistical learning pertains to the data analysis in this treatment, but the field of machine learning goes well beyond statistics and into algorithmic complexity of computational methods.

In business and finance, machine learning is used to search through huge amounts of data to find structure and pattern, and this is called *data mining.* In engineering, these methods are developed for *pattern recognition,* a term for classifying images into predetermined groups based on the study of statistical classification rules that statisticians refer to as *discriminant analysis.* In electrical engineering, specifically, the study of *signal processing* uses statistical learning techniques to analyze signals from sounds, radar, or other monitoring devices and convert them into digital data for easier statistical analysis.

Techniques called *neural networks* were so named because they were thought to imitate the way the human brain works. Analogous to neurons, connections between processing elements are generated dynamically in a learning system based on a large database of examples. In fact, most neural network algorithms are based on statistical learning techniques, especially nonparametric ones.

In this chapter, we will only present a brief exposition of classification and statistical learning that can be used in machine learning, discriminant analysis, pattern recognition, neural networks, and data mining. Nonparametric methods now play a vital role in statistical learning. As computing power has progressed through the years, researchers have taken on bigger and more complex problems.

Nonparametric Statistics with Applications to Science and Engineering with R, Second Edition.
Paul Kvam, Brani Vidakovic, and Seong-joon Kim.
© 2023 John Wiley & Sons, Inc. Published 2023 by John Wiley & Sons, Inc.

An increasing number of these problems cannot be properly summarized using parametric models.

This research area has a large and growing knowledge base that cannot be justly summarized in this book chapter. For students who are interested in reading more about statistical learning methods, both parametric and nonparametric, we suggest starting with the seminal textbook *The Elements of Statistical Learning* by Hastie, Tibshirani, and Friedman (2017). *An Introduction to Statistical Learning: With Applications in R* by James et al. (2019) represents an accessible primer for less experienced students. Efron and Hastie (2019) provide a helpful overview of just how pervasive and consequential these methods have been inside and outside of academia. Arnold, Kane, and Lewis (2019) offer an instructive text with helpful R code to aid in the implementation of common statistical learning algorithms.

17.1 Discriminant Analysis

Discriminant analysis is the statistical name for categorical prediction. The goal is to predict a categorical response variable, G, from one or more predictor variables, x. For example, if there is a partition of k groups $\mathcal{G} = (G_1, \dots, G_k)$, we want to find the probability that any particular observation x belongs to group $G_j, j = 1, \dots, k$ and then use this information to classify it in one group or the other. This is called *supervised classification* or *supervised learning* because the structure of the categorical response is known, and the problem is to find out in which group each observation belongs. *Unsupervised classification*, or *unsupervised learning* on the other hand, aims to find out how many relevant classes there are and then to characterize them.

One can view this simply as a categorical extension to prediction for simple regression: using a set of data of the form (x_1, g_1), ..., (x_n, g_n), we want to devise a rule to classify future observations $x_{n+1}, \dots, x_{n+m}$.

17.1.1 Bias Versus Variance

Recall that a loss function measures the discrepancy between the data responses and what the proposed model predicts for response values, given the corresponding set of inputs. For continuous response values y with inputs x, we are most familiar with squared error loss

$$L(y, f) = (y - f(x))^2.$$

We want to find the predictive function f that minimizes the *expected loss*, $\mathbb{E}[L(y, f)]$, where the expectation averages over all possible response values. With

| High variance | High variance | Low variance | Low variance |
| high bias | low bias | high bias | Low bias |

Figure 17.1   Targets illustrating the difference between model bias and variance.

the observed data set, we can estimate this as (Figure 17.1)

$$\mathcal{E}_f = \frac{1}{n}\sum_{i=1}^{n} L\left(y_i, \hat{f}(x_i)\right).$$

The function that minimizes the squared error is the conditional mean $\mathbb{E}(Y|X = x)$, and the expected squared errors $\mathbb{E}(Y - f(Y))^2$ consist of two parts: *variance* and the square of the *bias*. If the classifier is based on a global rule, such as linear regression, it is simple and rigid but at least stable. It has little variance but, by overlooking important nuances of the data, can be highly biased. A classifier that fits the model locally fails to garner information from as many observations and is more unstable. It has larger variance, but its adaptability to the detailed characteristics of the data ensures it has less bias. Compared with traditional statistical classification methods, most nonparametric classifiers tend to be less stable (more variable) but highly adaptable (less bias).

17.1.2   Cross-Validation

If you carry the egg basket do not dance.

Ambede[1] proverb

Obviously, the more local model will report less error than a global model, so instead of finding a model that simply minimizes error for the data set, it is better to put aside some of the data to test the model fit independently. The part of the data used to form the estimated model is called the *training sample*. The reserved group of data is the *test sample*.

The idea of using a training sample to develop a decision rule is paramount to empirical classification. Using the test sample to judge the method constructed

---

1   Bantu peoples group of sub-Saharan Africa.

from the training data is called *cross-validation*. Because data are often sparse and hard to come by, some methods use the training set to both develop the rule and to measure its misclassification rate (or error rate) as well. See the jackknife and bootstrap methods described in Chapter 15, for example.

17.1.3 Bayesian Decision Theory

There are two kinds of loss functions commonly used for categorical responses: a zero-one loss and cross-entropy loss. The zero-one loss merely counts the number of misclassified observations. Cross-entropy, on the other hand, uses the estimated class probabilities $\hat{p}_i(x) = \hat{P}(g \in G_i|x))$, and we minimize $\mathbb{E}(-2\ln\hat{p}_i(X))$.

By using zero-one loss, the estimator that minimizes risk classifies the observation to the most probable class, given the input $P(G|X)$. Because this is based on Bayesian rule of probability, this is called the *Bayesian classifier*. However, if $P(X|G_i)$ represents the distribution of observations from population G_i, it might be assumed we know a prior probability $P(G_i)$ that represents the probability any particular observation comes from population G_i. Furthermore, optimal decisions might depend on particular consequences of misclassification, which can be represented in cost variables; for example, $c_{ij} = $ cost of classifying an observation from population G_i into population G_j.

For example, if $k = 2$, the *Bayesian decision rule* that minimizes the expected cost (c_{ij}) is to classify x into G_1 if

$$\frac{P(x|G_1)}{P(x|G_2)} > \frac{(c_{21} - c_{22})P(G_2)}{(c_{12} - c_{11})P(G_1)}$$

and otherwise classify the observation into G_2.

Cross-entropy has an advantage over zero-one loss because of its continuity; in regression trees, for example, classifiers found via optimization techniques are easier to use if the loss function is differentiable.

17.2 Linear Classification Models

In terms of bias versus variance, a linear classification model represents a strict global model with potential for bias, but low variance that makes the classifier more stable. For example, if a categorical response depends on two ordinal inputs on the (x, y) axis, a linear classifier will draw a straight line somewhere on the graph to best separate the two groups.

The first linear rule developed was based on assuming the underlying distribution of inputs was normally distributed with different means for the different populations. If we assume further that the distributions have an identical covariance structure ($X_i \sim \mathcal{N}(\mu_i, \Sigma)$) and the unknown parameters have MLEs $\hat{\mu}_i$ and $\hat{\Sigma}$, then the discrimination function reduces to

$$x\hat{\Sigma}^{-1}\left(x_1 - x_2\right)' - \frac{1}{2}\left(x_1 + x_2\right)\hat{\Sigma}^{-1}\left(x_1 - x_2\right) > \delta \qquad (17.1)$$

for some value δ, which is a function of cost. This is called *Fisher's linear discrimination function* (LDF) because with the equal variance assumption, the rule is linear in x. The LDF was developed using normal distributions, but this linear rule can also be derived using a minimal squared-error approach. This is true, you can recall, for estimating parameters in multiple linear regression as well.

If the variances are not the same, the optimization procedure is repeated with extra MLEs for the covariance matrices, and the rule is quadratic in the inputs and hence called a *quadratic discriminant function* (QDF). Because the linear rule is overly simplistic for some examples, quadratic classification rules are used to extend the linear rule by including squared values of the predictors. With k predictors in the model, this begets $\binom{k+1}{2}$ additional parameters to estimate. So many parameters in the model can cause obvious problems, even in large data sets.

There have been several studies that have looked into the quality of linear and quadratic classifiers. While these rules work well if the normality assumptions are valid, the performance can be pretty lousy if they are not. There are numerous studies on the LDF and QDF robustness; for example, see Moore (1973), Marks and Dunn (1974), and Randles et al. (1978).

17.2.1 Logistic Regression as Classifier

The simple zero-one loss function makes sense in the categorical classification problem. If we relied on the squared-error loss (and outputs labeled with zeroes and ones), the estimate for g is not necessarily in $[0, 1]$, and even if the large sample properties of the procedure are satisfactory, it will be hard to take such results seriously.

One of the simplest models in the regression framework is the logistic regression model, which serves as a bridge between simple linear regression and statistical classification. Logistic regression, discussed in Chapter 12 in the context of generalized linear models (GLM), applies the linear model to binary response variables, relying on a *link function* that will allow the linear model to adequately describe probabilities for binary outcomes. Below we will use a simple illustration of how it can be used as a classifier. For a more comprehensive instruction on logistic regression and other models for ordinal data, Agresti's book *Categorical Data Analysis* serves as an excellent basis.

If we start with the simplest case where $k = 2$ groups, we can arbitrarily assign $g_i = 0$ or $g_i = 1$ for categories G_0 and G_1. This means we are modeling a binary response function based on the measurements on x. If we restrict our attention to a linear model $P(g = 1|x) = x'\beta$, we will be saddled with an unrefined model that can estimate probability with a value outside [0,1]. To avoid this problem, consider transformations of the linear model such as the following:

(i) *logit*: $p(x) = P(g = 1|x) = \exp(x'\beta)/[1 + \exp(x'\beta)]$, so $x'\beta$ is estimating $\ln[p(x)/(1 - p(x))]$ that has its range on $\mathbb{R}$.

(ii) *probit*: $P(g = 1|x) = \Phi(x'\beta)$, where Φ is the standard normal CDF. In this case $x'\beta$ is estimating $\Phi^{-1}(p(x))$.

(iii) *log-log*: $p(x) = 1 - \exp(\exp(x'\beta))$ so that $x'\beta$ is estimating $\ln[-\ln(1 - p(x))]$ on $\mathbb{R}$.

Because the logit transformation is symmetric and has relation to the natural parameter in the GLM context, it is generally the default transformation in this group of three. We focus on the logit link and seek to maximize the likelihood

$$L(\beta) = \prod_{i=1}^{n} p_i(x)^{g_i}(1 - p_i(x))^{1-g_i},$$

in terms of $p(x) = 1 - \exp(\exp(x'\beta))$ to estimate β and therefore obtain MLEs for $p(x) = P(g = 1|x)$. This likelihood is rather well behaved and can be maximized in a straightforward manner. We use the R function `glm` to perform a logistic regression in the example below.

Example 17.1   *Computer Programmers*
(Kutner, Nachtsheim, and Neter, 1996) A study of 25 computer programmers aims to predict task success based on the programmers' months of work experience:

```
> x <- c(14, 29, 6, 25, 18, 4, 18, 12, 22, 6, 30, 11, 30, 5, 20, 13,
+    9, 32, 24, 13, 19, 4, 28, 22, 8);
> y <- c(0, 0, 0, 1, 1, 0, 0, 0, 1, 0, 1, 0, 1, 0, 1, 0, 0, 1, 0, 1,
+    0, 0, 1, 1, 1);
> fit <- glm(y~x, family="binomial")
> fit
Call:  glm(formula = y ~ x, family = "binomial")

Coefficients:
(Intercept)           x
   -3.0597       0.1615

Degrees of Freedom: 24 Total (i.e. Null);  23 Residual
Null Deviance:        34.3
Residual Deviance: 25.42        AIC: 29.42
> # "summary(fit)" <- try this to obtain more information
>
> predict(fit,list(x=14),type="response")
         1
0.3102624
```

Here $\beta = (\beta_0, \beta_1)$ and $\hat{\beta} = (-3.0597, 0.1615)$. The estimated logistic regression function is

$$\hat{p} = \frac{e^{-3.0597+0.1615x}}{1 + e^{-3.0597+0.1615x}}.$$

For example, in the case $x_1 = 14$, we have $\hat{p}_1 = 0.31$; i.e. we estimate that there is a 31% chance a programmer with 14 months' experience will successfully complete the project.

In the logistic regression model, if we use $\hat{p}$ as a criterion for classifying observations, the regression serves as a simple linear classification model. If misclassification penalties are the same for each category, $\hat{p} = 1/2$ will be the classifier boundary. For asymmetric loss, the relative costs of the misclassification errors will determine an optimal threshold.

Example 17.2   Fisher's Iris Data
(Fisher's Iris Data) To illustrate this technique, we use Fisher's iris data (Fisher 1936), which is commonly used to show off classification methods. The iris data set contains physical measurements of 150 flowers – 50 for each of three types of iris (Virginica, Versicolor, and Setosa). Iris flowers have three petals and three outer petal-like sepals. Figure 17.2a shows a plot of petal length versus width for Versicolor (circles) and Virginica (plus signs) along with the line that best linearly categorizes them. How is this line determined?

From the logistic function $x'\beta = \ln(p/(1-p))$, $p = 1/2$ represents an observation that is half-way between the Virginica iris and the Versicolor iris. Observations with values of $p < 0.5$ are classified to be Versicolor while

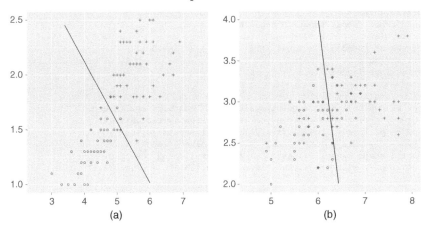

Figure 17.2 Two types of iris classified according to (a) petal length versus petal width and (b) sepal length versus sepal width. Versicolor = o, Virginica = +.

those with $p > 0.5$ are classified as Virginica. At $p = 1/2$, $x'\beta = \ln(p/(1 - p)) = 0$, and the line is defined by $\beta_0 + \beta_1 x_1 + \beta_2 x_2 = 0$, which in this case equates to $x_2 = (45.272 - 5.755 x_1)/10.447$. This line is drawn in Figure 17.2a:

```
> iris2 <- iris[-which(iris$Species=="setosa"),]
> fit <- glm(factor(Species)~Petal.Length+Petal.Width,data=iris2,
+ family="binomial")
> fit
Call:  glm(formula = factor(Species) ~ Petal.Length + Petal.Width,
     family = "binomial", data = iris2)

Coefficients:
(Intercept)   Petal.Length    Petal.Width
    -45.272          5.755         10.447

Degrees of Freedom: 99 Total (i.e. Null);   97 Residual
Null Deviance:       138.6
Residual Deviance: 20.56        AIC: 26.56
>
> x <- seq(3,7,by=0.1)
> p <- ggplot() + geom_point(aes(x=Petal.Length,y=Petal.Width,
+ group=Species,shape=Species,col=Species),data=iris2,size=2)
> p <- p + geom_line(aes(x=x,y=(45.272-5.775*x)/10.447))
> p <- p + scale_colour_manual(values=c("blue","red")) +
+ scale_shape_manual(values=c(1,3)) + theme(legend.position="none")
> p <- p + xlab("") + ylab("") + xlim(c(2.5,7.5)) + ylim(c(1,2.5))
> print(p)
>
> fit2 <- glm(factor(Species)~Sepal.Length+Sepal.Width,data=iris2,
+ family="binomial")
> fit2
Call:  glm(formula = factor(Species) ~ Sepal.Length + Sepal.Width,
family = "binomial", data = iris2)

Coefficients:
(Intercept)   Sepal.Length    Sepal.Width
   -13.0460         1.9024         0.4047
Degrees of Freedom: 99 Total (i.e. Null);   97 Residual
Null Deviance:       138.6
Residual Deviance: 110.3        AIC: 116.3
>
> x <- seq(4,8,by=0.1)
> p2 <- ggplot() + geom_point(aes(x=Sepal.Length,y=Sepal.Width,
+ group=Species,shape=Species,col=Species),data=iris2,size=2)
> p2 <- p2 + geom_line(aes(x=x,y=(13.046-1.9024*x)/0.4047))
> p2 <- p2+scale_colour_manual(values=c("blue","red")) +
+ scale_shape_manual(values=c(1,3)) + theme(legend.position="none")
> p2 <- p2 + xlab("") + ylab("") + xlim(c(4.5,8)) + ylim(c(2,4))
> print(p2)
```

While this example provides a spiffy illustration of linear classification, most populations are not so easily differentiated, and a linear rule can seem overly simplified and crude. Figure 17.2b shows a similar plot of sepal width versus length. The iris types are not so easily distinguished, and the linear classification does not help us in this example.

In Section 17.3, we will look at "nonparametric" classifying methods that can be used to construct a more flexible, nonlinear classifier.

17.3 Nearest Neighbor Classification

Recall from Chapter 13, nearest-neighbor methods can be used to create nonparametric regressions by determining the regression curve at x based on explanatory variables x_i that are considered closest to x. We will call this a k-nearest neighbor classifier if it considers the k closest points to x (using a majority vote) when constructing the rule at that point.

If we allow k to increase, the estimator eventually uses all of the data to fit each local response, so the rule is a global one. This leads to a simpler model with low variance. However, if the assumptions of the simple model are wrong, high bias will cause the expected mean squared error to explode. On the other hand, if we let k go down to one, the classifier will create minute neighborhoods around each observed x_i, revealing nothing from the data that a plot of the data has not already shown us. This is highly suspect as well.

The best model is likely to be somewhere in between these two extremes. As we allow k to increase, we will receive more smoothness in the classification boundary and more stability in the estimator. With small k, we will have a more jagged classification rule, but the rule will be able to identify more interesting nuances of the data. If we use a loss function to judge which is best, the 1-nearest-neighbor model will fit best, because there is no penalty for over-fitting. Once we identify each estimated category (conditional on X) as the observed category in the data, there will be no error to report.

In this case, it will help to split the data into a training sample and a test sample. Even with the loss function, the idea of local fitting works well with large samples. In fact, as the input sample size n gets larger, the k-nearest neighbor estimator will be consistent as long as $k/n \to 0$. That is, it will achieve the goals we wanted without the strong model assumptions that come with parametric classification. There is an extra problem using the nonparametric technique, however. If the dimension of X is somewhat large, the amount of data needed to achieve a satisfactory answer from the nearest neighbor grows exponentially.

17.3.1 The Curse of Dimensionality

The *curse of dimensionality*, termed by Bellman (1961), describes the property of data to become sparse if the dimension of the sample space increases. For example, imagine the denseness of a data set with 100 observations distributed uniformly on the unit square. To achieve the same denseness in a 10-dimensional unit hypercube, we would require 10^{20} observations.

This is a significant problem for nonparametric classification problems including nearest-neighbor classifiers and neural networks. As the dimension of inputs increase, the observations in the training set become relatively sparse. These procedures based on a large number of parameters help to handle complex problems but must be considered inappropriate for most small- or medium-sized data sets. In those cases, the linear methods may seem overly simplistic or even crude but still preferable to nearest neighbor methods.

17.3.2 Constructing the Nearest-Neighbor Classifier

The classification rule is based on the ratio of the nearest-neighbor density estimator. That is, if x is from population G, then $P(x|G) \approx$ (proportion of observations in the neighborhood around x)/(volume of the neighborhood). To classify x, select the population corresponding to the largest value of

$$\frac{P(G_i)P(x|G_i)}{\sum_j P(G_j)P(x|G_j)}, \quad i = 1, \dots, k.$$

This simplifies to the nearest-neighbor rule; if the neighborhood around x is defined to be the closest r observations, x is classified into the population that is most frequently represented in that neighborhood.

Figure 17.3 shows the output derived from the R example below. Fifty randomly generated points are classified into one of two groups in v in a partially random way. The nearest neighbor plots reflect three different smoothing conditions of $k = 11$, 5, and 1. As k gets smaller, the classifier acts more locally, and the rule appears more jagged:

```
> library(kknn)
> x <- matrix(runif(200),nrow=100);
> group <- round(0.3*runif(100)+0.3*x[,1]+0.4*x[,2]);
> dat <- data.frame(x1=x[,1],x2=x[,2],group=as.factor(group))
> newdat <- data.frame(expand.grid(seq(0,1,by=0.01),seq(0,1,by=0.01)))
> colnames(newdat) <- c("x1","x2");
>
> fit <- kknn(group ~ x1+x2,train=dat,test=newdat,k=4)
> fit.predict <- fitted(fit);
> p <- ggplot(aes(x=x1,y=x2,group=group,shape=group,col=group),data=dat)
> p <- p + geom_point(size=3)+theme(legend.position="none")
> print(p)
>
> nn.plot <- function(k){
+ fit <- fitted(kknn(group ~ x1+x2,train=dat,test=newdat,k=k));
+ dat2 <- newdat[which(fit==1),];
+ p <- ggplot(aes(x=x1,y=x2),data=dat2) + geom_point(pch=20)
+ p <- p + xlab("") + ylab("")
+ print(p)
+ }
> nn.plot(11)
> nn.plot(5)
> nn.plot(1)
```

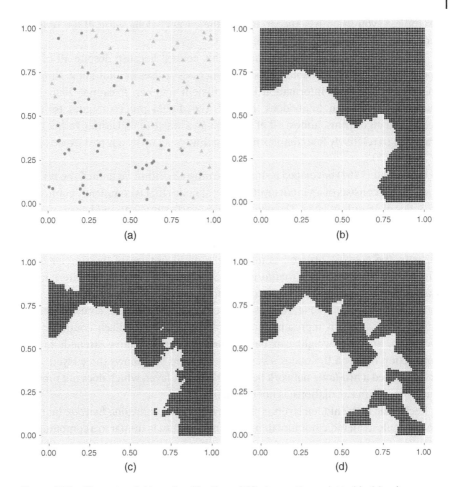

Figure 17.3 Nearest-neighbor classification of 50 observations plotted in (a) using neighborhood sizes of (b) 11, (c) 5, and (d) 1.

17.4 Neural Networks

Despite what your detractors say, you have a remarkable brain. Even with the increasing speed of computer processing, the much slower human brain has surprising ability to sort through gobs of information, disseminate some of its peculiarities, and make a correct classification often several times faster than a computer. When a familiar face appears to you around a street corner, your brain has several processes working in parallel to identify this person you see, using past experience to gauge your expectation (you might not believe your eyes, for example, if you saw Elvis appear around the corner) along with all the sensory data from what you see, hear, or even smell.

The computer is at a disadvantage in this contest because despite all of the speed and memory available, the static processes it uses cannot parse through the same amount of information in an efficient manner. It cannot adapt and learn as the human brain does. Instead, the digital processor goes through sequential algorithms, almost all of them being a waste of CPU time, rather than traversing a relatively few complex neural pathways set up by our past experiences (Figure 17.3).

Rosenblatt (1962) developed a simple learning algorithm he named the *perceptron*, which consists of an input layer of several nodes that is completely connected to nodes of an output layer. The perceptron is overly simplistic and has numerous shortcomings, but it also represents the first neural network. By extending this to a two-step network that includes a *hidden layer* of nodes between the inputs and outputs, the network overcomes most of the disadvantages of the simpler map. Figure 17.4 shows a simple *feed-forward* neural network, that is, the information travels in the direction from input to output.

The square nodes in Figure 17.4 represent neurons, and the connections (or *edges*) between them represent the synapses of the brain. Each connection is weighted, and this weight can be interpreted as the relative strength in the connection between the nodes. Even though Figure 17.4 shows three layers, this is considered a *two-layer* network because the input layer, which does not process data or perform calculations, is not counted.

Each node in the hidden layers is characterized by an *activation function* that can be as simple as an indicator function (the binary output is similar to a computer) or

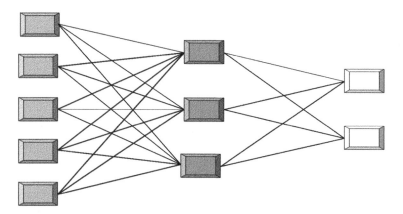

Figure 17.4 Basic structure of feed-forward neural network.

have more complex nonlinear forms. A simple activation function would represent a node that would react when the weighted input surpassed some fixed threshold.

The neural network essentially looks at repeated examples (or input observations) and recalls patterns appearing in the inputs along with each subsequent response. We want to train the network to find this relationship between inputs and outputs using supervised learning. A key in training the network is to find the weights to go along with the activation functions that lead to supervised learning. To determine weights, we use a *back-propagation* algorithm.

17.4.1 Back-Propagation

Before the neural network experiences any input data, the weights for the nodes are essentially random (noninformative). So at this point, the network functions like the scattered brain of a college freshman who has celebrated his first weekend on campus by drinking way too much beer.

The feed-forward neural network is represented by

$$
\begin{array}{ccccc}
n_I & \Longrightarrow & n_H & \Longrightarrow & n_O \\
\text{input nodes} & & \text{hidden nodes} & & \text{output nodes}
\end{array}.
$$

With an input vector $X = (x_1, \ldots, x_{n_I})$, each of the n_I input node codes the data and "fires" a signal across the edges to the hidden nodes. At each of the n_H hidden nodes, this message takes the form of a weighted linear combination from each attribute:

$$
\mathcal{H}_j = A\left(\alpha_{0j} + \alpha_{1j}x_1 + \cdots + \alpha_{n_Ij}x_{n_I}\right), \quad j = 1, \ldots, n_H \tag{17.2}
$$

where A is the activation function which is usually chosen to be the *sigmoid function*

$$
A(x) = \frac{1}{1 + e^{-x}}.
$$

We will discuss why A is chosen to be a sigmoid later. In the next step, the n_H hidden nodes fire this nonlinear outcome of the activation function to the output nodes, each translating the signals as a linear combination

$$
\mathcal{O}_k = \beta_0 + \beta_1\mathcal{H}_1 + \cdots + \beta_{n_H}\mathcal{H}_{nH}, \quad k = 1, \ldots, n_O. \tag{17.3}
$$

Each output node is a function of the inputs, and through the steps of the neural network, each node is also a function of the weights α and β. If we observe $X_l = (x_{1l}, \ldots, x_{n_Il})$ with output $g_l(k)$ for $k = 1, \ldots, n_O$, we use the same kind of transformation used in logistic regression:

$$
\hat{g}_l(k) = \frac{e^{\mathcal{H}_k}}{e^{\mathcal{H}_1} + e^{\mathcal{H}_2} + \cdots + e^{\mathcal{H}_{n_o}}}, \quad k = 1, \ldots, n_O.
$$

For the training data $\{(X_1, g_1), \ldots, (X_n, g_n)\}$, the classification is compared with the observation's known group, which is then *back-propagated* across the network, and the network responds (learns) by adjusting weights in the cases an error in classification occurs. The loss function associated with misclassification can be squared errors, such as

$$SSQ(\alpha, \beta) = \sum_{l=1}^{n} \sum_{k=1}^{n_O} (g_l(k) - \hat{g}_l(k))^2, \tag{17.4}$$

where $g_l(k)$ is the actual response of the input X_l for output node k and $\hat{g}_l(k)$ is the estimated response.

Now we look how those weights are changed in this back-propagation. To minimize the squared error SSQ in (17.4) with respect to weights α and β from both layers of the neural net, we can take partial derivatives (with respect to weight) to find the direction the weights should go to decrease the error. However, there are a lot of parameters to estimate: α_{ij} with $1 \leq i \leq n_I$ and $1 \leq j \leq n_H$ and β_{jk} with $1 \leq j \leq n_H$ and $1 \leq k \leq n_O$. It is not helpful to think of this as a parameter set, as if they have their own intrinsic value. If you do, the network looks terribly over-parameterized and unnecessarily complicated. Remember that α and β are artificial, and our focus is on the n predicted outcomes instead of estimated parameters. We will do this iteratively using *batch learning* by updating the network after the entire data set is entered.

Actually, finding the global minimum of SSQ with respect to α and β will lead to over-fitting the model, that is, the answer will not represent the true underlying process because it is blindly mimicking every idiosyncrasy of the data. The gradient is expressed here with a constant γ called the *learning rate*

$$\Delta\alpha_{ij} = \gamma \sum_{l=1}^{n} \frac{\partial(\sum_{k=1}^{n_O} (g_l(k) - \hat{g}_l(k))^2)}{\partial\alpha_{ij}} \tag{17.5}$$

$$\Delta\beta_{jk} = \gamma \sum_{l=1}^{n} \frac{\partial(\sum_{k=1}^{n_O} (g_l(k) - \hat{g}_l(k))^2)}{\partial\beta_{jk}} \tag{17.6}$$

and is solved iteratively with the following back-propagation equations (see Chapter 11 of Hastie et al. (2017)) via error variables a and b:

$$a_{il} = \left[\frac{\partial A(t)}{\partial(t)}\right]_{t=\alpha_i' X_l} \sum_{l=1}^{n} \beta_{jk} b_{jl}. \tag{17.7}$$

Obviously, the activation function A must be differentiable. Note that if $A(x)$ is chosen as a binary function such as $I(x \geq 0)$, we end up with a regular linear model from (17.2). The sigmoid function when scaled as $A_c(x) = A(cx)$ will look like $I(x \geq 0)$ as $c \to \infty$, but the function also has a well-behaved derivative.

In the first step, we use current values of α and β to predict outputs from (17.2) and (17.3). In the next step we compute errors b from the output layer and use (17.7) to compute a from the hidden layer. Instead of batch processing, updates to the gradient can be made sequentially after each observation. In this case, γ is not constant and should decrease to zero as the iterations are repeated (this is why it is called the learning rate).

The hidden layer of the network, along with the nonlinear activation function, gives it the flexibility to learn by creating convex regions for classification that need not be linearly separable like the more simple linear rules require. One can introduce another hidden layer that in effect can allow nonconvex regions (by combining convex regions together). Applications exist with even more hidden layers, but two hidden layers should be ample for almost every nonlinear classification problem that fits into the neural network framework.

17.4.2 Implementing the Neural Network

Implementing the steps above into a computer algorithm is not simple, nor is it free from potential errors. One popular method for processing through the back-propagation algorithm uses six steps:

1. Assign random values to the weights.
2. Input the first pattern to get outputs to the hidden layer $(\mathcal{H}_1, \ldots, \mathcal{H}_{n_H})$ and output layer $(\hat{g}(1), \ldots, \hat{g}(k))$.
3. Compute the output errors b.
4. Compute the hidden layer errors a as a function of b.
5. Update the weights using (17.5).
6. Repeat the steps for the next observation.

Computing a neural network from scratch would be challenging for many of us, even if we have a good programming background. In R, there are a few packages that can be used for classification: AMORE, nnet, and neuralnet. In AMORE, the TAO-robust back-propagation learning algorithm is implemented, and nnet package provides a function to train feed-forward neural network with a single hidden layer using traditional back-propagation algorithm. In neuralnet, the resilient back-propagation algorithm is used to build a neural network with multiple hidden layers.

17.4.3 Projection Pursuit

The technique of projection pursuit is similar to that of neural networks, as both employ a nonlinear function that is applied only to linear combinations of the input. While the neural network is relatively fixed with a set number of hidden layer nodes (and hence a fixed number of parameters), projection pursuit seems more nonparametric because it uses unspecified functions in its transformations. We will start with a basic model:

$$g(X) = \sum_{i=1}^{n_P} \psi\left(\theta_i' X\right),\qquad(17.8)$$

where n_P represents the number of unknown parameter vectors $(\theta_1, \dots, \theta_{n_p})$.

Note that $\theta_i' X$ is the projection of X onto the vector θ_i. If we pursue a value of θ_i that makes this projection effective, it seems logical enough to call this projection pursuit. The idea of using a linear combination of inputs to uncover structure in the data was first suggested by Kruskal (1969). Friedman and Stuetzle (1981) derived a more formal projection pursuit regression using a multistep algorithm:

1. Define $\tau_i^{(0)} = g_i$.
2. Maximize the standardized squared errors

$$SSQ^{(j)} = 1 - \frac{\sum_{i=1}^{n}\left(\tau_i^{(j-1)} - \hat{g}^{(j-1)}(\hat{w}^{(j)'}x_i)\right)^2}{\sum_{i=1}^{n}\left(\tau_i^{(j-1)}\right)^2}\qquad(17.9)$$

over weights $\hat{w}^{(j)}$ (under the constraint that $\hat{w}^{(j)'}1 = 1$) and $\hat{g}^{(j-1)}$.
3. Update τ with $\tau_i^{(j)} = \tau_i^{(j-1)} - \hat{g}^{(j-1)}(\hat{w}^{(j)'}x_i)$.
4. Repeat the first step k times until $SSQ^{(k)} \le \delta$ for some fixed $\delta > 0$.

Once the algorithm finishes, it essentially has given up trying to find other projections, and we complete the projection pursuit estimator as

$$\hat{g}(x) = \sum_{j=1}^{n_P} \hat{g}^{(j)}(\hat{w}^{(j)'}x).\qquad(17.10)$$

17.5 Binary Classification Trees

Binary trees offer a graphical and logical basis for empirical classification. Decisions are made sequentially through a route of branches on a tree – every time a choice is made, the route is split into two directions. Observations that are collected at the same endpoint (node) are classified into the same population. Those junctures on the route where the split is made are *nonterminal nodes*, and *terminal nodes* denote all the different endpoints where a classification of the tree. These endpoints are also called the leaves of the tree, and the starting node is called the root.

With the training set $(x_1, g_1), \dots, (x_n, g_n)$, where x is a vector of m components, splits are based on a single variable of x, possibly a linear combination. This leads to decision rules that are fairly easy to interpret and explain, so binary trees are popular for disseminating information to a broad audience. The phases of of tree construction include the following:

- Deciding whether to make the node a terminal node.
- Selection of splits in a nonterminal node
- Assigning classification rule at terminal nodes.

This is the essential approach of *classification and regression trees* (CART). The goal is to produce a simple and effective classification tree without an excess number of nodes.

If we have k populations $G_1, \ldots, G_k$, we will use the frequencies found in the training data to estimate population frequency in the same way we constructed nearest-neighbor classification rules: the proportion of observations in training set from the ith population $= P(G_i) = n_i/n$. Suppose there are $n_i(r)$ observations from G_i that reach node r. The probability of such an observation reaching node r is estimated as

$$P_i(t) = P(G_i)P(\text{reach node } r \mid G_i) = \frac{n_i}{n} \times \frac{n_i(r)}{n_i} = \frac{n_i(r)}{n}.$$

We want to construct a perfectly pure split where we can isolate one or some of the populations into a single node that can be a terminal node (or at least split more easily into one during a later split). Figure 17.5 illustrates a perfect split of node r. This, of course, is not always possible. This quality measure of a split is defined in an impurity index function

$$\mathcal{I}(r) = \psi\left(P_1(r), \ldots, P_k(r)\right),$$

where ψ is nonnegative and symmetric in its arguments, maximized at $(1/k, \ldots, 1/k)$, and minimized at any k-vector that has a one and $k - 1$ zeroes.

Several different methods of impurity have been defined for constructing trees. The three most popular impurity measures are cross-entropy, Gini impurity, and misclassification impurity:

1. **Cross-entropy:** $\mathcal{I}(r) = -\sum_{i:P_i(r)>0} P_i(r) \ln[P_i(r)]$.
2. **Gini:** $\mathcal{I}(r) = -\sum_{i\neq j} P_i(r)P_j(r)$.
3. **Misclassification:** $\mathcal{I}(r) = 1 - \max_j P_j(r)$

Figure 17.5 Purifying a tree by splitting.

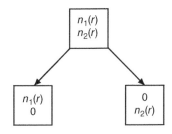

The misclassification impurity represents the minimum probability that the training set observations would be (empirically) misclassified at node r. The Gini measure and cross-entropy measure have an analytical advantage over the discrete impurity measure by being differentiable. We will focus on the most popular index of the three, which is the cross-entropy impurity.

By splitting a node, we will reduce the impurity to

$$q(L)\mathcal{I}(r_L) + q(R)\mathcal{I}(r_R),$$

where $q(R)$ is the proportion of observations that go to node r_R and $q(L)$ is the proportion of observations that go to node r_L. Constructed this way, the binary tree is a *recursive* classifier.

Let Q be a potential split for the input vector x. If $x = (x_1, \ldots, x_m)$, $Q = \{x_i > x_0\}$ would be a valid split if x_i is ordinal, or $Q = \{x_i \in S\}$ if x_i is categorical and S is a subset of possible categorical outcomes for x_i. In either case, the split creates two additional nodes for the binary response of the data to Q. For the first split, we find the split Q_1 that will minimize the impurity measure the most. The second split will be chosen to be the Q_2 that minimizes the impurity from one of the two nodes created by Q_1.

Suppose we are the middle of constructing a binary classification tree T that has a set of terminal nodes $\mathcal{R}$. With

$$P(\text{reach node } r) = P(r) = \sum P_i(r),$$

suppose the current impurity function is

$$\mathcal{I}_T = \sum_{r \in \mathcal{R}} \mathcal{I}(r)P(r).$$

At the next stage, then, we split the node that will most greatly decrease $\mathcal{I}_T$.

Example 17.3   The following made-up example was used in Elsner, Lehmiller, and Kimberlain (1996) to illustrate a case for which linear classification models fail and binary classification trees perform well. Hurricanes are categorized according to season as "tropical only" or "baroclinically influenced." Also these are classified according to location (longitude, latitude), and Figure 17.6a shows that no linear rule can separate the two categories without a great amount of misclassification. The average latitude of origin for tropical-only hurricanes is 18.8°N, compared with 29.1°N for baroclinically influenced storms. The baroclinically influenced hurricane season extends from mid-May to December, while the tropical-only season is largely confined to the months of August through October.

For this problem, simple splits are considered, and the ones that minimize impurity are Q_1 with longitude ≥ 67.75 and Q_2 with longitude ≤ 62.5 *(see homework)*.

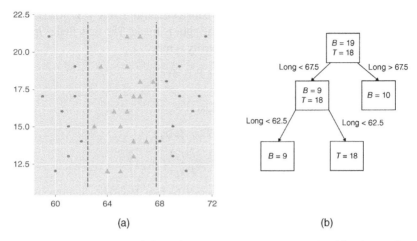

(a)   (b)

Figure 17.6 (a) Location of 37 tropical (circles) and other (plus signs) hurricanes from Elsner et al. (1996). (b) Corresponding separating tree.

In this case, the tree perfectly separates the two types of storms with two splits and three terminal nodes in Figure 17.6b:

```
> long <- c(59.00,59.50,60.00,60.50,61.00,61.00,61.50,61.50,62.00,
+ 63.00,63.50,64.00,64.50,65.00,65.00,65.00,65.50,65.50,65.50,
+ 66.00,66.00,66.00,66.50,66.50,66.50,67.00,67.50,68.00,68.50,
+ 69.00,69.00,69.50,69.50,70.00,70.50,71.00,71.50);
> lat <- c(17.00,21.00,12.00,16.00,13.00,15.00,17.00,19.00,14.00,
+ 15.00,19.00,12.00,16.00,12.00,15.00,17.00,16.00,19.00,21.00,
+ 13.00,14.00,17.00,17.00,18.00,21.00,14.00,18.00,14.00,18.00,
+ 13.00,15.00,17.00,19.00,12.00,16.00,17.00,21.00);
> trop <- c(0,0,0,0,0,0,0,0,0,1,1,1,1,1,1,1,1,1,1,1,1,1,1,1,1,1,
+        0,0,0,0,0,0,0,0,0,0);
> dat <- data.frame(long=long,lat=lat,trop=as.factor(trop));
>
> p <- ggplot() + geom_point(aes(x=long,y=lat,group=trop,shape=trop,
+ col=trop),data=dat,size=4) + geom_vline(xintercept=c(62.5,67.75),lty=2)
> p <- p + theme(legend.position="none") + xlab("") + ylab("")
> print(p)
```

17.5.1 Growing the Tree

So far we have not decided how many splits will be used in the final tree; we have only determined which splits should take place first. In constructing a binary classification tree, it is standard to grow a tree that is initially too large and to then prune it back, forming a sequence of subtrees. This approach works well; if one of the splits made in the tree appears to have no value, it might be worth saving if there exists below it an effective split.

In this case we define a branch to be a split direction that begins at a node and includes all the subsequent nodes in the direction of that split (called a subtree or descendants). For example, suppose we consider splitting tree T at node r and T_r represents the classification tree after the split is made. The new nodes made under r will be denoted r_R and r_L. The impurity is now

$$\mathcal{I}_{T_r} = \sum_{s \in T, s \neq r} \mathcal{I}_T(s)P(s) + P(r_R)\mathcal{I}_T(r_R) + P(r_L)\mathcal{I}(r_L).$$

The change in impurity caused by the split is

$$\Delta \mathcal{I}_{T_r}(r) = P(r)\mathcal{I}_T(r) - P(r_R)\mathcal{I}_T(r_R) - P(r_L)\mathcal{I}_T(r_L)$$

$$= P(r)\left(\mathcal{I}_T(r) - \frac{P(r_R)}{P(r_R)}\mathcal{I}_T(r_R) - \frac{P(r_L)}{P(r_L)}\mathcal{I}_T(r_L)\right).$$

Again, let $\mathcal{R}$ be the set of all terminal nodes of the tree. If we consider the potential differences for any particular split Q, say, $\Delta \mathcal{I}_{T_r}(r; Q)$, then the next split should be chosen by finding the terminal node r and split Q corresponding to

$$\max_{r \in \mathcal{R}}\left(P(r)\left(\max_Q \Delta \mathcal{I}_{T_r}(r; Q)\right)\right).$$

To prevent the tree from splitting too much, we will have a fixed threshold level $\tau > 0$ so that splitting must stop once the change no longer exceeds τ. We classify each terminal node according to majority vote: observations in terminal node r are classified into the population i with the highest $n_i(r)$. With this simple rule, the misclassification rate for observations arriving at node r is estimated as $1 - P_i(r)$.

17.5.2 Pruning the Tree

With a tree constructed using only a threshold value to prevent overgrowth, a large set of training data may yield a tree with an abundance of branches and terminal nodes. If τ is small enough, the tree will fit the data locally, similar to how a 1-nearest neighbor overfits a model. If τ is too large, the tree will stop growing prematurely, and we might fail to find some interesting features of the data. The best method is to grow the tree a bit too much and then prune back unnecessary branches.

To make this efficable, there must be a penalty function $\zeta_T = \zeta_T(|\mathcal{R}|)$ for adding extra terminal nodes, where $|\mathcal{R}|$ is the cardinality, or number of terminal nodes of $\mathcal{R}$. We define our cost function to be a combination of misclassification error and penalty for over-fitting:

$$C(T) = L_T + \zeta_T,$$

where

$$L_T = \sum_{r \in \mathcal{R}} P(r)\left(1 - \max_j P_j(r)\right) \equiv \sum_{r \in \mathcal{R}} L_T(r).$$

This is called the *cost-complexity* pruning algorithm in Breiman et al. (1984). Using this rule, we will always find a subtree of T that minimizes $C(T)$. If we allow $\zeta_T \to 0$, the subtree is just the original tree T, and if we allow $\zeta_T \to \infty$, the subtree is a single node that does not split at all. If we increase ζ_T from 0, we will get a sequence of subtrees, each one being nested in the previous one.

In deciding whether or not to prune a branch of the tree at node r, we will compare $C(T)$ of the tree with the new cost that would result from removing the branches under node r. L_T will necessarily increase, while ζ_T will decrease as the number of terminal nodes in the tree decreases.

Let T_r be the branch under node r, so the tree remaining after cutting branch T_r (we will call this $T_{(r)}$) is nested in T, i.e. $T_{(r)} \subset T$. The set of terminal nodes in the branch T_r is denoted $\mathcal{R}_r$. If another branch at node s is pruned, we will denote the remaining subtree as $T_{(r,s)} \subset T_{(r)} \subset T$. Now,

$$C(T_r) = \sum_{s \in \mathcal{R}_r} L_{T_r}(s) + \zeta_{T_r}$$

is equal to $C(T)$ if ζ_T is set to

$$h(r) = \frac{L_T - \sum_{s \in \mathcal{R}_r} L_{T_r}(s)}{|\mathcal{R}_r|}.$$

Using this approach, we want to trim the node r that minimizes $h(r)$. Obviously, only nonterminal nodes $r \in \mathcal{R}^C$ because terminal nodes have no branches. If we repeat this procedure after recomputing $h(r)$ for the resulting subtree, this pruning will create another sequence of nested trees

$$T \supset T_{(r_1)} \supset T_{(r_1, r_2)} \supset \cdots \supset r_0,$$

where r_0 is the first node of the original tree T. Each subtree has an associated cost $(C(T), C(T_{r_1}), \ldots, C(r_0))$ that can be used to determine at what point the pruning should finish. The problem with this procedure is that the misclassification probability is based only on the training data.

A better estimator can be constructed by cross-validation. If we divide the training data into ν subsets $S_1, \ldots, S_\nu$, we can form ν artificial training sets as

$$S_{(j)} \equiv \bigcup_{i \neq j} S_i$$

and constructing a binary classification tree based on each of the ν sets $S_{(1)}, \ldots, S_{(\nu)}$. This type of cross-validation is analogous to the jackknife "leave-one-out" resampling procedure. If we let $L^{(j)}$ be the estimated misclassification probability based on the subtree chosen in the jth step of the cross-validation (i.e. leaving out S_j) and let $\zeta^{(j)}$ be the corresponding penalty function, then

$$L^{CV} \equiv \frac{1}{n} \sum_{j=1}^{\nu} L^{(j)}$$

provides a bona fide estimator for misclassification error. The corresponding penalty function for L^{CV} is estimated as the geometric mean of the penalty functions in the cross-validation. That is,

$$\zeta^{CV} = \sqrt[\nu]{\prod_{j=1}^{\nu} \zeta^{(j)}}.$$

To perform a binary tree search in R, a number of packages can be used to fit data. The ideas in the CART book are implemented in `tree` and `rpart` packages. Package `party` fits a decision tree using a recursive partitioning algorithm in a conditional inference framework. The Quinlan's C5.0 algorithm for building decision trees, which is popular in the machine learning community, is implemented in `C50` package. We will present R examples using `rpart` package as it implements the recursive partitioning method described in this chapter. A function `rpart` creates a decision tree based on an input data and modeling formula. Several options are available to control tree growth, tree pruning, and misclassification costs. The function `prune` is pruning a fitted tree to the desired size. For example, if `T` is the output of a `rpart` function that needs to be decided how depth of the tree to retain, the `prune(T, cp=XX)` generates a pruned tree of T (Figure 17.7):

```
> library(rpart)
> library(rpart.plot)
> fit <- rpart(Species ~ Sepal.Length+Sepal.Width, data=iris,
+ control=rpart.control(cp=0.0,minbucket=4))
> rpart.plot(fit,type=4,extra=2,cex=0.7)
```

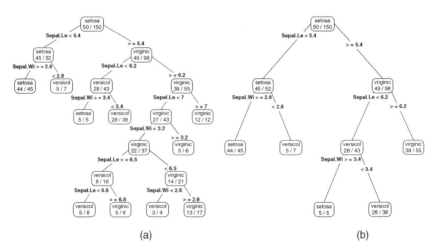

(a)   (b)

Figure 17.7 Binary tree classification applied to Fisher's iris data using (a) less pruning and (b) more pruning.

```
> printcp(fit)
Classification tree:
rpart(formula = Species ~ Sepal.Length + Sepal.Width, data = iris,
control = rpart.control(cp = 0, minbucket = 4))

Variables actually used in tree construction:
[1] Sepal.Length Sepal.Width

Root node error: 100/150 = 0.66667

n= 150

        CP nsplit rel error xerror       xstd
1 0.440      0      1.00    1.18 0.050173
2 0.180      1      0.56    0.63 0.060448
3 0.050      2      0.38    0.39 0.053722
4 0.040      3      0.33    0.39 0.053722
5 0.008      4      0.29    0.31 0.049592
6 0.000      9      0.25    0.41 0.054583
> # the cross-validation error is minimum at cp=0.008
>
> fit2<-prune(fit,cp=0.008)
> rpart.plot(fit2,type=4,extra=2,cex=0.7)
```

17.5.3 General Tree Classifiers

CARTs can be conveniently divided to five different families.

(i) *The CART family*: Simple versions of CART have been emphasized in this chapter. This method is characterized by its use of two branches from each nonterminal node. Cross-validation and pruning are used to determine size of tree. Response variable can be quantitative or nominal. Predictor variables can be nominal or ordinal, and continuous predictors are supported. *Motivation* is statistical prediction.

(ii) *The CLS family*: These include ID3, originally developed by Quinlan (1979), and off-shoots such as CLS and C4.5. For this method, the number of branches equals the number of categories of the predictor. Only nominal response and predictor variables are supported in early versions, so continuous inputs had to be binned. However, the latest version of C4.5 supports ordinal predictors. *Motivation* is concept learning.

(iii) *The AID family*: Methods include AID, THAID, CHAID, MAID, XAID, FIRM, and TREEDISC. The number of branches varies from two to the number of categories of the predictor. Statistical significance tests (with multiplicity adjustments in the later versions) are used to determine the size of tree. AID, MAID, and XAID are for quantitative responses. THAID, CHAID, and TREEDISC are for nominal responses, although the version of CHAID from Statistical Innovations, distributed by SPSS, can handle a quantitative categorical response. FIRM comes in two varieties for categorical or continuous response. Predictors can be nominal or ordinal, and there is usually provision for a missing-value category. Some versions can handle continuous predictors; others cannot. *Motivation* is detecting complex statistical relationships.

(iv) *Linear combinations*: Methods include OC1 and SE-Trees. The number of branches varies from two to the number of categories of predictor. *Motivation* is detecting linear statistical relationships combined to concept learning.

(v) *Hybrid models*: IND is one example. It combines CART and C4, as well as Bayesian and minimum encoding methods. Knowledge Seeker combines methods from CHAID and ID3 with a novel multiplicity adjustment. *Motivation* is combining methods from other families to find optimal algorithm.

17.6 Exercises

17.1 Create a simple nearest-neighbor program using R. It should input a training set of data in $m + 1$ columns; one column should contain the population identifier $1, \ldots, k$, and the others contain the input vectors that can have length m. Along with this training set, also input another m column matrix representing the classification set. The output should contain n, m, k, and the classifications for the input set.

17.2 For the Example 17.3, show the optimal splits, using the cross-entropy measure, in terms of intervals {longitude $\geq l_0$} and {latitude $\geq l_1$}.

17.3 In this exercise the goal is to discriminate between observations coming from two different normal populations, using logistic regression.
Simulate a training data set, $\{(X_i, Y_i), i = 1, \ldots, n\}$, (take n even) as follows: For the first half of data, X_i, $i = 1, \ldots, n/2$ are sampled from the standard normal distribution and $Y_i = 0$, $i = 1, \ldots, n/2$. For the second half, X_i, $i = n/2 + 1, \ldots, n$ are sampled from normal distribution with mean 2 and variance 1, while $Y_i = 1$, $i = n/2 + 1, \ldots, n$. Fit the logistic regression to this data: $\hat{P}(Y = 1) = f(X)$.
Simulate a validation set $\{(X_j^*, Y_j^*), j = 1, \ldots, m\}$ the same way, and classify these new Y_j^*'s as 0 or 1 depending whether $f(X_j^*) < 0.5$ or ≥ 0.5 :
(a) Calculate the error of this logistic regression classifier:
$$L_n(m) = \frac{1}{m} \sum_{j=1}^{m} \mathbf{1} \left(\mathbf{1}(f(X_j^*) > 0.5) \neq Y_j^* \right).$$

In your simulations use $n = 60, 200$, and 2000 and $m = 100$.
(b) Can the error $L_n(m)$ be made arbitrarily small by increasing n?

| RELEVANT R FUNCTIONS AND DATA SETS IN THIS CHAPTER |

R functions: glm, kknn, rpart, rpart.plot, prune
R package: kknn, rpart, rpart.plot

References

Arnold, T., Kane, M., and Lewis, B. W. (2019), *A Computational Approach to Statistical Learning*, Chapman and Hall/CRC.

Bellman, R. E. (1961), *Adaptive Control Processes*, Princeton, NJ: Princeton University Press.

Breiman, L., Friedman, J., Olshen, R., and Stone, C. (1984), *Classification and Regression Trees*, Belmont, CA: Wadsworth.

Efron, B, and Hastie, T. (2019), *Computer Age Statistical Inference*, London: Cambridge University Press.

Elsner, J. B., Lehmiller, G. S., and Kimberlain, T. B. (1996), "Objective Classification of Atlantic Basin Hurricanes," *Journal of Climate*, 9, 2880–2889.

Fisher, R. A. (1936), "The Use of Multiple Measurements in Taxonomic Problems," *Annals of Eugenics*, 7, 179–188.

Friedman, J., and Stuetzle, W. (1981), "Projection Pursuit Regression," *Journal of the American Statistical Association*, 76, 817–823.

Hastie, T., Tibshirani, R., and Friedman, J. (2017), *The Elements of Statistical Learning*, Second Edition, New York: Springer-Verlag.

James, G., Witten, D., Hastie, T., and Tibshirani, R. (2019), *An Introduction to Statistical Learning: with Applications in R*, Second Edition, New York: Springer-Verlag.

Kutner, M. A., Nachtsheim, C. J., and Neter, J. (1996), *Applied Linear Regression Models*, Fourth Edition, Chicago: Irwin.

Kruskal, J. (1969), "Toward a Practical Method which Helps Uncover the Structure of a Set of Multivariate Observations by Finding the Linear Transformation which Optimizes a New Index of Condensation," in *Statistical Computation*, New York: Academic Press, pp. 427–440.

Marks, S., and Dunn, O. (1974), "Discriminant Functions when Covariance Matrices are Unequal," *Journal of the American Statistical Association*, 69, 555–559.

Moore, D. H. (1973), "Evaluation of Five Discrimination Procedures for Binary Variables," *Journal of the American Statistical Association*, 68, 399–404.

Quinlan, J. R. (1979), "Discovering Rules from Large Collections of Examples: A Case Study," in *Expert Systems in the Microelectronics Age*, Ed. D. Michie, Edinburgh: Edinburgh University Press.

Randles, R. H., Broffitt, J. D., Ramberg, J. S., and Hogg, R. V. (1978), "Generalized Linear and Quadratic Discriminant Functions Using Robust Estimates," *Journal of the American Statistical Association*, 73, 564–568.

Rosenblatt, R. (1962), *Principles of Neurodynamics: Perceptrons and the Theory of Brain Mechanisms*, Washington, DC: Spartan.

18

Nonparametric Bayes

*Bayesian (bey'-zhuhn) n. **1**. Result of breeding a statistician with a clergyman to produce the much sought honest statistician.*

Anonymous

This chapter is about nonparametric Bayesian inference. Understanding the computational machinery needed for non-conjugate Bayesian analysis in this chapter can be quite challenging, and it is beyond the scope of this text. Instead, we will use specialized software, WinBUGS, to implement complex Bayesian models in a user-friendly manner. Some applications of WinBUGS have been discussed in Chapter 4, and an overview of WinBUGS is given in the Appendix B.

Our purpose is to explore the useful applications of the nonparametric side of Bayesian inference. At first glance, the term *nonparametric Bayes* might seem like an oxymoron; after all, Bayesian analysis is all about introducing prior distributions on parameters. Actually, nonparametric Bayes is often seen as a synonym for Bayesian models with process priors on the spaces of densities and functions. Dirichlet process (DP) priors are the most popular choice. However, many other Bayesian methods are nonparametric in spirit. In addition to DP priors, Bayesian formulations of contingency tables and Bayesian models on the coefficients in atomic decompositions of functions will be discussed later in this chapter.

18.1 Dirichlet Processes

The central idea of traditional nonparametric Bayesian analysis is to draw inference on an unknown distribution function. This leads to models on function

Nonparametric Statistics with Applications to Science and Engineering with R, Second Edition. Paul Kvam, Brani Vidakovic, and Seong-joon Kim.
© 2023 John Wiley & Sons, Inc. Published 2023 by John Wiley & Sons, Inc.

spaces, so that the Bayesian nonparametric approach to modeling requires a dramatic shift in methodology. In fact, a commonly used technical definition of nonparametric Bayesian models involves infinitely many parameters, as mentioned in Chapter 10.

Results from Bayesian inference are comparable with classical nonparametric inference, such as density and function estimation, estimation of mixtures, and smoothing. There are two main groups of nonparametric Bayes methodologies: (i) methods that involve prior/posterior analysis on distribution spaces and (ii) methods in which standard Bayesian analysis is performed on a vast number of parameters, such as atomic decompositions of functions and densities. Although these two methodologies can be presented in a unified way (see Mueller and Quintana, 2004), because of simplicity we present them separately.

Recall a Dirichlet random variable can be constructed from gamma random variables. If $X_1, \dots, X_n$ are i.i.d. $Gamma(a_i, 1)$, then for $Y_i = X_i / \sum_{j=1}^{n} X_j$, the vector $(Y_1, \dots, Y_n)$ has Dirichlet $Dir(a_1, \dots, a_n)$ distribution. The Dirichlet distribution represents a multivariate extension of the beta distribution: $Dir(a_1, a_2) \equiv Be(a_1, a_2)$. Also, from Chapter 2, $\mathbb{E}Y_i = a_i / \sum_{j=1}^{n} a_j$, $\mathbb{E}Y_i^2 = a_i(a_i + 1) / \sum_{j=1}^{n} a_j(1 + \sum_{j=1}^{n} a_j)$, and $\mathbb{E}(Y_i Y_j) = a_i a_j / \sum_{j=1}^{n} a_j(1 + \sum_{j=1}^{n} a_j)$.

The DP, with precursors in the work of Freedman (1963) and Fabius (1964), was formally developed by Ferguson (1973, 1974). It is the first prior developed for spaces of distribution functions. The DP is, formally, a probability measure (distribution) on the space of probability measures (distributions) defined on a common probability space $\mathcal{X}$. Hence, a realization of DP is a random distribution function.

The DP is characterized by two parameters: (i) Q_0, a specific probability measure on $\mathcal{X}$ (or equivalently, G_0 a specified distribution function on $\mathcal{X}$) and (ii) α, a positive scalar parameter.

Definition 18.1 (Ferguson, 1973) The DP generates random probability measures (random distributions) Q on $\mathcal{X}$ such that for any finite partition $B_1, \dots, B_k$ of $\mathcal{X}$,

$$(Q(B_1), \dots, Q(B_k)) \sim Dir(\alpha Q_0(B_1), \dots, \alpha Q_0(B_k)),$$

where, $Q(B_i)$ (a random variable) and $Q_0(B_i)$ (a constant) denote the probability of set B_i under Q and Q_0, respectively. Thus, for any B,

$$Q(B) \sim Be(\alpha Q_0(B), \alpha(1 - Q_0(B)))$$

and

$$\mathbb{E}(Q(B)) = Q_0(B), \quad \mathbb{V}ar(Q(B)) = \frac{Q_0(B)(1 - Q_0(B))}{\alpha + 1}.$$

The probability measure Q_0 plays the role of the center of the DP, while α can be viewed as a *precision* parameter. Large α implies small variability of DP about its center Q_0.

The above can be expressed in terms of CDFs, rather than in terms of probabilities. For $B = (-\infty, x]$ the probability $Q(B) = Q((-\infty, x]) = G(x)$ is a distribution function. As a result, we can write

$$G(x) \sim Be(\alpha G_0(x), \alpha(1 - G_0(x)))$$

and

$$\mathbb{E}(G(x)) = G_0(x)), \quad \mathbb{V}\mathrm{ar}(G(x)) = \frac{G_0(x)(1 - G_0(x))}{\alpha + 1}.$$

The notation $G \sim DP(\alpha G_0)$ indicates that the DP prior is placed on the distribution G.

Example 18.1 Let $G \sim DP(\alpha G_0)$ and $x_1 < x_2 < \cdots < x_n$ are arbitrary real numbers from the support of G. Then

$$\left(G(x_1), G(x_2) - G(x_1), \ldots, G(x_n) - G(x_{n-1})\right) \sim$$
$$Dir(\alpha G_0(x_1), \alpha(G_0(x_2) - G_0(x_1)), \ldots, \alpha(G_0(x_n) - G_0(x_{n-1}))), \tag{18.1}$$

which suggests a way to generate a realization of density from DP at discrete points.

If $(d_1, \ldots, d_n)$ is a draw from (18.1), then $(d_1, d_1 + d_2, \ldots, \sum_{i=1}^{n} d_i)$ is a draw from $(G(x_1), G(x_2), \ldots, G(x_n))$. The R script below generates 15 draws from $DP(\alpha G_0)$ for the base CDF $G_0 \equiv Be(2,2)$ and the precision parameter $\alpha = 20$. In Figure 18.1 the base CDF $Be(2,2)$ is shown as a dotted line. Fifteen random CDFs from $DP(20, Be(2,2))$ are scattered around the base CDF:

```
> library(MCMCpack)
> n <- 30; # generate random CDF's at 30 equispaced points
> a <- 2;  # a, b are parameters of theme
> b <- 2;  # BASE distribution G_0 = Beta(2,2)
>
> alpha = 20;
> # The precision parameter alpha = 20 describes
> # scattering about the BASE distribution.
> # Higher alpha, less variability.
> # -------------
>
> x <- seq(0,1,length=n);
> # The equispaced points at which
> # random CDF's are evaluated.
>
> y <- pbeta(x,a,b);
> # find CDF's of BASE
> par <- c(y[1], diff(y));
>
```

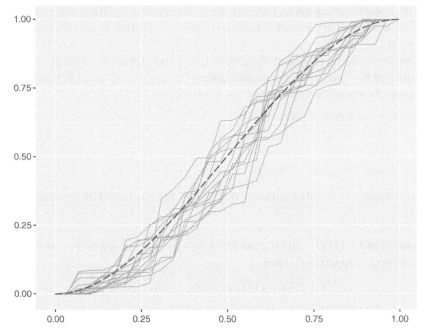

Figure 18.1 The base CDF *Be*(2,2) is shown as a dotted line. Fifteen random CDFs from DP(20, *Be*(2,2)) are scattered around the base CDF.

```
> yy <- rdirichlet(15,alpha*par);
> yy2 <- apply(yy,1,cumsum);
> yy2 <- data.frame(x=rep(x,15),y=as.vector(yy2),
+ group=as.vector(sapply(1:15,rep,times=length(par))));
> ggplot()+geom_line(aes(x=x,y=y,group=group),data=yy2,lwd=0.7
+,col="darkgray") + geom_line(aes(x=x,y=y),lty=2,lwd=1,col="red")
```

An alternative definition of DP, due to Sethuraman and Tiwari (1982) and Sethuraman (1994), is known as the *stick-breaking algorithm*.

Definition 18.2 Let $U_i \sim Be(1, \alpha)$, $i = 1, 2, \ldots$ and $V_i \sim G_0, i = 1, 2, \ldots$ be two independent sequences of i.i.d. random variables. Define weights $\omega_1 = U_1$ and $\omega_i = U_i \prod_{j=1}^{i-1}(1 - U_j), i > 1$. Then,

$$G = \sum_{k=1}^{\infty} \omega_k \delta(V_k) \sim DP(\alpha G_0),$$

where $\delta(V_k)$ is a point mass at V_k.

The distribution G is discrete, as a countable mixture of point masses, and from this definition one can see that with probability 1 only discrete distributions fall in the support of DP. The name stick breaking comes from the fact that $\sum \omega_i = 1$ with probability 1, that is, the unity is broken on infinitely many random weights. The Definition 18.2 suggests another way to generate approximately from a given DP.

Let $G_K = \sum_{k=1}^{K} \omega_k \delta(V_k)$ where the weights $\omega_1, \dots, \omega_{K-1}$ are as in Definition 18.2 and the last weight ω_K is modified as $1 - \omega_1 - \cdots - \omega_{K-1}$, so that the sum of K weights is 1. In practical applications, K is selected so that $\left(1 - (\alpha/(1+\alpha))^K\right)$ is small.

18.1.1 Updating Dirichlet Process Priors

The critical step in any Bayesian inference is the transition from the prior to the posterior, that is, updating a prior when data are available. If $Y_1, Y_2, \dots, Y_n$ is a random sample from G and G has Dirichlet prior $DP(\alpha G_0)$, the posterior remains Dirichlet, $G|Y_1, \dots, Y_n \sim DP\left(\alpha^* G_0^*\right)$, with $\alpha^* = \alpha + n$, and

$$G_0^*(t) = \frac{\alpha}{\alpha + n} G_0(t) + \frac{n}{\alpha + n}\left(\frac{1}{n}\sum_{i=1}^{n} I(Y_i \leq t \leq \infty)\right). \tag{18.2}$$

Notice that the DP prior and the EDF constitute a *conjugate pair* because the posterior is also a DP. The posterior estimate of distribution is $E(G|Y_1, \dots, Y_n) = G_0^*(t)$ that is, as we saw in several examples with conjugate priors, a weighted average of the "prior mean" and the maximum likelihood estimator (the EDF).

Example 18.2 In the spirit of classical nonparametrics, the problem of estimating the CDF at a fixed value x has a simple nonparametric Bayesian solution. Suppose the sample $X_1, \dots, X_n \sim F$ is observed and that one is interested in estimating $F(x)$. Also, suppose the $F(x)$ is assigned a DP prior with a center F_0 and a small precision parameter α. The posterior distribution for $F(x)$ is

$$Be(\alpha F_0(x) + \ell_x, \alpha(1 - F_0(x)) + n - \ell_x)$$

where ℓ_x is the number of observations in the sample smaller than or equal to x. As $\alpha \to 0$, the posterior tends to a $Be(\ell_x, n - \ell_x)$. This limiting posterior is often called *noninformative*. By inspecting the $Be(\ell_x, n - \ell_x)$ distribution, or generating from it, one can find a posterior probability region for the CDF at any value x. Note that the posterior expectation of $F(x)$ is equal to the classical estimator ℓ_x/n, which makes sense because the prior is noninformative. (See Figure 18.2)

Example 18.3 Hartsfield–Jackson airport. The underground train at Hartsfield–Jackson airport arrives at its starting station every four minutes. The number of people Y entering a single car of the train is random variable with a Poisson distribution,

$$Y|\lambda \sim \mathcal{P}(\lambda).$$

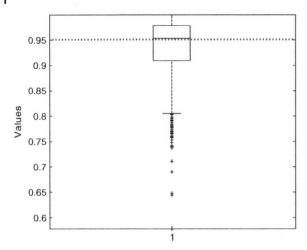

Figure 18.2 For a sample $n = 15$ Beta(2,2) observations, a boxplot of "noninformative" posterior realizations of $P(X \le 1)$ is shown. Exact value $F(1)$ for Beta(2,2) is shown as dotted line.

A sample of size $N = 20$ for Y is obtained below:

9	7	7	8	8	11	8	7	5	7
13	5	7	14	4	6	18	9	8	10

The prior on λ is *any* discrete distribution supported on integers $[1, 17]$:

$$\lambda | P \sim Discr\left((1,2,\dots,17), P = (p_1, p_2, \dots, p_{17})\right),$$

where $\sum_i p_i = 1$. The hyperprior on probabilities P is Dirichlet:

$$P \sim Dir(\alpha G_0(1), \alpha G_0(2), \dots, \alpha G_0(17)).$$

We can assume that the prior on λ is a DP with

$$G_0 = [1,1,1,2,2,3,3,4,4,5,6,5,4,3,2,1,1]/48$$

and $\alpha = 48$. We are interested in posterior inference on the rate parameter λ:

```
model
{
for (i in 1:N)
    {
    y[i] ~ dpois(lambda)
    }
    lambda ~ dcat(P[])
    P[1:bins] ~ ddirch(alphaG0[])
}
#data
list(bins=17, alphaG0=c(1,1,1,2,2,3,3,4,4,5,6,5,4,3,2,1,1),
  y=c(9,7,7,8,8,11,8,7,5,7,13,5,7,14,4,6,18,9,8,10),  N=20
```

```
)
#inits
list(lambda=12,
P=c(0,0,0,0,0,0,0,0,0,0,0,1,0,0,0,0,0))
)
```

The summary posterior statistics were found directly from within WinBUGS:

Node	Mean	sd	MC error	2.5%	Median	97.5%
λ	8.634	0.6687	0.003232	8	9	10
P[1]	0.02034	0.01982	8.556E−5	5.413E−4	0.01445	0.07282
P[2]	0.02038	0.01995	78.219E−5	5.374E−4	0.01423	0.07391
P[3]	0.02046	0.02004	8.752E−5	5.245E−4	0.01434	0.07456
P[4]	0.04075	0.028	1.179E−4	0.004988	0.03454	0.1113
P[5]	0.04103	0.028	1.237E−4	0.005249	0.03507	0.1107
P[6]	0.06142	0.03419	1.575E−4	0.01316	0.05536	0.143
P[7]	0.06171	0.03406	1.586E−4	0.01313	0.05573	0.1427
P[8]	0.09012	0.04161	1.981E−4	0.02637	0.08438	0.1859
P[9]	0.09134	0.04163	1.956E−4	0.02676	0.08578	0.1866
P[10]	0.1035	0.04329	1.85E−4	0.03516	0.09774	0.2022
P[11]	0.1226	0.04663	2.278E−4	0.04698	0.1175	0.2276
P[12]	0.1019	0.04284	1.811E−4	0.03496	0.09649	0.1994
P[13]	0.08173	0.03874	1.71E−4	0.02326	0.07608	0.1718
P[14]	0.06118	0.03396	1.585E−4	0.01288	0.05512	0.1426
P[15]	0.04085	0.02795	1.336E−4	0.005309	0.03477	0.1106
P[16]	0.02032	0.01996	9.549E−5	5.317E−4	0.01419	0.07444
P[17]	0.02044	0.01986	8.487E−5	5.475E−4	0.01445	0.07347

The main parameter of interest is the arrival rate, λ. The posterior mean of λ is 8.634. The median is nine passengers every four minutes. Either number could be justified as an estimate of the passenger arrival rate per four-minute interval. WinBUGS provides an easy way to save the simulated parameter values, in order, to a text file. This then enables the data to be easily imported into another environment, such as **R** or MATLAB, for data analysis and graphing. In this example, R was used to provide the histograms for λ and p_{10}. The histograms in Figure 18.3 illustrate that λ is pretty much confined to the five integers 7, 8, 9, 10, and 11, with the mode 9.

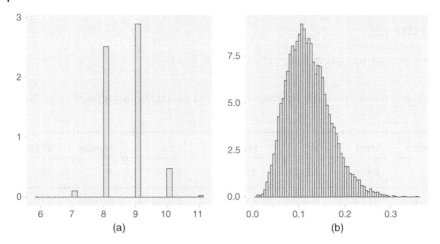

Figure 18.3 Histograms of 40 000 samples from (a) posterior of lambda and (b) posterior of p[10].

18.1.2 Generalized Dirichlet Processes

Some popular NP Bayesian models employ a mixture of DPs. The motivation for such models is their extraordinary modeling flexibility. Let $X_1, X_2, \ldots, X_n$ be the observations modeled as

$$X_i|\theta_i \sim Bin(n_i, \theta_i),$$
$$\theta_i|F \sim F, \quad i = 1, \ldots, n \tag{18.3}$$
$$F \sim Dir(\alpha).$$

If α assigns mass to every open interval on [0,1] then the support of the distributions on F is the class of *all* distributions on [0,1]. This model allows for pooling information across the samples. For example, observation X_i will have an effect on the posterior distribution of θ_j, $j \neq i$, via the hierarchical stage of the model involving the common DP.

The model (18.4) is used extensively in the applications of Bayesian nonparametrics. For example, Berry and Christensen (1979) use the model for the quality of welding material submitted to a naval shipyard, implying an interest in posterior distributions of θ_i. Liu (1996) uses the model for results of flicks of thumbtacks and focuses on distribution of $\theta_{n+1}|X_1, \ldots, X_n$. MacEarchern, Clyde, and Liu (1999) discuss estimation of the posterior predictive $X_{n+1}|X_1, \ldots, X_n$, and some other posterior functionals.

The DP is the most popular nonparametric Bayes model in the literature (for a recent review, see MacEachern and Mueller, 2000). However, limiting the prior to discrete distributions may not be appropriate for some applications.

A simple extension to remove the constraint of discrete measures is to use a convoluted DP:

$$X|F \sim F$$

$$F(x) = \int f(x|\theta)dG(\theta),$$

$$G \sim DP(\alpha G_0).$$

This model is called *Dirichlet process mixture* (DPM), because the mixing is done by the DP. Posterior inference for DMP models is based on MCMC posterior simulation. Most approaches proceed by introducing latent variables θ as

$$X_i|\theta_i \sim f(x|\theta_i), \quad \theta_i|G \sim G \text{ and } G \sim DP(\alpha G_0).$$

Efficient MCMC simulation for general mixture of Dirichlet process (MDP) models is discussed, among others, in Escobar (1994), Escobar and West (1995), Bush and MacEachern (1996), and MacEachern and Mueller (1998). Using a Gaussian kernel, $f(x|\mu, \Sigma) \propto \exp\{(x - \mu)'\Sigma(x - \mu)/2\}$, and mixing with respect to $\theta = (\mu, \Sigma)$, a density estimate resembling traditional kernel density estimation is obtained. Such approaches have been studied in Lo (1984) and Escobar and West (1995).

A related generalization of DPs is the *mixture of Dirichlet processes* (MDP). The MDP is defined as a DP with a center CDF that depends on random θ,

$$F \sim DP(\alpha G_\theta)$$

$$\theta \sim \pi(\theta).$$

Antoniak (1974) explored theoretical properties of MDP's and obtained posterior distribution for θ.

18.2 Bayesian Contingency Tables and Categorical Models

In contingency tables, the cell counts N_{ij} can be modeled as realizations from a count distribution, such as multinomial $\mathcal{M}n(n, p_{ij})$ or Poisson $\mathcal{P}(\lambda_{ij})$. The hypothesis of interest is independence of row and column factors, $H_0 : p_{ij} = a_i b_j$, where a_i and b_j are marginal probabilities of levels of two factors satisfying $\sum_i a_i = \sum_j b_j = 1$.

The expected cell count for the multinomial distribution is $\mathbb{E}N_{ij} = np_{ij}$. Under H_0, this equals $na_i b_j$, so by taking the logarithm on both sides, one obtains

$$\log \mathbb{E}N_{ij} = \log n + \log a_i + \log b_j$$
$$= \text{const} + \alpha_i + \beta_j,$$

for some parameters α_i and β_j. This shows that testing the model for additivity in parameters α and β is equivalent to testing the original independence hypothesis H_0. For the Poisson counts, the situation is analogous; one uses $\log \lambda_{ij} = \text{const} + \alpha_i + \beta_j$.

Example 18.4   Activities of Dolphin Groups Revisited. We revisit the dolphin's activity example from p. 175. Groups of dolphins were observed off the coast of Iceland, and the table providing group counts is given below. The counts are listed according to the time of the day and the main activity of the dolphin group. The hypothesis of interest is independence of the type of activity from the time of the day:

	Traveling	Feeding	Socializing
Morning	6	28	38
Noon	6	4	5
Afternoon	14	0	9
Evening	13	56	10

The WinBUGS program implementing the additive model is quite simple. We assume the cell counts are assumed distributed Poisson and the logarithm of intensity (expectation) is represented in an additive manner. The model parts (intercept, α_i, and β_j) are assigned normal priors with mean zero and precision parameter xi. The precision parameter is given a gamma prior with mean 1 and variance 10. In addition to the model parameters, the WinBUGS program will calculate the deviance and chi-square statistics that measure goodness of fit for this model:

```
model {
    for (i in 1:nrow) {
        for (j in 1:ncol) {
        groups[i,j]  ~ dpois(lambda[i,j])
        log(lambda[i,j]) <-  c + alpha[i] + beta[j]
        } }

        c ~ dnorm(0, xi)
        for (i in 1:nrow) { alpha[i] ~ dnorm(0, xi)  }
        for (j in 1:ncol) {  beta[j] ~ dnorm(0, xi)   }
        xi ~ dgamma(0.01, 0.01)

    for (i in 1:nrow) {
        for (j in 1:ncol) {
            devG[i,j] <- groups[i,j] * log((groups[i,j]+0.5)/
            (lambda[i,j]+0.5))-(groups[i,j]-lambda[i,j]);
```

```
        devX[i,j] <- (groups[i,j]-lambda[i,j])
                   *(groups[i,j]-lambda[i,j])/lambda[i,j];} }
    G2 <- 2 * sum( devG[,]);
    X2 <- sum( devX[,])}
```

The data are imported as

```
list(nrow=4, ncol=3,
        groups = structure(
            .Data = c( 6, 28, 38, 6, 4, 5,
                14, 0, 9, 13, 56, 10),.Dim=c(4,3))))
```

and initial parameters are

```
list(xi=0.1, c = 0, alpha=c(0,0,0,0), beta=c(0,0,0))
```

The following output gives Bayesian estimators of the parameters and measures of fit. This additive model conforms poorly to the observations; under the hypothesis of independence, the test statistic is χ^2 with $3 \times 4 - 6 = 6$ degrees of freedom, and the observed value $X^2 = 77.73$ has a p-value (1-chi2cdf(77.73, 6)) that is essentially zero:

Node	Mean	sd	MC error	2.5%	Median	97.5%
c	1.514	0.7393	0.03152	−0.02262	1.536	2.961
alpha[1]	1.028	0.5658	0.0215	−0.07829	1.025	2.185
alpha[2]	−0.5182	0.5894	0.02072	−1.695	−0.5166	0.6532
alpha[3]	−0.1105	0.5793	0.02108	−1.259	−0.1113	1.068
alpha[4]	1.121	0.5656	0.02158	0.02059	1.117	2.277
beta[1]	0.1314	0.6478	0.02492	−1.134	0.1101	1.507
beta[2]	0.9439	0.6427	0.02516	−0.3026	0.9201	2.308
beta[3]	0.5924	0.6451	0.02512	−0.6616	0.5687	1.951
c	1.514	0.7393	0.03152	−0.02262	1.536	2.961
G2	77.8	3.452	0.01548	73.07	77.16	86.2
X2	77.73	9.871	0.03737	64.32	75.85	102.2

Example 18.5   Cæsarean Section Infections Revisited. We now consider the Bayesian solution to the Cæsarean section birth problem from p. 254. The model for probability of infection in a birth by Cæsarean section was given in terms of the *logit* link as

$$\log \frac{P(\text{infection})}{P(\text{no infection})} = \beta_0 + \beta_1 \text{ noplan} + \beta_2 \text{ riskfac} + \beta_3 \text{ antibio}.$$

The WinBUGS program provided below implements the model in which the number of infections is $Bin(n, p)$ with p connected to covariates noplan riskfac and antibio via the logit link. Priors on coefficients in the linear predictor are set to be a vague Gaussian (small precision parameter):

```
model{
  for(i in 1:N){
    inf[i] ~ dbin(p[i],total[i])
    logit(p[i]) <- beta0 + beta1*noplan[i] +
            beta2*riskfac[i] + beta3*antibio[i]
  }
  beta0 ~dnorm(0, 0.00001)
  beta1 ~dnorm(0, 0.00001)
  beta2 ~dnorm(0, 0.00001)
  beta3 ~dnorm(0, 0.00001)
}
#DATA
list( inf=c(1, 11, 0, 0, 28, 23, 8, 0),
      total = c(18, 98, 2, 0, 58, 26, 40, 9),
      noplan = c(0,1,0,1,0,1,0,1),
      riskfac = c(1,1, 0, 0, 1,1, 0, 0),
      antibio =c(1,1,1,1,0,0,0,0), N=8)
#INITS
list(beta0 =0, beta1=0,
     beta2=0, beta3=0)
```

The Bayesian estimates of the parameters $\beta_0 - \beta_3$ are given in the WinBUGS output below:

Node	Mean	sd	MC error	2.5%	Median	97.5%
beta0	−1.962	0.4283	0.004451	−2.861	−1.941	−1.183
beta1	1.115	0.4323	0.003004	0.29	1.106	1.988
beta2	2.101	0.4691	0.004843	1.225	2.084	3.066
beta3	−3.339	0.4896	0.003262	−4.338	−3.324	−2.418

Note that Bayesian estimators are close to the estimators obtained in the frequentist solution in Chapter 12 $(\beta_0, \beta_1, \beta_2, \beta_3) = (-1.89, 1.07, 2.03, -3.25)$ and that in addition to the posterior means, posterior medians, and 95% credible sets for the parameters are provided. WinBUGS can provide various posterior location and precision measures. From the table, the 95% credible set for β_0 is $[-2.861, -1.183]$.

18.3 Bayesian Inference in Infinitely Dimensional Nonparametric Problems

Earlier in the book we argued that many statistical procedures classified as nonparametric are, in fact, infinitely parametric. Examples include wavelet regression, orthogonal series density estimators, and nonparametric MLEs (Chapter 10). To estimate such functions, we rely on shrinkage, tapering, or truncation of coefficient estimators in a potentially infinite expansion class. (Chencov's orthogonal series density estimators, Fourier and wavelet shrinkage, and related.) The benefits of shrinkage estimation in statistics were first explored in the mid-1950s by C. Stein. In the 1970s and 1980s, many statisticians were active in research on statistical properties of classical and Bayesian shrinkage estimators.

Bayesian methods have become popular in shrinkage estimation because Bayesian rules are, in general, "shrinkers." Most Bayesian rules shrink large coefficients slightly, whereas small ones are more heavily shrunk. Furthermore, interest for Bayesian methods is boosted by the possibility of incorporating prior information about the function to model wavelet coefficients in a realistic way.

Wavelet transformations W are applied to noisy measurements $y_i = f_i + \varepsilon_i$, $i = 1, \ldots, n$, or, in vector notation, $\boldsymbol{y} = \boldsymbol{f} + \boldsymbol{\varepsilon}$. The linearity of W implies that the transformed vector $\boldsymbol{d} = W(\boldsymbol{y})$ is the sum of the transformed signal $\boldsymbol{\theta} = W(\boldsymbol{f})$ and the transformed noise $\boldsymbol{\eta} = W(\boldsymbol{\varepsilon})$. Furthermore, the orthogonality of W implies that ε_i, i.i.d. normal $\mathcal{N}(0, \sigma^2)$ components of the noise vector $\boldsymbol{\varepsilon}$, are transformed into components of $\boldsymbol{\eta}$ with the same distribution.

Bayesian methods are applied in the wavelet domain, that is, after the wavelet transformation has been applied and the model $d_i \sim \mathcal{N}(\theta_i, \sigma^2), i = 1, \ldots, n$ has been obtained. We can model coefficient by coefficient because wavelets decorrelate and d_i's are approximately independent.

Therefore we concentrate just on a single typical wavelet coefficient and one model: $d = \theta + \varepsilon$. Bayesian methods are applied to estimate the location parameter θ. As θ's correspond to the function to be estimated, back-transforming an estimated vector θ will give the estimator of the function.

18.3.1 BAMS Wavelet Shrinkage

Bayesian adaptive multiscale shrinkage, or BAMS for short, is a simple efficient shrinkage in which the shrinkage rule is a Bayesian rule for properly selected prior and hyperparameters of the prior. Starting with $[d|\theta, \sigma^2] \sim \mathcal{N}(\theta, \sigma^2)$ and the prior $\sigma^2 \sim \mathcal{E}(\mu)$, $\mu > 0$, with density $f(\sigma^2|\mu) = \mu e^{-\mu\sigma^2}$, we obtain the marginal likelihood

$$d|\theta \sim \mathcal{DE}\left(\theta, \sqrt{2\mu}\right), \quad \text{with density } f(d|\theta) = \frac{1}{2}\sqrt{2\mu}e^{-\sqrt{2\mu}|d-\theta|}.$$

If the prior on θ is a mixture of a point mass δ_0 at zero and a double-exponential distribution,

$$\theta|\varepsilon \sim \varepsilon\delta_0 + (1-\varepsilon)D\mathcal{E}(0,\tau), \tag{18.4}$$

then the posterior mean of θ (from Bayes rule) is

$$\delta^*(d) = \frac{(1-\varepsilon)\,m(d)\,\delta(d)}{(1-\varepsilon)\,m(d) + \varepsilon\,f(d|0)}, \tag{18.5}$$

where

$$m(d) = \frac{\frac{1}{\tau}e^{-\tau|d|} - \frac{1}{\sqrt{2\mu}}e^{-\sqrt{2\mu}|d|}}{2/\tau^2 - 1/\mu}, \tag{18.6}$$

and

$$\delta(d) = \frac{(1/\tau^2 - 1/(2\mu))de^{-\tau|d|}/\tau + \left(e^{-\sqrt{2\mu}|d|} - e^{-\tau|d|}\right)/(\mu\tau^2)}{(1/\tau^2 - 1/(2\mu))\left(e^{-\tau|d|}/\tau - e^{-\sqrt{2\mu}|d|}/\sqrt{2\mu}\right)}. \tag{18.7}$$

As evident from Figure 18.4, the Bayesian rule (18.5) falls between comparable hard- and soft-thresholding rules. To apply the shrinkage in (18.5) on a specific problem, the hyperparameters μ, τ, and ε have to be specified. A default choice for the parameters is suggested in Vidakovic and Ruggeri (2001); see also Antoniadis, Bigot, and Sapatinas (2001) for a comparative study of many shrinkage rules, including BAMS.

Figure 18.5(a) shows a noisy `doppler` function of size $n = 1024$, where the signal-to-noise ratio (defined as a ratio of variances of signal and noise) is 7. Figure 18.5b shows the smoothed function by BAMS. The graphs are based on default values for the hyperparameters.

Example 18.6 Bayesian Wavelet Shrinkage in WinBUGS. Because of the decorrelating property of wavelet transforms, the wavelet coefficients are modeled

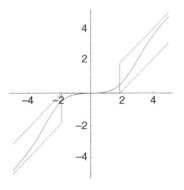

Figure 18.4 Bayesian rule (18.7) and comparable hard and soft thresholding rules.

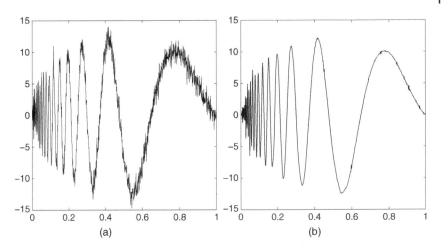

Figure 18.5 (a) A noisy `doppler` signal (SNR=7, $n = 1024$, noise variance $\sigma^2 = 1$). (b) Signal reconstructed by BAMS.

independently. A selected coefficient d is assumed to be normal $d \sim \mathcal{N}(\theta, \xi)$ where θ is the coefficient corresponding to the underlying signal in data and ξ is the precision, reciprocal of variance. The signal component θ is modeled as a mixture of two double-exponential distributions with zero mean and different precisions, because WinBUGS will not allow a point mass prior. The precision of one part of the mixture is large (so the variance is small) indicating coefficients that could be ignored as negligible. The corresponding precision of the second part is small (so the variance is large) indicating important coefficients of possibly large magnitude. The densities in the prior mixture are taken in proportion $p : (1 - p)$ where p is Bernoulli. For all other parameters and hyperparameters, appropriate prior distributions are adopted.

We are interested in the posterior means for θ. Here is the WinBUGS implementation of the described model acting on some imaginary wavelet coefficients ranging from -50 to 50, as an illustration. Figure 18.6 shows the Bayesian rule. Note a desirable shape close to that of the thresholding rules:

```
model{
    for (j in 1:N){
        DD[j] ~ dnorm(theta[j], tau);
        theta[j] <- p[j] * mu1[j] + (1-p[j]) * mu2[j];
        mu1[j] ~ ddexp(0, tau1);
        mu2[j] ~ ddexp(0, tau2);
        p[j] ~ dbern(r);
    }
```

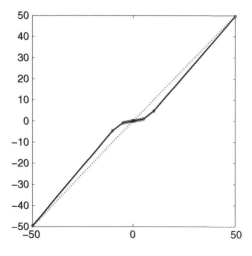

Figure 18.6 Approximation of Bayesian shrinkage rule calculated by WinBUGS.

```
    r ~ dbeta(1,10);
    tau ~ dgamma(0.5, 0.5);
    tau1 ~ dgamma(0.005, 0.5);
    tau2 ~dgamma(0.5, 0.005);
}
# DATA
list( DD=c(-50, -10, -5,-4,-3,-2,-1,-0.5, -0.1, 0,
        0.1, 0.5, 1, 2,3,4,5, 10, 50), N=19);
# INITS
list(tau=1, tau1=0.1, tau2=10);
```

18.4 Exercises

18.1 Show that in the DP Definition 18.2, $\mathbb{E}\left(\sum_{i=1}^{T} \omega_i\right) = 1 - [\alpha/(1-\alpha)]^T$.

18.2 Let $\mu = \int_{-\infty}^{\infty} y \, dG(y)$ and let G be a random CDF with Dirichlet process prior $DP(\alpha G_0)$. Let **y** be a sample of size n from G. Using (18.2), show that

$$\mathbb{E}(\mu|\mathbf{y}) = \frac{\alpha}{\alpha + n}\mathbb{E}\mu + \frac{n}{\alpha + n}\bar{y}.$$

In other words, show that the expected posterior mean is a weighted average of the expected prior mean and the sample mean.

18.3 Redo Exercise 9.13, where the results for 148 survey responses are broken down by program choice and by race. Test the fit of the properly set additive Bayesian model. Use WinBUGS for model analysis.

18.4 Show that $m(d)$ and $\delta(d)$ from (18.6) and (18.7) are marginal distributions and the Bayesian rule for the model is

$$d|\theta \sim \mathcal{DE}\left(\theta, \sqrt{2\mu}\right), \quad \theta \sim \mathcal{DE}(0, \tau),$$

where μ and τ are the hyperparameters.

18.5 This is an open-ended question. Select a data set with noise present in it (a noisy signal), transform the data to the wavelet domain, apply shrinkage on wavelet coefficients by the Bayesian procedure described below, and back-transform the shrunk coefficients to the domain of original data.

  (i) Prove that for $[d|\theta] \sim \mathcal{N}(\theta, 1)$, $[\theta|\tau^2] \sim \mathcal{N}(0, \tau^2)$, and $\tau^2 \sim (\tau^2)^{-3/4}$, the posterior is unimodal at 0 if $0 < d^2 < 2$ and bimodal otherwise with the second mode

$$\delta(d) = \left(1 - \frac{1 - \sqrt{1 - 2/d^2}}{2}\right)d.$$

  (ii) Generalize to $[d|\theta] \sim \mathcal{N}(\theta, \sigma^2)$, σ^2 known, and apply *the larger mode shrinkage*. Is this shrinkage of the thresholding type?

  (iii) Use the approximation $(1 - u)^\alpha \sim (1 - \alpha u)$ for u small to argue that the largest mode shrinkage is close to a James–Stein-type rule $\delta^*(d) = \left(1 - \frac{1}{2d^2}\right)_+ d$, where $(f)_+ = \max\{0, f\}$.

18.6 Chipman, Kolaczyk, and McCulloch (1997) propose the following model for Bayesian wavelet shrinkage (ABWS) in which we give in a simplified form:

$$d|\theta \sim \mathcal{N}(\theta, \sigma^2).$$

The prior on θ is defined as a mixture of two normals with a hyperprior on the mixing proportion:

$$\theta|\gamma \sim \gamma \mathcal{N}(0, (c\tau)^2) + (1 - \gamma)\mathcal{N}(0, \tau^2),$$

$$\gamma \sim Bin(1, p).$$

Variance σ^2 is considered known, and $c \gg 1$.

  (i) Show that the Bayesian rule (posterior expectation) for θ has the explicit form of

$$\delta(d) = \left[P(\gamma = 1|d)\frac{(c\tau)^2}{\sigma^2 + (c\tau)^2} + P(\gamma = 0|d)\frac{\tau^2}{\sigma^2 + \tau^2}\right]d,$$

where

$$P(\gamma = 1|d) = \frac{p\pi(d|\gamma = 1)}{p\pi(d|\gamma = 1) + (1 - p)\pi(d|\gamma = 0)}$$

and $\pi(d|\gamma = 1)$ and $\pi(d|\gamma = 0)$ are densities of $\mathcal{N}(0, \sigma^2 + (c\tau)^2)$ and $\mathcal{N}(0, \sigma^2 + \tau^2)$ distributions, respectively, evaluated at d.

(ii) Plot the Bayesian rule from (i) for selected values of parameters and hyperparameters $(\sigma^2, \tau^2, \gamma, c)$ so that the shape of the rule is reminiscent of thresholding.

RELEVANT R FUNCTIONS AND DATA SETS IN THIS CHAPTER

R function: rdirichlet
R package: MCMCpack

dolphins.txt, hartsfields.txt, shrinkage.txt

References

Antoniadis, A., Bigot, J., and Sapatinas, T. (2001), "Wavelet Estimators in Nonparametric Regression: A Comparative Simulation Study," *Journal of Statistical Software*, 6, 1–83.

Antoniak, C. E. (1974), "Mixtures of Dirichlet Processes with Applications to Bayesian Nonparametric Problems," *Annals of Statistics*, 2, 1152–1174.

Berry, D. A., and Christensen, R. (1979), " Empirical Bayes Estimation of a Binomial Parameter Via Mixtures of Dirichlet Processes," *Annals of Statistics*, 7, 558–568.

Bush, C. A., and MacEachern, S. N. (1996), "A Semi-Parametric Bayesian Model for Randomized Block Designs," *Biometrika*, 83, 275–286.

Chipman, H. A., Kolaczyk, E. D., and McCulloch, R. E. (1997), "Adaptive Bayesian Wavelet Shrinkage," *Journal of American Statistical Association*, 92, 1413–1421.

Escobar, M. D. (1994), "Estimating Normal Means with a Dirichlet Process Prior," *Journal of American Statistical Association*, 89, 268–277.

Escobar, M. D., and West, M. (1995), "Bayesian Density Estimation and Inference Using Mixtures," *Journal of American Statistical Association*, 90, 577–588.

Fabius, J. (1964), "Asymptotic Behavior of Bayes' Estimates," *Annals of Mathematical Statistics*, 35, 846–856.

Ferguson, T. S. (1973), "A Bayesian Analysis of Some Nonparametric Problems," *Annals of Statistics*, 1, 209–230.

Ferguson, T. S. (1974), "Prior Distributions on Spaces of Probability Measures," *Annals of Statistics*, 2, 615–629.

Freedman, D. A. (1963), "On the Asymptotic Behavior of Bayes' Estimates in the Discrete Case," *Annals of Mathematical Statistics*, 34, 1386–1403.

Liu, J. S. (1996), "Nonparametric Hierarchical Bayes via Sequential Imputations," *Annals of Statistics*, 24, 911–930.

Lo, A. Y. (1984), "On a Class of Bayesian Nonparametric Estimates, I. Density Estimates," *Annals of Statistics*, 12, 351–357.

MacEachern, S. N., and Mueller, P. (1998), "Estimating Mixture of Dirichlet Process Models," *Journal of Computational and Graphical Statistics*, 7, 223–238.

MacEachern, S. N., and Mueller, P. (2000), "Efficient MCMC Schemes for Robust Model Extensions Using Encompassing Dirichlet Process Mixture Models," in *Robust Bayesian Analysis*, Eds. F. Ruggeri and D. Rios-Insua, New York: Springer-Verlag.

MacEachern, S. N., Clyde, M., and Liu, J. S. (1999), "Sequential Importance Sampling for Nonparametric Bayes Models: The Next Generation," *Canadian Journal of Statistics*, 27, 251–267.

Mueller, P., and Quintana, F. A. (2004), "Nonparametric Bayesian Data Analysis," *Statistical Science*, 19, 95–110.

Sethuraman, J. (1994), "A Constructive Definition of Dirichlet Priors," *Statistica Sinica*, 4, 639–650.

Sethuraman, J., and Tiwari, R. C. (1982), "Convergence of Dirichlet Measures and the Interpretation of their Parameter," in *Statistical Decision Theory and Related Topics III*, Eds. S. Gupta and J. O. Berger, New York: Springer-Verlag, Volume 2, pp. 305–315.

Vidakovic, B., and Ruggeri, F. (2001), "BAMS Method: Theory and Simulations," *Sankhyā, Series B*, 63, 234–249.

Appendix A

WinBUGS

BUGS and WᵢₙBUGS are distributed freely and are the result of many years of development by a team of statisticians and programmers at the Medical Research Council Biostatistics Research Unit in Cambridge (BUGS and WinBUGS) and recently by a team at University of Helsinki (OpenBUGS); see the project pages:

> `http://www.mrc-bsu.cam.ac.uk/bugs/`

and

> `http://mathstat.helsinki.fi/openbugs/`.

Models are represented by a flexible language, and there is also a graphical feature, DOODLEBUGS, which allows users to specify their models as directed graphs. For complex models the DOODLEBUGS can be very useful. As of May 2007, the latest version of WinBUGS is 1.4.1, and OpenBUGS 3.0.

A.1 Using WinBUGS

We start the introduction to WinBUGS with a simple regression example. Consider the model

$$y_i | \mu_i, \tau \sim \mathcal{N}(\mu_i, \tau), \quad i = 1, \dots, n,$$
$$\mu_i = \alpha + \beta(x_i - \bar{x}),$$
$$\alpha \sim \mathcal{N}(0, 10^{-4}),$$
$$\beta \sim \mathcal{N}(0, 10^{-4}),$$
$$\tau \sim \mathcal{G}amma(0.001, 0.001).$$

The scale in normal distributions here is parameterized in terms of a *precision* parameter τ that is the reciprocal of variance, $\tau = 1/\sigma^2$. Natural distributions for the precision parameters are gamma, and small values of the precision reflect

Nonparametric Statistics with Applications to Science and Engineering with R, Second Edition.
Paul Kvam, Brani Vidakovic, and Seong-joon Kim.

the flatness (noninformativeness) of the priors. The parameters α and β are less correlated if predictors $x_i - \bar{x}$ are used instead of x_i. Assume that (x,y)-pairs $(1, 1), (2, 3), (3, 3), (4, 3)$, and $(5, 5)$ are observed.

Estimators in classical least-squares regression of y on $x - \bar{x}$ are given in the following table:

```
Coef       LSEstimate   SE Coef      t        p
ALPHA          3.0000    0.3266    9.19   0.003
BETA           0.8000    0.2309    3.46   0.041
S = 0.730297   R-Sq =  80.0%   R-Sq(adj) = 73.3%
```

How about Bayesian estimators? We will find the estimators by MCMC calculations as means on the simulated posteriors. Assume that the initial values of parameters are $\alpha_0 = 0.1$, $\beta_0 = 0.6$, and $\tau = 1$. Start BUGS, and input the following code in [**File > New**]:

```
# A simple regression
model{
   for (i in 1:N) {
   Y[i] ~ dnorm(mu[i],tau);
   mu[i] <- alpha + beta * (x[i] - x.bar);
   }
   x.bar <- mean(x[]);
   alpha ~ dnorm(0, 0.0001);
   beta ~ dnorm(0, 0.0001);
   tau ~ dgamma(0.001, 0.001);
   sigma <- 1.0/sqrt(tau);
}
#-----------------------------
#these are observations
list( x=c(1,2,3,4,5), Y=c(1,3,3,3,5), N=5);
#-----------------------------
#the initial values
list(alpha = 0.1, beta = 0.6, tau = 1);
```

Next, put the cursor at an arbitrary position within the scope of model that delimited by wiggly brackets. Select the **Model** menu and open **Specification.** The **Specification Tool** window will pop out. If your model is highlighted, you may **check model** in the specification tool window. If the model is correct, the response on the lower bar of the BUGS window should be as follows: **model is syntactically correct**. Next, highlight the "list" statement in the data part of your code. In the Specification Tool window, select **load data**. If the data are in correct format, you should receive response on the bottom bar of BUGS window: **data**

loaded. You will need to compile your model on order to activate **inits** buttons. Select **compile** in the Specification Tool window. The response should be **model compiled**, and the buttons **load inits** and **gen inits** become active. Finally, highlight the "list" statement in the initials part of your code, and in the Specification Tool window select **load inits**. The response should be as follows: **model is initialized**, and this finishes reading in the model. If the response is **initial values loaded but this or other chain contains uninitialized variables,** click on the **gen inits** button. The response should be as follows: **initial values generated, model initialized**.

Now, you are ready to burn in some simulations and at the same time check that the program is working. In the **Model** menu, choose **Update …**, and open **Update Tool** to check if your model updates.

From the **Inference** menu, open **Samples …** A window titled **Sample Monitor Tool** will pop out. In the **node** subwindow, input the names of the variables you want to monitor. In this case, the variables are `alpha, beta,` and `tau`. If you correctly input the variable, the **set** button becomes active, and you should set the variable. Do this for all three variables of interest. In fact, `sigma` as transformation of `tau` is available, as well.

Now choose `alpha` from the subwindow in **Sample Monitor Tool**. All of the buttons (**clear, set, trace, history, density, stats, coda, quantiles, bgr diag, auto cor**) are now active. Return to **Update Tool**, and select the desired number of simulations, say, 10 000, in the **updates** subwindow. Press the **update** button.

Return to **Sample Monitor Tool**, and check **trace** for the part of MC trace for α, **history** for the complete trace, **density** for a density estimator of α, etc (see Figure A.1) For example, pressing **stats** button will produce something similar to Table A.1:

	mean	sd	MCerror	val2.5pc	median	val97.5pc	start	sample
alpha	3.003	0.549	0.003614	1.977	3.004	4.057	10 000	20 001

The mean 3.003 is the Bayesian estimator (as the mean from the sample from the posterior for α.) There are two precision outputs, `sd` and `MCerror`. The former is an estimator of the standard deviation of the posterior and can be improved by increasing the sample size but not the number of simulations. The later one is the error of simulation and can be improved by additional simulations. The 95% credible set is bounded by `val2.5pc` and `val97.5pc`, which are the 0.025 and 0.975 (empirical) quantiles from the posterior. The empirical median of the posterior is given by `median`. The outputs `start` and `sample` show the starting index for the simulations (after burn-in) and the available number of simulations.

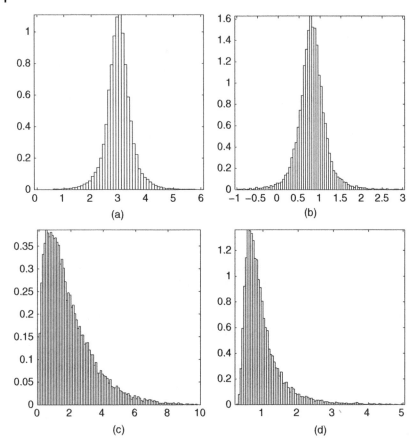

Figure A.1 Traces of the four parameters from simple example: (a) α, (b) β, (c) τ, and (d) σ from WinBUGS. Data are plotted in MATLAB after being exported from WinBUGS.

For all parameters, a comparative table is

	Mean	sd	MCerror	val2.5pc	Median	val97.5pc	Start	Sample
Alpha	3.003	0.549	0.003614	1.977	3.004	4.057	10 000	20 001
Beta	0.7994	0.3768	0.002897	0.07088	0.7988	1.534	10 000	20 001
Tau	1.875	1.521	0.01574	0.1399	1.471	5.851	10 000	20 001
Sigma	1.006	0.7153	0.009742	0.4134	0.8244	2.674	10 000	20 001

If you want to save the trace for α in a file and process it in MATLAB, say, select **coda**, and the data window will open with an information window as well. Keep the data window active and select **Save As** from the **File** menu. Save the αs in `alphas.txt` where it will be ready to be imported to MATLAB.

Table A.1 Built-in functions in WinBUGS.

BUGS code	Function
abs(y)	$|y|$
cloglog(y)	$\ln(-\ln(1-y))$
cos(y)	$\cos(y)$
equals(y, z)	1 if $y = z$; 0 otherwise
exp(y)	$\exp(y)$
inprod(y, z)	$\sum_i y_i z_i$
inverse(y)	y^{-1} for symmetric positive-definite matrix y
log(y)	$\ln(y)$
logfact(y)	$\ln(y!)$
loggam(y)	$\ln(\Gamma(y))$
logit(y)	$\ln(y/(1-y))$
max(y, z)	y if $y > z$; y otherwise
mean(y)	$n^{-1}\sum_i y_i, \ n = dim(y)$
min(y, z)	y if $y < z$; z otherwise
phi(y)	Standard normal CDF $\Phi(y)$
pow(y, z)	y^z
sin(y)	$\sin(y)$
sqrt(y)	$\sqrt{y}$
rank(v, s)	Number of components of v less than or equal to v_s
ranked(v, s)	The sth smallest component of v
round(y)	Nearest integer to y
sd(y)	Standard deviation of components of y
step(y)	1 if $y \geq 0$; 0 otherwise
sum(y)	$\sum_i y_i$
trunc(y)	Greatest integer less than or equal to y

Kevin Murphy leads the project for communication between WinBUGS and MATLAB:

http://www.cs.ubc.ca/ ~murphyk/Software/MATBUGS/matbugs.html.

His suite MATBUGS, maintained by several researchers, communicates with WinBUGS directly from MATLAB.

A.2 Built-in Functions and Common Distributions in BUGS

This section contains two tables: one with the list of built-in functions and the second with the list of available distributions.

Table A.2 Built-in distributions with BUGS names and their parametrizations.

Distribution	BUGS code	Density
Bernoulli	x ~ dbern(p)	$p^x(1-p)^{1-x}$, $x = 0, 1$; $0 \le p \le 1$
Binomial	x ~ dbin(p, n)	$\binom{n}{x} p^x(1-p)^{n-x}$, $x = 0,\ldots,n$; $0 \le p \le 1$
Categorical	x ~ dcat(p[])	$p[x]$, $x = 1, 2, \ldots, \dim(p)$
Poisson	x ~ dpois(lambda)	$\frac{\lambda^x}{x!}\exp\{-\lambda\}$, $x = 0, 1, 2, \ldots$, $\lambda > 0$
Beta	x ~ dbeta(a,b)	$\frac{1}{B(a,b)}x^{a-1}(1-x)^{b-1}$, $0 = x \le 1$, $a, b > -1$
Chi-square	x ~ dchisqr(k)	$\frac{x^{k/1-1}\exp(-x/2)}{2^{k/2}\Gamma(k/2)}$, $x \ge 0$, $k > 0$
Double exponential	x ~ ddexp(mu, tau)	$\frac{\tau}{2}\exp\{-\tau|x-\mu|\}, x \in R$, $\tau > 0$, $\mu \in R$
Exponential	x ~ dexp(lambda)	$\lambda\exp\{-\lambda x\}$, $x \ge 0$, $\lambda \ge 0$
Flat	x ~ dflat()	Constant; not a proper density
Gamma	x ~ dgamma(a, b)	$\frac{b^a x^{a-1}}{\Gamma(a)}\exp(-bx)$, $x, a, b > 0$
Normal	x ~ dnorm(mu, tau)	$\sqrt{\tau/(2\pi)}\exp\{-\frac{\tau}{2}(x-\mu)^2\}$, $x, \mu \in R, \tau > 0$
Pareto	x ~ dpar)alpha,c)	$\alpha c^\alpha x^{-(\alpha+1)}$, $x > c$
Student-t	x ~ dt(mu, tau, k)	$\frac{\Gamma((k+1)/2)}{\Gamma(k/2)}\sqrt{\frac{\tau}{k\pi}}\left[1+\frac{\tau}{k}(x-\mu)^2\right]^{-(k+1)/2}$, $x \in R, k \ge 2$
Uniform	x ~ dunif(a, b)	$\frac{1}{b-a}$, $a \le x \le b$
Weibull	x ~ dweib(v, lambda)	$v\lambda x^{v-1}\exp\{-\lambda x^v\}$, $x, v, \lambda > 0$,
Multinomial	x[] ~ dmulti(p[], N)	$\frac{(\sum_i x_i)!}{\prod_i x_i!}\prod_i p_i^{x_i}$, $\sum_i x_i = N, 0 < p_i < 1, \sum_i p_i = 1$
Dirichlet	p[] ~ ddirch(alpha[])	$\frac{\Gamma(\sum_i a_i)}{\prod_i \Gamma(a_i)}\prod_i p_i^{a_i-1}$, $0 < p_i < 1, \sum_i p_i = 1$
Multivariate normal	x[] ~ dmnorm(mu[], T[,])	$(2\pi)^{-d/2}|T|^{1/2}\exp\{-1/2(x-\mu)'T(x-\mu)\}$, $x \in R^d$
Multivariate Student-t	x[] ~ dmt(mu[], T[,], k)	$\frac{\Gamma((k+d)/2)}{\Gamma(k/2)}\frac{|T|^{1/2}}{k^{d/2}\pi^{d/2}}\left[1+\frac{1}{k}(x-\mu)'T(x-\mu)\right]^{-(k+d)/2}$, $x \in R^d, k \ge 2$
Wishart	x[,] ~ dwish(R[,], k)	$|R|^{k/2}|x|^{(k-p-1)/2}\exp\{-1/2Tr(Rx)\}$

The first-time WinBUGS user may be disappointed by the selection of built-in functions – the set is minimal but sufficient. The full list of distributions in WinBUGS can be found in **Help** > **WinBUGS User Manual** under **The_BUGS_language: _stochastic_nodes** > **Distributions**. BUGS also allows for construction of distributions for which are not in default list. In Table A.2, a list of important continuous and discrete distributions, with their BUGS syntax and parametrization, is provided. BUGS has the capability to define custom distributions, both as likelihood and as a prior, via the so-called *zero-Poisson device*.

Appendix B

R Coding

I've got a lot to say, I've got a lot to say,
I've got a lot to say, I've got a lot to say.
I can't remember now, I can't remember now,
I can't remember now, I can't remember now

The Ramones

B.1 Programming in R

This appendix provides little more than a reference for R-commands that are used frequently or applied in this textbook. We recommend you program using RStudio. If you have not yet installed R on your computer, you can download it (for free) at

```
http://www.r-project.org/.
```

Once you have R installed, you can also install RStudio at

```
http://www.rstudio.org/.
```

By creating R notebooks with RStudio, users can conveniently create programs and subprograms that can be run separately (in "chunks"). Figure B.1 displays the interactive environment for an RStudio session. The four windows shown have specific uses in programming:

- **Main Program Window** in the upper left corner: where you can type code (in chunks) and save it. You can separate subprograms into chunks that can generate output as well. This is the main window of use.
- **Console** in the lower left corner: for typing commands and generating output. This is what you see if you run R in the command line (without the aid of RStudio).

Nonparametric Statistics with Applications to Science and Engineering with R, Second Edition.
Paul Kvam, Brani Vidakovic, and Seong-joon Kim.

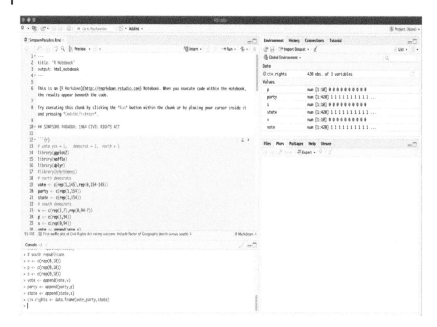

Figure B.1 RStudio console features four boxes: Source Editor, Console, Workspace Browser (and History), and Plots (and Files, Packages, Help). Source: Rstudio, Inc.

- **Environment/History** in the upper right: environment shows current objects, variables, and history records from commands that run in console.
- **Files/Plots/Packages/Help** in the lower right: several menu tabs for help in seeing/storing plots, looking up commands (under the Help tab). The Packages tab shows you all of the R-packages downloaded to your program (or cloud-space) and allows you to upload them for use with a simple click.

In the book, we frequently use graphics provided by Tidyverse, which is a refined collection of R-packages. Consider inst+alling the package via

```
install.packages("tidyverse").
```

We rely on one of the libraries (ggplot2) extensively for the book's graphics, although those commands are not necessary for producing simple plots, including scatterplots, box plots, density plots, etc.

B.1.1 Vectors

It is helpful to remember R has the propensity to treat all variables as lists or vectors. The frequently used command c(...) joins arguments into a vector (make sure not to assign this letter to a variables name!). Parentheses after a variable name indicate a function. For example, `sqrt(x1)` finds the square root of all values in the vector `x1`. Squared brackets (e.g. `x1[7]`) indicate elements of a vector or list.

B.1.2 Missing Values

Missing data are encoded using the symbol `NA`. You can test to see if a variable `x1` has missing values using function `is.na(x1)`, and it will respond with a logical output `TRUE` or `FALSE`. If you want R to ignore missing values while performing a command, you can add an optional argument to the function. For example, you can compute the median of a list using

```
median(x1, na.rm = TRUE).
```

B.1.3 Logical Arguments

We cover examples in the Section B.3 that include logical tests of variables. The operator symbols for testing whether A is equal to B are A==B, for example. Note that there are two equal signs, not one. Greater, less, and not equal are >, <, !=. To apply and/or logical operators, use A & B for **and** and A | B for **or**.

This appendix will focus mainly on functions used in the text. We will not cover how to create and apply functions in this summary.

B.2 Basics of R

If you are able, we recommend you obtain an RStudio account (for small computing tasks, you may be able to do this for free). This allows you to rely on cloud computing instead of downloading the software on your computer. In either case, you can set up directories and folders to store your work.

R uses a default folder on your computer to look for input and store output from the program. This folder is called your working directory. Find your working directory type

```
> getwd()
```

and find that folder on your computer. Before you start generating your own files, set your working directory to where you want you output to be stored (and to where you want R to look for your input files):

```
> setwd("/Users/SomeFolder")
```

There are many helpful subroutines in R that are part of installed packages or libraries. To see what packages are installed on your computer at any time, type

```
> library()
```

For example, if you want to install a package called "MASS" on your computer, type

```
> install.packages("MASS")
```

This allows programs and subroutines from the MASS package to be available to you, but to use them, you need to load the installed package using the library command:

```
> library(MASS)
```

If you are using RStudio, you can also upload any installed packages via the Package menu tab in the lower right corner. You can find out what is inside this package using library(help = "MASS").

There are many useful packages out there, and we just picked MASS as one to illustrate. MASS also contains interesting data sets that are uploaded into variables. For example, if you type

```
> summary(Rabbit)
```

you will find that Rabbit is a data frame with 60 observations and five variables. The summary command is useful in finding out what is in a vector, matrix, or data frame without having to type out the whole thing. You could also just type Rabbit to have R echo all 5×60 components of the data frame.

Alternative to the summary command, if you just want to see a few lines of the data (each line represents an observation with multiple measurements), you can type

```
> head(Rabbit,n=10)
```

and this will print out the first 10 observations of the data set.

B.3   R Commands

If you are novice at coding in R, you can try the following simple commands that vary from simple calculations to logical tests and list sorting. Some comments are included:

```
> 4*5 + 2^3 - exp(1)    # R can be used as a simple calculator
[1] 25.28172

> rep(2,10)  # repeat the number 2 ten times
[1] 2 2 2 2 2 2 2 2 2 2

> rep(c(1,2,3),4) # repeat the list c(1,2,3) four times
[1] 1 2 3 1 2 3 1 2 3 1 2 3

> seq(from=10, to=40, by=5)  # create sequence of numbers
[1] 10 15 20 25 30 35 40
```

```
> x1 <- c(6,1,30,9,12,2,21,5)   # assign list to variable x1
> x1   # print x1 to see output
[1]   6   1 30   9 12   2 21   5

> length(x1)   # number of items in list x1
[1] 8

>  x1[4]   # the 4th element of x1
[1] 9

> x1[-1]   #x1 without the 1st element
[1]   1 30   9 12   2 21   5

> sort(x1)   # order list from smallest to largest
[1]   1   2   5   6   9 12 21 30

> rank(x1)   # compute rank of each element in list
[1] 4 1 8 5 6 2 7 3

> c(min(x1),mean(x1),median(x1),sd(x1),max(x1))
[1]   1.00000 10.75000   7.50000 10.05343 30.00000

> x1 == 30   # R logical test: == means test to see if equal
[1] FALSE FALSE   TRUE FALSE FALSE FALSE FALSE FALSE

> (x1 == 30) + 0   # turns Boolean output into 0s and 1s
[1] 0 0 1 0 0 0 0 0

> sum(x1<10)   # logical test counts number of elements less than 10
[1] 5

> x1[x1>10]   # picks out elements of x1 that are bigger than 10
[1] 30 12 21

> x1*10   # * does component-wise multiplication
[1]   60   10 300   90 120   20 210   50

> x1^2   # all functions are applied component-wise
[1]   36     1 900   81 144     4 441   25

> x1*x1   # even vector multiplication is component-wise
[1]   36     1 900   81 144     4 441   25

> sample(x1)   # randomly re-arrage elements of list
[1] 30   1 12 21   9   6   2   5

> sample(x1,replace=TRUE)   # randomly draw with replacement
[1]   9   1 21   2 12   9 21   9

> range(x1)   # smallest and largest elements in list
[1]   1 30

> x1[[9]]<-50  # add 9th element to list
> x1
[1]   6   1 30   9 12   2 21   5 50
```

B.4   R for Statistics

Table B.1 contains basic summaries of data and rudimentary graphical summaries (including histograms, scatterplots, and box plots). As an example, we will utilize a preexisting data set named `rivers` that is a vector of 141 observations representing the length of 141 major rivers in North America. The graphical output is contained in Figure B.2:

```
> c(length(rivers),mean(rivers),median(rivers),sd(rivers))
   [1] 141.0000 591.1844 425.0000 493.8708
> mean(rivers) > median(rivers)
   [1] TRUE
>boxplot(rivers)
>hist(rivers)
```

Table B.1   Built-in functions in R.

R code	Function description
mean(x)	Computes the mean of the variable x
median(x)	Computes the median of the variable x
sd(x)	Computes the standard deviation of the variable x
IQR(x)	Computer the IQR of the variable x
summary(x)	Computes the five-number summary and the mean of the variable x
cor(x,y)	Computes the correlation coefficient
cumsum(x)	Cumulative sum of vector x
hist(x)	Creates a histogram for the variable x
boxplot(x)	Creates a boxplot for the variable x
boxplot(y~x)	Creates side-by-side boxplots
stem(x)	Creates a stem plot for the variable x
plot(y~x)	Creates a scatterplot of y versus x
abline(lm(y~x))	Adds regression line to plot
lines(lowess(x,y))	Adds LOWESS line (x,y) to scatter plot
table(x)	Summarizes count data
merge(a,b)	Merges two data frames
read.csv(file)	Reads .csv file in as a data frame
write.csv(file,x) Saves data frame x as .csv file	
nrow(x)	Number of rows for list or data frame
subset(x, condition)	Returns only rows of data frame satisfying condition

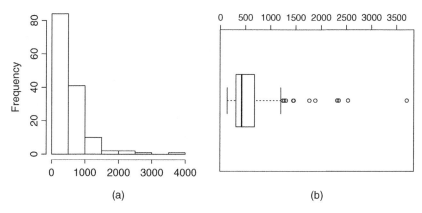

Figure B.2 Graphical summary of lengths for 141 rivers in North America using (a) histogram and (b) box plot.

Figure B.3 Histogram using ggplot.

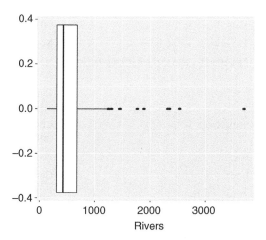

Graphics in the textbook use the Tidyverse packages, especially `ggplot2`, so we present graphs like Figure B.3.

Table B.2 contains several R functions used in the textbook for nonparametric data analysis. Please refer to the index of R commands to aid you in finding specific R functions.

In Table B.3, we list distributions that are commonly used in statistics. Several of these distributions were summarized in Chapter 2.

As an example, we use the R code to illustrate the normal density plotted in Figure B.4:

```
x <- seq(0,10,0.01)
y <- dnorm(x,5,1.5)
w <- dnorm(x,2.5,0.5)
```

Table B.2 Statistics functions in R.

R code	Function description
prop.test	Test of one or two proportions
t.test	Parametric *t*-test for one or two samples
binom.test(x, n, p)	Binomial test for proportions
conover.test	Nonparametric test of two variances
wilcoxon.signed(x)	Wilcoxon signed rank test
wilcox.test(x, y)	Wilcoxon rank sum test
KMcdfSM	Kaplan–Meier estimator
fligner.test(x, g)	Fligner–Killeen test of equal variances; g is a grouping variable
ks.test(x,y)	Kolmogorov–Smirnov test
kruskal.test(x, g)	Kruskal–Wallis rank sum test; g is a grouping variable
friedman.test(y)	Nonparametric Friedman test for correlation
friedman.pairwise.comparison	Nonparametric multiple comparisons for ANOVA
cor.test(x,y)	Correlation test plus CI for several measures of association (r, rho, tau)
tablerxc	Chi-square test for independence in categorical analysis
boot	Bootstrap procedure
boot.ci	Generate bootstrap confidence intervals
runs.test	Chi-square runs test for sequence of binary variables
mantel.haenszel	Mantel–Haenszel test

```
          # y, w: different normal densities
   norm <- data.frame(x,y,w)
         # install.packages("ggplot2")
   ggplot(norm,aes(x = x, y = y)) +
       geom_ribbon(aes(ymin=0,ymax=y),alpha=0.2) +
       geom_ribbon(aes(ymin=0,ymax=w),alpha=0.5)
```

This code created two plots for two different normal densities: $N(5, 1.5^2)$ and $N(2.5, 0.5^2)$. The first argument in dnorm specifies the value of the random variable X (in this case a large vector of arguments). The second argument is the mean (μ) of the normal distribution, and the third is the standard deviation (σ).

Table B.3 Probability functions in R.

Distribution	Functions			
Beta	pbeta	qbeta	dbeta	rbeta
Binomial	pbinom	qbinom	dbinom	rbinom
Cauchy	pcauchy	qcauchy	dcauchy	rcauchy
Chi-square	pchisq	qchisq	dchisq	rchisq
Exponential	pexp	qexp	dexp	rexp
F	pf	qf	df	rf
Gamma	pgamma	qgamma	dgamma	rgamma
Geometric	pgeom	qgeom	dgeom	rgeom
Hypergeometric	phyper	qhyper	dhyper	rhyper
Logistic	plogis	qlogis	dlogis	rlogis
Log normal	plnorm	qlnorm	dlnorm	rlnorm
Negative binomial	pnbinom	qnbinom	dnbinom	rnbinom
Normal	pnorm	qnorm	dnorm	rnorm
Poisson	ppois	qpois	dpois	rpois
Student *t*	pt	qt	dt	rt
Uniform	punif	qunif	dunif	runif
Weibull	pweibull	qweibull	dweibull	rweibull

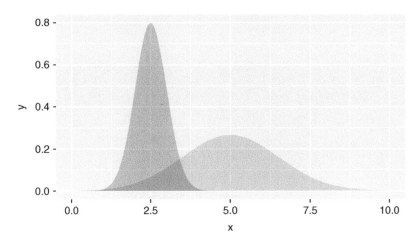

Figure B.4 Plot of two normal densities.

R Index

Nonparametric Statistics with Applications to Science and Engineering with R, Second Edition.
Paul Kvam, Brani Vidakovic, and Seong-joon Kim.
© 2023 John Wiley & Sons, Inc. Published 2023 by John Wiley & Sons, Inc.

Author Index

a

Adams, D. 213
Agresti, A. 42, 167, 347
Altman, N. S. 267, 277
Anderson, T. W. 96
Anscombe, F. J. 51, 55, 244
Antoniadis, A. 294, 382
Antoniak, C. E. 377
Arbuthnot, J. 129
Arnold, B. 73
Arnold, T. 344
Arvin, D. V. 135

b

Babbage, C. 55
Bai, Z. 81
Baines, L. 223
Baines, M. J. 277
Balakrishnan, N. 73
Balmukand, B. 328
Bayes, T. 51
Bellman, R. E. 351
Benford, F. 172
Berger, J. O. 62
Berry, D. A. 376
Best, N. G. 65
Bickel, P. J. 189
Bigot, J. 294, 382

Birnbaum, Z. W. 89
Bradley, J. V. 2
Breiman, L. 363
Broffitt, J.D. 347
Brown, J. S. 199
Buddha 87
Bush, C. A. 377

c

Carter, W. C. 215
Casella, G. 1, 45, 65
Charles, J. A. 319
Chen, M.-H. 65
Chen, Z. 81, 114
Chernick, M. R. 321
Christensen, R. 376
Cleveland, W. 267
Clopper, C. J. 41
Clyde, M. 376
Cochran, W. G. 181
Congdon, P. 65
Conover, W. J. 2, 144, 161
Cox, D. R. 212
Cox, G. 163
Cramér, H. 98
Crowder, M. J. 205
Crowley, J. 327
Cummings, T. L. 192

Nonparametric Statistics with Applications to Science and Engineering with R, Second Edition.
Paul Kvam, Brani Vidakovic, and Seong-joon Kim.
© 2023 John Wiley & Sons, Inc. Published 2023 by John Wiley & Sons, Inc.

Subject Index

Nonparametric Statistics with Applications to Science and Engineering with R, Second Edition.
Paul Kvam, Brani Vidakovic, and Seong-joon Kim.
© 2023 John Wiley & Sons, Inc. Published 2023 by John Wiley & Sons, Inc.

Printed and bound by CPI Group (UK) Ltd, Croydon, CR0 4YY

16/04/2025

14658365-0002